AF753270

CUEVAS MEDEK EXERCISE THERAPY

RAMON J. CUEVAS, PT
CME therapy creator

Since 1972

CUEVAS MEDEK EXERCISE

INTERNATIONAL CENTER:

5757 Kennedy Av. 5th. Floor, suite 504,

Las Condes, Santiago, Chile.

Phone:	(56-2) 246-0609
Email:	cme@cuevasmedek.com, cmeinternational@gmail.com
Website:	www.cuevasmedek.com
Facebook:	Cuevas Medek Exercise
Twitter:	@cme_therapy

ISBN 978-956-345-919-7

Printed by Lulu.com under the knowledge of the Cuevas Medek Exercise International Center, Chile.

"LIKE ARROWS IN THE HAND OF A WARRIOR,
SO ARE THE CHILDREN OF YOUTH."

Tehillim, psalm 127

ACKNOWLEDGEMENT

Love and appreciation to my wife Dalal and children Alexis, Andrés, Christian and Claudia for the endless support in my professional endeavor.

Special thanks and love to my son Alexis, who is following my path on the CME road.

Thanks to all those parents and physicians in Venezuela, who gave me the necessary reliance and support in the early seventies, which allowed me to develop this new approach of Physical Therapy for motor impaired children.

Thanks to all the parents around the world, who gave me the privilege of helping their children as a CME practitioner.

Thanks to the Heller and Miller families from Toronto for giving a big push to the promotion of CME therapy world wide.

Thanks to the enlightened and dedicated Physical and Occupational Therapists who learned from me, and recognized the potential of CME therapy for their patient's sake.

Thanks to the Yanofsky family who did so much to promote CME therapy around the world.

Thanks to a generous man from Long Island (Z.W.), who gave me the concrete means to achieve this CME project.

Special thanks for the person who helped me directly with the different stages of this project.

In Memoriam
Pablo Antonio Massis.

1945 - 2005

Beloved member of my family, always supportive and generous who told me a few hours before going forever: "Ramon, finish the book..."

Dedicated to the ONE who inspired me,
and to my parents.

TABLE OF CONTENTS

DISCLAIMER

This manual is presented to the general public to promote the existence of CUEVAS MEDEK EXERCISE therapy (CME), former MEDEK therapy. The author does not intend by any means to instigate or suggest to the reader to use the assessment format following the inserted information, nor to use these exercises described in this manual with any particular motor delayed child.

The reader must be aware that any child suffering motor delay will first need to be diagnosed by a physician, who should refer the child to the appropriate motor stimulation program.

Only therapists in possession of a CMECEP registered certificate are allowed to apply CME therapy.

There are no world wide MEDEK CENTER approved branches, other than Chilean Home office of CUEVAS MEDEK EXERCISE.

MEDEK and CME are intellectual registered property by Ramon Cuevas Unique founder of CUEVAS MEDEK EXERCISE since 1972.

The author
Ramón Cuevas

PREFACE

It is a pleasure to write the foreword for Cuevas Medek Exercises, a manual reflecting the considerable accomplishment of Ramon Cuevas's lifelong research and clinical experience in treating infants and children with motor delay. Ramon has devoted his entire career to this challenging field. The unique CME approach described in the text is the culmination of 30 years of inspiration, accumulated knowledge, hard work, and dedication. As a pediatric orthopedist, I have been intrigued by CME therapy since I first encountered it 10 years ago. Ramon has successfully used CME to treat patients with spasticity, hypotonia, ataxia, Down's syndrome, and recently, myelomeningocele.

The goal of CME is to produce functional motor control in young children with motor delay so that they might attain the highest level of independence in sitting, standing, and walking. Among the most difficult challenges, facing children with motor delay are lack of independence, and secondary contractures leading to structural trunk and limb deformities that often require multiple corrective surgeries. Unfortunately, due to the underlying abnormal muscle tone, the deformities eventually recur and repeated surgeries are required for these patients. CME therapy has prevented the need for these major surgical procedures by improving the child's muscle tone and by increasing flexibility and mobility.

CME exercises gradually expose the child to anti-gravity motor-postural challenges that facilitate the learning of new anti-gravity control reactions. The focus on facilitating the patient's self-produced actions to environmental challenges likely leads to neural connection rearrangement in the child's brain as long as the neurons in the brain retain some form of neural plasticity. This degree of concentration on self-produced motor actions to anti-gravity stimulation, a main feature of CME therapy, is rare among interventions that target motor function but is a common feature of successful ones.

By integrating the components of antigravity posture, range of motion, weight bearing, and stretching in the functional exercises, Ramon has been able to improve the agility, flexibility, balance, and coordination of the children he treats. Early intervention is important, as muscles weaken from lack of use, making treatment more difficult as time goes on. Also important for success is the participation of the parents or caregiver, who, after careful instruction, continues with the child's treatment exercises at home.

The CME manual is well organized and begins methodically with the assessment protocol. Accurate assessment is the basis of CME therapy, and allows the examiner to quantify the delay in the child's development. After scoring the child's motor ability, a treatment plan can be tailored, and the child's progress can be followed objectively using the same assessment protocol. The manual provides help for elaborate and complicated problems, but also allows for simplicity in dealing with less complex situations.

CME therapy offers us a window into the exciting future of management of the complex range of children's developmental disabilities. As such, this manual is essential reading for all therapists who work with motor delayed infants and children and a valuable resource for pediatricians, pediatric neurologists, pediatric physiatrists, pediatric orthopedists, and all professionals interested in this subject.

Dr. Azriel Benaroya *, New York, July 2007.

INTRODUCCION

When a child is affected by developmental motor delay, our role as physical therapists is to provide the parents an efficient approach for motor therapy to promote the recovery of the child's motor functions.

There exists a wide spectrum of information in the infant rehabilitation field about different approaches of motor intervention, which among them share a common goal: to improve developmental motor function. However, the difference between the various physical therapy approaches stems from the theoretical frame work supporting each particular type of intervention. These methods should be in the memory of any contemporary physical therapist and other professionals working in the pediatric rehabilitation field. It is part of our responsibility as therapists to be informed and to be able to answer the questions that parents pose related to their child's particular situation.

Furthermore, it is very important to explain to the parents, who place their confidence and hope of their child's recovery in our hands, the exact aims and means of the physical therapy approach that we intend to use on their child.

CME therapy proposed an original assessment protocol in which the automatic motor reactions within the first 16 months of age are the core of the evaluation; in addition to muscle-skeletal conditions. Subsequently, with the direct participation of the parents, two to three absent motor responses are chosen as short term goals which must be achieved within the first eight weeks of therapy.

The parents are also advised that CME does not possess "any scientific proof" supporting the efficacy of the approach compared to non therapy intervention, to conventional therapy or to any other existing motor therapy approach. However, there is ample evidence of the remarkable progress made by children treated with CME therapy from early ages or older than 6 or 10 years old, compared to children treated by other therapies or methods.

Ramon's quote: "In 35 years of continous practice, the lack of scientific proof regarding CME therapy results, has never been an obstacle for the parents to put their children in my hands. Only the solid progress experienced by the children treated with CME, has spread the reputation of this approach".

The basic assumption of Cuevas Medek Exercise therapy, is that the child's antigravity motor functions are triggered from the moment of birth and reach their peak with the independent gait around the average age of 14 months. Early walkers made reach their peak at 10 months and late normal at 14-16 months, even as late as 18 months. Motor control stems from genetic encoded information induced to be manifested as a consequence of the aging process. The aging process results in maturity and maturity allows the child to take gradual control of posture and movements.

In different moments of life, a wide variety of noxious situations can affect the brain, which is the depositary of the codes for motor evolution. Consecuently there will be a negative effect on the maturity process which ceases to be active. As result the child stops evolving. When damage has occurred the aging process pushes the child's evolution ahead, however, the natural aging process is insufficient to lead the child to motor independence. Intervention of a high quality motor therapy is imperative to lead child's motor evolution.

The CME therapy theoretical frame work is based on one premise: ***the brain's recovery potential can produce better responses, if the child is gradually exposed to experience motor-postural challenges***. It is the core idea of CME therapy: "to provoke the central nervous system to produce new functional reactions related to antigravity postural control, gradually providing the least amount possible of external support".

The rationale behind this postulate, is that the laws of standard developmental process does not apply for developmentally delayed children. The special child can not be expected to evolve in the same sequence as average children. When the brain is affected, the child can only express the "aging maturity" in a random way and shaped with the dominant muscle tone abnormality.

As long as the aging maturity process (or recovery potential) does exist, we can expect significant progress, particularly if the therapy begins between the first four to six months of age.

CME therapy proposes a trial period of eight weeks of daily treatment in order to prove that the therapy program can help the child to progress. The main goal for these 8 weeks of treatment is to achieve at least three of the goals established at the CME assessment. By using these three goals as a criteria, parents will have a concrete method to avaluate the effectiveness of the CME approach for their child. ***CME is the only physical therapy approach committed to this standards of effectiveness.***

This manual explains the history of CME, the origin of the principles, the context in which it began developing, the particular characteristics of the approach, the assessment format and treatement protocol. There are 99 exercises displayed, chosen from approximately 600 that compose the current total amount of exercises. Each exercise represents a particular biomechanical challenge for the child.

Each exercise demands an active response from the child and always the choice of the exercise is directly related to the child's reaction potential. The "art" portion of CME therapy depends on the hability of the CME practitioner to choose and apply the optimal sequence of exercises during the therapy session, in order to "provoke" new spontaneous postural-functional reactions. The "science" portion of CME resides in the new responses emerging from the immature brain.

CME DEFINITION

CME, the acronym for Cuevas Medek Exercise, is a physical therapy approach for infants suffering abnormal developmental motor evolution caused by a known or unknown non-degenerative syndrome affecting the Central Nervous System.

This therapy can be applied to children from the age of 3 months until they achieve and control independent walking. Because the CME practitioner needs to expose the infant to the influence of gravity, by providing progressive distal support, the use of this therapy may be limited by the child's size and weight.

This approach was created and developed by Ramon Cuevas, Chilean physical therapist, during the early seventies in Caracas, Venezuela.

The fundamental assumption of CME is that developmentally compromised infants need to reinforce their natural recovery potential. This underlying property of the Central Nervous System continues to propel the developmental process even after the sequelae are established.

The natural reaction of the "recovery potential" alone, cannot lead the compromised child out of the abnormal motor developmental situation. Therefore it is crucial to begin motor therapy at the same time the first signs of delayed motor function appear.

The final level of motor independence achieved by the child will depend on the following:

1) The early recognition of the motor developmental abnormality.
2) The existence of brain's recovery potential.
3) The timely application of a reliable approach of motor therapy.
4) The constant execution of appropiate exercise until achieve total motor control.

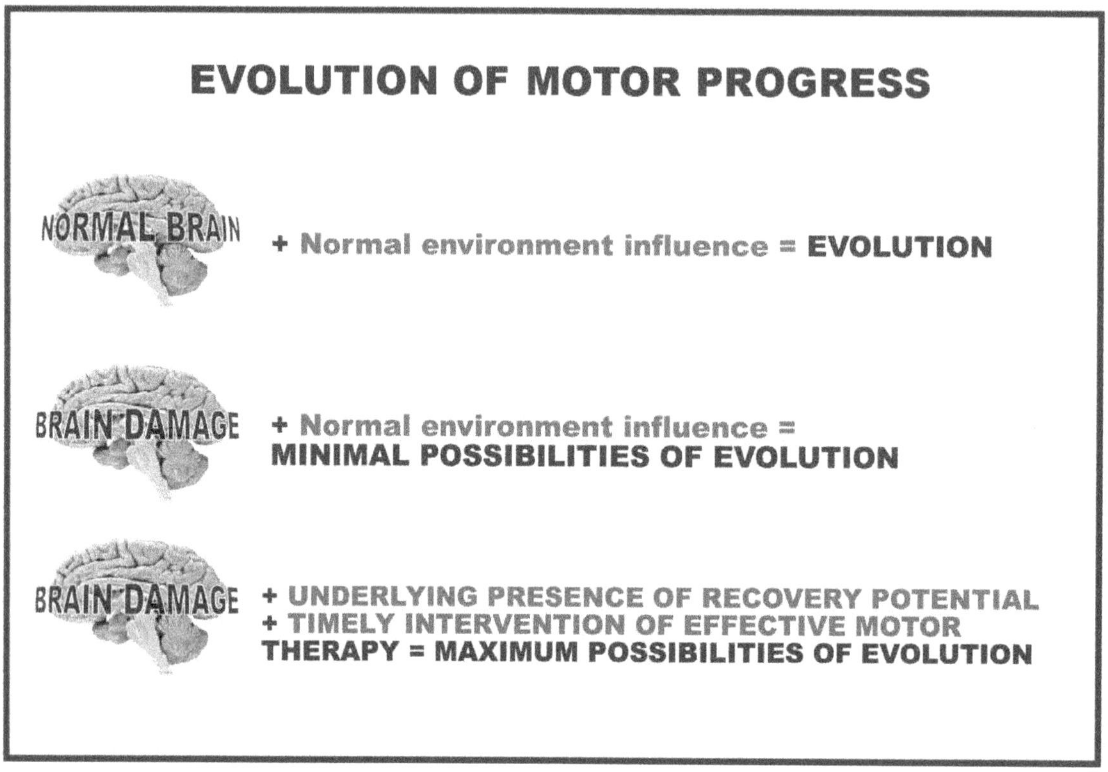

Figure 1, Evolution of motor development.

Ample data and literature are available in the infant rehabilitation field regarding the theoretical basis of motor control, models of how motor evolution occurs, and the relationship between neuro-sciences and developmental evolution, etc. Excellent examples include the books "Motor Control by A. Shumway & Woollacott, and Neuro-Developmental Treatment Approach by J. Howle". Furthermore, there are several web sites that provide resource information on these topics. One very useful site is the National Institute for Neurological Disorder and Stroke (www.ninds.com). The NINDS site presents and explains several topics related to child development.

For this reason the reader will not find any chapters dedicated to the theory about the developmental evolution of children, since this subject is treated in a peerless way by the referred sources.

From the CME point of view, it is imperative to mention Cerebral Palsy (CP) when analyzing the causes of infant developmental problems. There are three issues related to CP according to the NINDS website, which are very pertinent to discuss:

According to NINDS:
"70 percent of cerebral palsy cases correspond to unknown causes".

It is very important to be able to recognize the motor developmental abnormality on behalf of this huge amount of undiagnosed children in a timely fashion, in order to begin to provide motor therapy help as soon as possible. The early detection can be done according to CME postulates and assessment procedures, between 4 to 6 months of age, with high levels of certainty. Thrirty percent of CP cases are affected in their developmental evolution by known causes during the pre, peri and post natal period. The causes therefore are already known by the parents and medical team surrounding the infants, and in theory this 30 percent of children should receive motor treatment in a timely manner.

For the remaining 70 percent, the situation is more difficult. The absence of identified causes for developmental abnormalities places this group at a disadvantage for receiving timely motor stimulation, large because the immediate attention is focused on the quest for diagnosis. As a result, the motor therapy intervention is delayed while the search for a cause is ongoing. Regardless of the cause of the problem, there are no valid reasons to postpone the therapy intervention for children even if know history of neurological damage or high risk condition exists, if early signs of delayed postural-motor functions begin to appear.

It is fair to agree that on some occasions, infants with a normal medical history can show signs of motor developmental delay in the first 4 to 6 months of age, and subsequently, without any therapy, they reach the motor functions and can walk freely and normally between 14 to 18 months of age. In fact, those children never stop progressing slowly, but steadily in their motor functions. When the child is really affected, however the motor functions expected for any given age are absent and stay absent for the next month, and for the next, and for the next.

In no way should this be taken as an alarmist attitude, since as soon as the lack of function is detected, it will only be necessary to pay attention and monitor the altered function more closely for a period of four weeks. If the altered function does not improve in 4 weeks, then it becomes imperative to begin immediately with a good motor stimulation program. The reason to react quickly during the first six months of age is because during this period, all the basic motor functions already exist in a sort of organized master plan manifested mainly by the ability of the child to take the weight of his body on his legs when his feet are placed in contact with a solid surface.

According to NINDS:

"Early signs of CP usually appear before 3 years of age, and the diagnosis is made on average, around 3 years of age".

Cerebral Palsy is a syndrome characterized by the interruption of the normal motor developmental evolution. Other developmental areas can be affected as well, so the child can show different degrees of mental impairment which means that the affected child does not posses the motor control that an average infant displays at each level of chronological age. For instance, at 4 months of age children maintain their head in full vertical control, at 6 months of age children can keep a self supported sitting position, at 8 to 9 months of age or earlier, infants can move from a lying to a sitting position, at 10 months or earlier they can go to a standing position by holding the bars of the crib, and the standard average infant begins to walk independently around the year of age.

These milestones can show some normal statistical variation of less than or more than 2 months, and remain in the range of normality. Furthermore, the least worrisome of the delayed motor functions is independent walking, so long as the progression of head control, trunk control and standing control is manifested into the average range.

The possibility of developmental motor delay beginning with ***"the absence of free walking"***, after showing normal head, trunk and standing control at the age of 10 months, is extremely rare. Except of course, when the brain damage is caused by head trauma, stroke, acute anoxia, acute infection (encephalitis, meningitis), or other unexpected causes affecting a child after having had a normal evolution.

Consequently, motor developmental alterations caused by Cerebral Palsy or another syndrome, cannot begin to appear in an abrupt manner in infants developing normally, unless an acute episode presents itself which is impossible to be ignored by the parents,

According to NINDS:

"Parents are often the first to suspect that their infant is not developing motor skills normally".

Parents do not want to discover problems with their children, but ***parental instinct*** always works. They are able to recognize the early signals of developmental delay threatening the evolution of the child. These signals are commonly depicted by mothers within the first six months of life. Many times the response received from the involved care-giver is: ***"don't worry, not all children develop at the same rate, he/she is just a little lazy, lets give it time and you will see how everything will soon be fine***". But, what will happen if the mother's concerns are for real?

After receiving this kind of advice the mother and the family enjoy a temporary relief of their concerns, so, they wait and hope that time will push ahead the child's maturity process. Unfortunately, in most cases, the mother's concerns were the early signals of a real problem, and for the oncoming months, the developmental delay symptoms will become obvious.

At this point, a few months have elapsed since the first delay signals were noticed, and now the gap between expected functions and chronological age is undeniably bigger than a few months ago.

Now, if the family is lucky, the child will immediately be referred to motor therapy and a wide variety of medical and laboratory tests, (MRI, EEG, CTCS, MRS, genetic screening, muscle biopsy, etc.) will be indicated in order to investigate the cause of the problem. Other children are not so lucky, they are sent to take special lab and medical examinations, prior to beginning treatment with a good motor therapy program. All the while, a few more months of precious recovery time elapses in the quest for diagnosis.

In one way or another, the fact is that the child has lost the opportunity to begin to bridge the gap between age and motor function at the optimal time, when the mother noticed the first abnormal signals from the child, and became worried and concerned.

Who was the winner on this bet?. Certainly not the infant, nor the family. Then, the next question is, what will be the greatest danger for the child if the mother's concerns are recognized and the child is immediately sent to motor therapy ?

There is no risk involved for the baby to be timely referred to an experienced physical or occupation therapist specialized in early motor treatment. The professionals will perform a complete developmental motor assessment, and will discuss the results of the evaluation with the parents, and if a delayed condition is corroborated, the child will immediately begin to receive the necessary motor treatment with the experienced and competent therapist.

The final conclusion is that the concepts of preventing major developmental delay and preventing further lack of maturity by not neglecting the necessary and timely motor stimulation should be important concerns for professionals working in the pediatric field.

CME BEGINNINGS

Ramon Cuevas, CME Therapy creator, received his PT degree, Summa Cum Laude, from the University of Chile, in November 1969 and moved to Venezuela in April 1970. There, he came in contact with the world of infant rehabilitation, as the person responsible for re-organizing the Physical Therapy Department of the private institute AVEPANE (Venezuelan Association of Parent and Friends of Exceptional Children). The population treated in the Institute, was mainly children between the ages of 13 and 8 years affected by Cerebral Palsy or Down's syndrome.

At the time, the concepts of ***"high risk"*** condition and ***"early stimulation"*** were not well known. Furthermore, the common policy in pediatric rehabilitation for treating a child diagnosed with "cerebral palsy" was based on two conditions:

1. ***The child should be older than 2 years of age to begin physical therapy treatment.***
2. ***The child should be able to "cooperate" with the rehabilitation process otherwise, they were not eligible for treatment.***

After two years of full time work, using an eclectic combination of exercises from the conventional approaches of physical therapy, the author recognized two important issues: First, they were not producing noticeable motor improvement in the children as result of the intervention; rather they were only helping the children to "repeat" the motor functions which they had already acquired, and second, he understood that to demand "cooperation" and "motivation" from the delayed child, was unfair and illusory requisites to perform the exercise program.

Trying to find some answers to this self-critical analysis, Ramon developed the premise that at early ages, the only hope to improve the evolution of the motor delayed child is to ***"provoke"*** the activation of the underlying recovery potential through the use of ***"specific dynamic exercises"***.

At the beginning of 1972, Ramon established the basic goal of his professional endeavor as a physical therapist: to create a method of exercises that would ***"provoke absent developmental postural-functional reactions regardless of the symptoms of damage and the level of awareness of the child"***. This premise was based on the observed biomechanical fact that in the sitting and standing exercises, ***"as more distal support is applied by the therapist, stronger functional antigravity reactions are provoked"***.

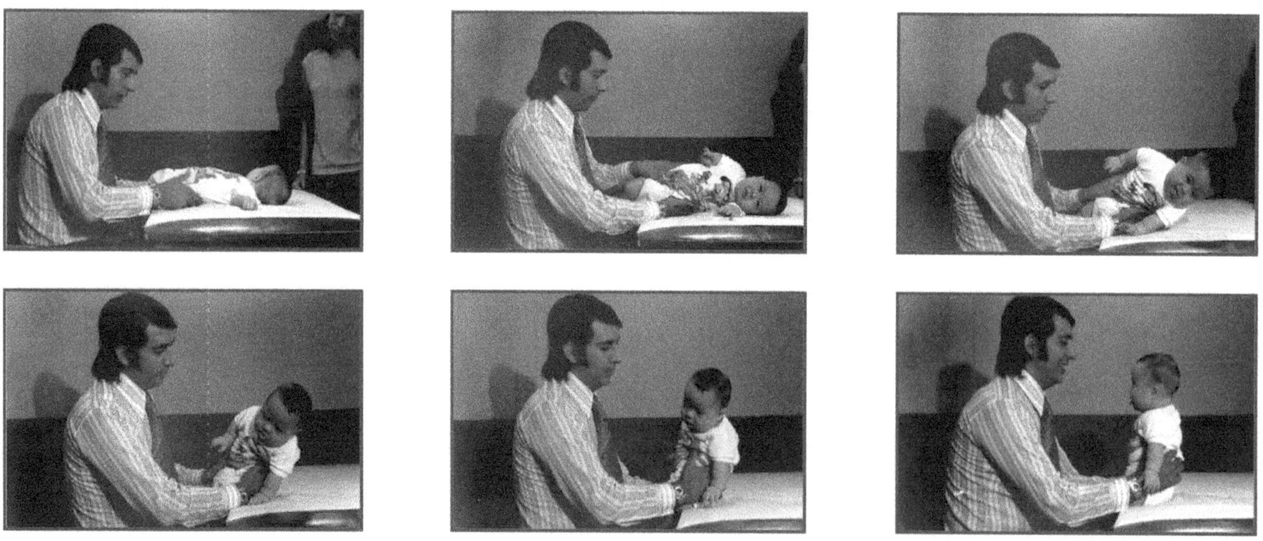

The sequence of pictures shows Ramon, doing the CME called Supine sitting by hips to a Down syndrome child of 6 ½ months old, in Caracas in 1972. The purpose of the exercises was to create an alternative way to provoke the sitting reaction instead of the classical maneuver of pulling the child up to sitting by holding both hands and / or pulling the child up to sitting by one hand while the other hand is contained against the table. These two maneuvers were part of the standard repertoire of conventional therapy to help the child to sit up. The new idea was to provoke the sitting reaction by distal input without indicating to the child's brain that the sitting begins by hand traction.

At that particular time, research teams in USA, were developing totally new concepts of premature baby handling. These new concepts, which were brought to the general public attention, were: "high risk condition" and "early stimulation", which means early developmental delay diagnosis follow by early stimulation intervention.

Nevertheless, the original North American early stimulation program did not promote physical therapy intervention as a priority in the program, which was centered totally on sensory stimulation input, by means of tactile, visual and auditory stimuli. At the forefront of intervention were Occupational Therapists, Nurses, and Psychologists.

However, in Venezuela, it was a different story. In 1973, the AVEPANE Institute opened the first "high risk and early stimulation" program, whose main goal was to incorporate physical therapy treatment as a basic component of the early stimulation intervention. Thanks to this policy, Ramon got the perfect opportunity to begin to use his original physical therapy approach on the dozens of developmental motor delayed children who started to flow in to the new early stimulation program open in the AVEPANE Institute. Ramon had the opportunity to work with a variety of diagnosis such as Down syndrome, unspecific hypotonic condition, high risk premature babies, motor retardation due to metabolic disorders, and cerebral palsy. All the cases were babies at an average of 3 to 4 months old at the initial motor assessment. This allowed Ramon to carefully apply and develop new exercises, under the constant supervision and support of the medical team and the parents.

Positive results began to appear after one year of closely monitored effort creating and applying the new exercises. The children began to improve in a clearly superior manner compared with the motor evolution obtained before with the same type of population using the conventional physical therapy. Based on the positive results of this experience through the undeniable better motor performance of the treated children during the year 1972, the AVEPANE early stimulation program was expanded in the beginning of 1973, from having two rooms and only one PT and one OT, to a new location with a staff of 12 professionals working full time.

Furthermore, the success of the new therapy was of such magnitude that by the end of 1974 it was necessary to create a new center in order to cope with the growing population of new cases coming to the early motor stimulation program. This new Institute was named INVEDIN (Spanish acronym that stands for Instituto Venezolano para el Desarrollo Integral del Niño), it was founded by the parents and the professional staff of the early stimulation program of the AVEPANE institute.

For the ensuing years, Ramon worked part time in the INVEDIN Institute. He began to invest long hours in his private practice and registered his physical therapy approach under the acronym MEDEM (Metodo Dinamico de Estimulacion Motora). In 1980, the method was renamed and registered as MEDEK, changing the last letter "M" (motoric) for the "K" (Kinetic), and also the first logo was created (fig. 2). This name and logo lasted until the year 1999 in which the author decided to incorporate his last name (Cuevas) into the new registered acronym and logo CME (fig. 3), which stands for Cuevas Medek Exercises.

Figure 2, first logo (1980).

Figure 3, new logo (1999).

Ramon has dedicated his professional life without interruption to treat patients and teaching CME therapy to other committed and dedicated physical, and occupational therapists working in the pediatric rehabilitation field. He still continues today to dedicate the best of his energy and expertise to treat patients and sharing his knowledge with those open minded therapist willing to learn CME. In the year 2000 the current Cuevas-Medek-Exercise Continuing Education Program was established, which is the program dedicated to promote the CME therapy concept by means of offering seminars and formal courses, in order to provide first quality information to those therapists interested in knowing from the original source.

CME therapy is in constant evolution and the creation of new exercises is still in progress due to the imperative need to provide the optimal exercise for each child in particular. For instance, the original assessment format has been modified along the last 5 years reducing the number of items in order to eliminate the non significant ones and to simplify the initial evaluation, without loosing accuracy in the final results which reveal the relation between the chronological age and the motor-functional level.

CME ASSESSMENT FORMAT

The current version of the CME evaluation format is a reviewed version of the original test developed in 1972. During 4 years (1976 to 1980), the assessment was under experimental scrutiny as a curricular component of the research project ARYET (Spanish acronym stand for High Risk and Early Stimulation). The ARYET project was sponsored by the Neumann Foundation and the Clinic Hospital of the Central University of Venezuela. The result of the project was published in Caracas in 1982 in a 3 volume edition, plus one volume specially dedicated to the CME assessment protocol.

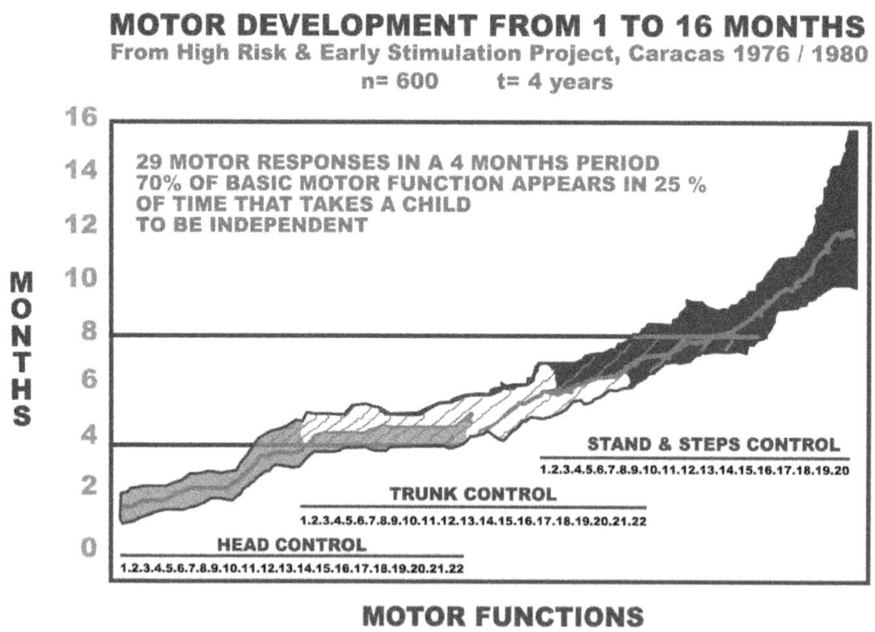

Figure 4, basic motor development STUDY / ARYET PROJECT.

The above diagram show us an ascending ribbon from left to right which illustrates the motor progress of a group of children evaluated 3 times per week during their first 16 months of life. This experimental project was developed in a time period of 3 years. The illustration is based in the simple "cloud of dots" technique, consisting in marking in millimeter paper the age at which each child responds positively to each motor item from 6 weeks old until he/she achieves independent walking. The curve shows major changes in width towards the last items of walking control that suggest a normal variability in the age for controlling free walking. The gray portion of the curve represents the head control items, with some children performing the final items of head control around 5 months old. The white section of the ribbon illustrates the trunk control responses with some children responding the first items before 5 months and the latest responses of independent sitting were obtained around 7 to 8 months of age. The upper blue portion of the ribbon represents the standing and walking responses. Here is a noticeable gap between the precocious walkers and the tardiest ones.

It is important to remark the overlapping of motor responses belonging to different conventional areas of motor control, and especially the narrow temporal span in which the final functions of head control, the total responses of trunk control and the 50% of standing control responses appear. All this happens from 4 to 8 months of age; in consequence this short time period concentrates the largest percentage on motor functions within the first 16 months of age. These findings suggest that the optimum moment for identifying a motor developmental abnormality regarding only motoric symptoms, should be between 4 to 8 months of age.

The first version of the motor test was composed of 86 items to explore the motor developmental evolution from 3 to 16 months of age, which is the standard period within which the child becomes able to control primary motor-postural functions such as sitting, standing and independent walking. The present the test has gradually been simplified to 41 items to measure motor development from the neonatal period until reaching independent walking.

C.M.E CUEVAS MEDEK EXERCISE

Note: *Only CME certified practitioners are allowed to apply this test.*

NAME ____ DATE___/_____/_____

DATE OF BIRTH____________________ AGE_______ DIAGNOSIS____________________

____ BY______________________________

Score 3 = Complete response / **Score 2** = Incomplete response / **Score 1** = Minimal response / **Score** = 0 No response

ITEMS OR MOTOR FUNCTION:	**Assessment**	**Control**
1. NECK EXTENSION IN PRONE POSITION		
2. HEAD VERTICAL TO FORWARD SLOPE		
3. HEAD VERTICAL TO BACKWARD SLOPE		
4. HEAD VERTICAL TO RIGHT SIDE SLOPE		
5. HEAD VERTICAL TO LEFT SIDE SLOPE		
6. NECK EXTENSION IN PRONE / HEAD OFF THE TABLE		
7. NECK LEFT SIDE FLEXION / HEAD OFF THE TABLE		
8. NECK RIGHT SIDE FLEXION / HEAD OFF THE TABLE		
9. NECK FLEXION IN SUPINE / HEAD OFF THE TABLE		
10. ARMS PROTECTIVE REACTION		
11. LEFT ARM SIDEWAYS PROTECTIVE REACTION		
12.RIGHT ARM SIDEWAYS PROTECTIVE REACTION		
13. ROLLING FROM PRONE TO SUPINE POSITION		
14. ROLLING FROM SUPINE TO PRONE POSITION		
15. SELF SUPPORTED SITTING POSTURE		
16. UNSUPPORTED SITTING POSTURE		
17. TRUNK VERTICAL TO BACKWARD SLOPE		
18. TRUNK VERTICAL TO FORWARD SLOPE		
19.TRUNK VERTICAL TO RIGHT SIDE SLOPE		
20. TRUNK VERTICAL TO LEFT SIDE SLOPE		
21. TRUNK EXTENSION		
22. WEIGHT BEARING		
23.INDEPENDENT SITTING FROM PRONE POSITION		
24. INDEPENDENT SITTING FROM SUPINE POSITION		

ITEMS OR MOTOR FUNCTION:	Assessment	Control
25. QUADRUPED LOCOMOTION		
26. LEANING / GRASPING STANDING POSTURE		
27. STANDING BY ONE HORIZONTAL FOREARM		
28.STANDING BY HANDS ON WALL		
29. KNEELING TO STANDING LEAN / GRASP		
30.KNEELING TO STANDING BY HANDS ON WALL		
31. LEANING / GRASPING SIDEWAYS STEPS		
32. STEPS BY ONE HAND / ARM HORIZONTAL		
33. STEPS BY HANDS ON WALL		
34. FREE STANDING POSTURE		
35. FREE STEPS		
36. FREE STANDING FROM THE FLOOR		
37. FREE BACKWARD STEPS		
38. FREE STEP OVER SMALL OBSTACLE		
39. FREE STEPS OUT OF 10 CM. BOX		
40. FREE STEPS DOWN FROM 10 CM. BOX		
41. FREE STEPS UP ON 10 CM. BOX		

Total score__________Divided by 7.6 = Functional age ______months versus chronological age of______months. FunctionalGap____________months.

CME Assumption: Low score in items 9 at 4 months of age, 15 at 7 months of age, and 26 at 10 months of age can be predictors of developmental motor impairment.

RECORD

PRENATAL______________________________

PERINATAL______________________________

POSTNATAL______________________________

MUSCLETONE______________________________

REFLEXES______________________________

SKELETAL CONDITION______________________________

INITIAL CME PROGRAM______________________________

Items 1 to 9 are indicators of head control maturity; items 10 to 12 are for measuring scapular control (arm control); items 13 and 14 are for measuring the primary horizontal mobility; items 15 to 25 are for measuring trunk control.

Item 22 in this group is for measuring the automatic reaction of ***weight bearing*** which is normally acquired since 4 months of age. Full trunk extension is a requisite to achieving weight bearing. Items 26 to 41 cover the standing and walking reactions expected in normal average children at 16 months of age.

The responses to each item can be qualified within a range of four possibilities: complete response equivalent to a 3, incomplete response equal to 2, minimal response equal to 1 and absence of a response equal to 0.

In the CME assessment, there are three significant items because the absence of response grade 3 or 2 on any of those functions is a clear indicator of motor delay condition when they are applied to the adequate chronological age. The item 9: ***"neck flexion-supine-head out of table"*** should be qualified as a 3 or a 2 for the age of four months, less than a 2 indicates motor delayed condition.

For item 15: ***"self supported sitting posture"*** a result of less than 2 at the age of 7 months will indicate a high possibility of motor delayed condition, and for item 26: ***"standing posture by horizontal or seized support"*** getting less than 2 at the age of 10 months indicates a high possibility of motor retardation.

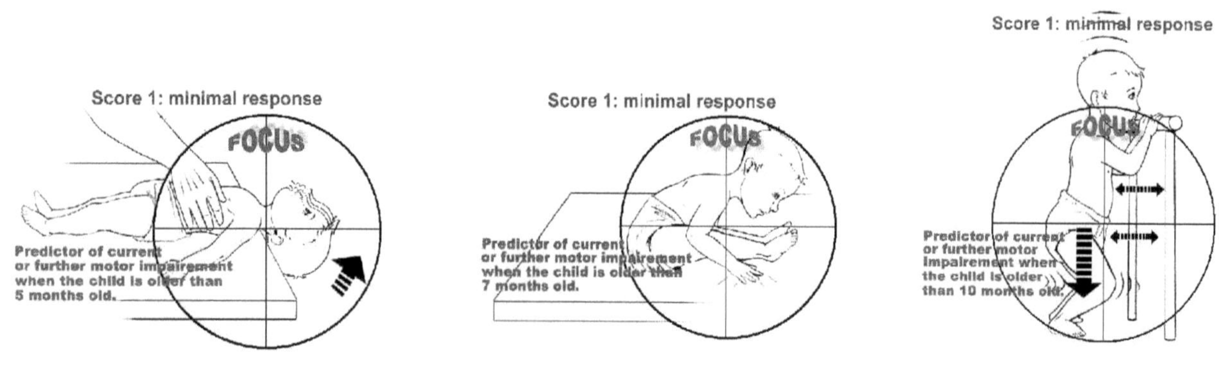

Figure 5, Immature response for the items 9, 15 and 26.

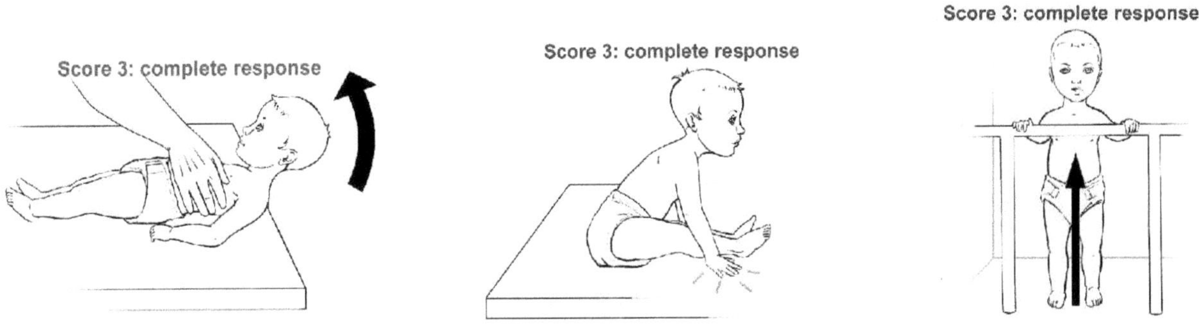

Figure 6, Mature reaction for the items 9, 15 and 26.

The main goal of the CME assessment is to unveil any abnormality of the developmental motor functions provided that the child is tested during the first year and a half of life.

To be able to classify a child as motor developmentally delayed, it is important to acknowledge that the average ages to achieve different motor functions vary from one child to another and by using a broad statement it is possible to say that motor normality can fluctuate within 3 categories:

A: Precocious population, walking around 9 to 11 months of age.
B: Average population, walking around 12 to 14 months.
C. Normal slow population, walking around 15 to 18 months.

It is not significant in being able to walk later than 12 / 14 months, if the head, trunk and standing control functions are aroused at the average time. No motor delayed condition begins at 12 months of age after normal evolution, without a clear episode of crisis of some nature, only progressive muscular diseases can gradually occur after years of normal motor maturity. After administering the assessment, it is important to take into consideration setting up the CME therapy program. The items which have a one and a zero, indicate the motor functions which require immediate attention.

In addition, the CME assessment results make it possible to calculate the current level of motor-postural performance of the child related to the chronological age, which is significant for the prognosis, only during the first 2 years of life. To obtain the motor maturity level, it is necessary to find the sum of the score assigned to each item and divide the total amount by the factor 7.6, and the final figure will represent the age level of motor-functional maturity of the child at the time of the initial evaluation.

CME ASSESSMENT PROTOCOL

All the motor responses are displayed in correlative sequences from 1 to 41, and only the positives scores of 3 (complete response), 2 (incomplete response) and 1 (minimal response) are illustrated, the score of 0 or absence of response is not represented.

ITEM 1: Neck extension in prone position

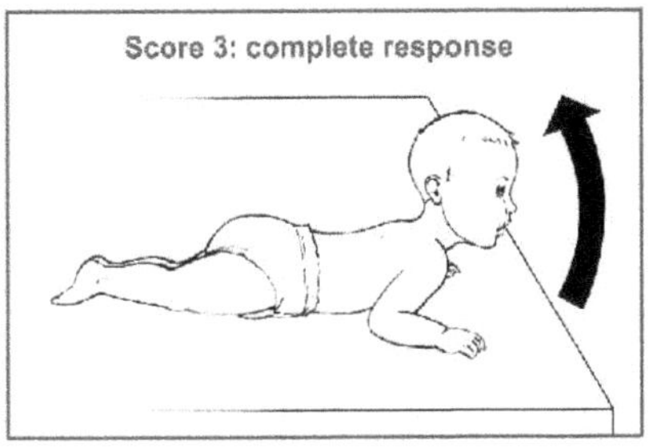

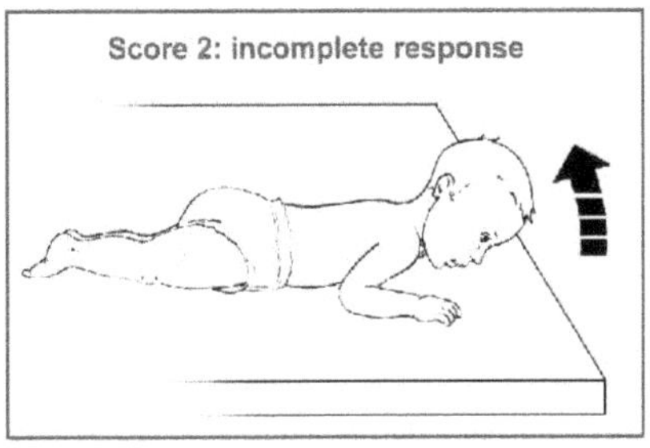

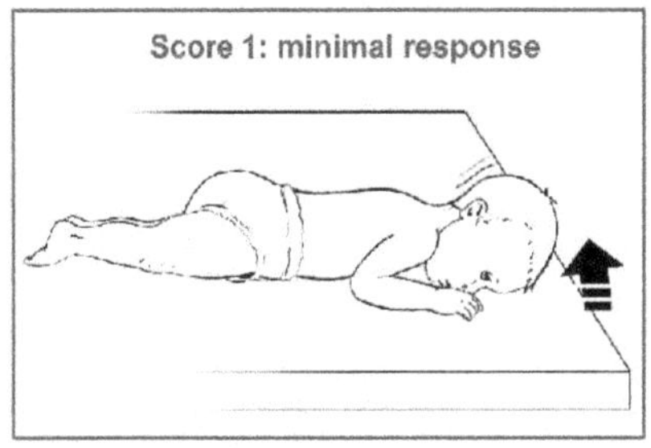

ITEM 2: Head vertical to forward slope

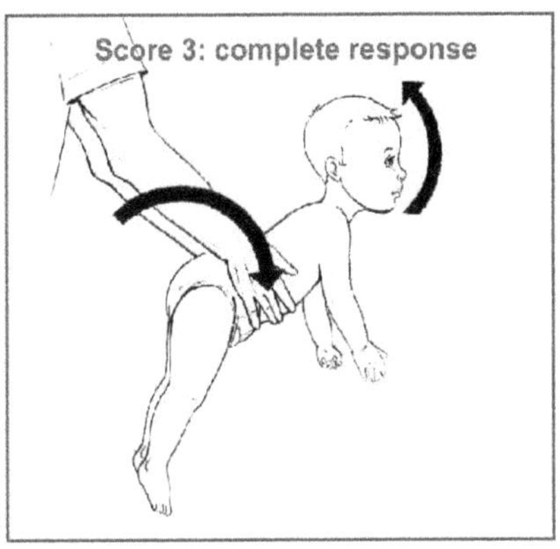

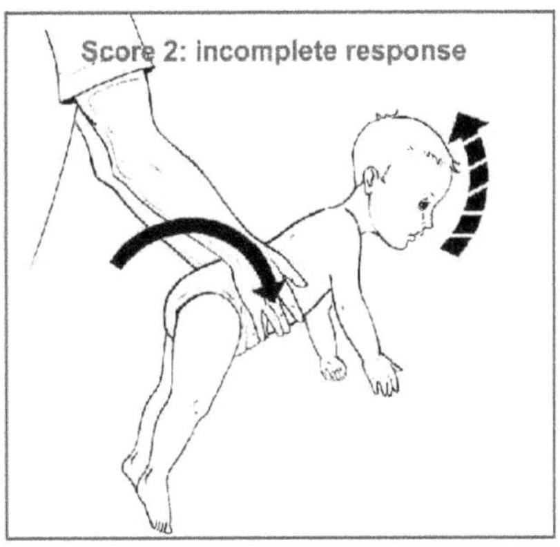

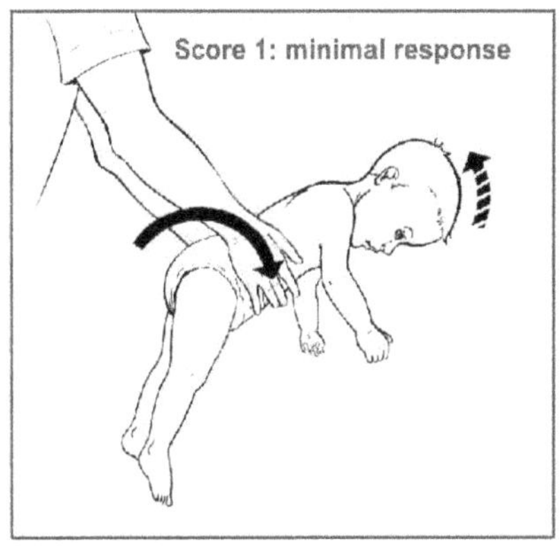

ITEM 3: Head vertical to backward slope

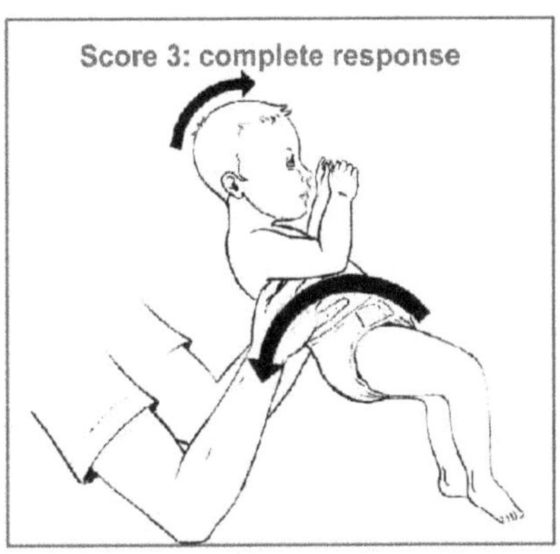

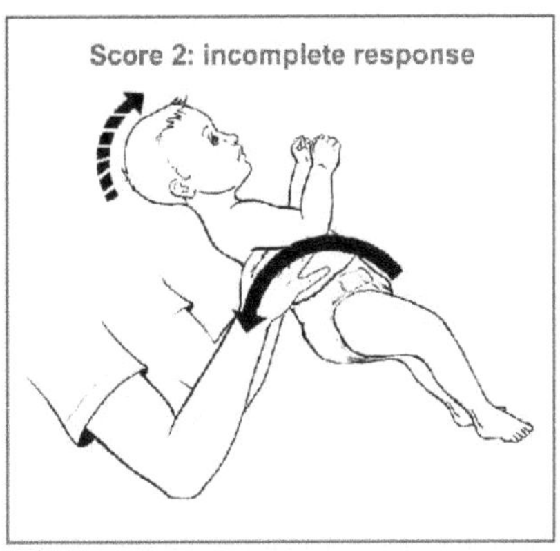

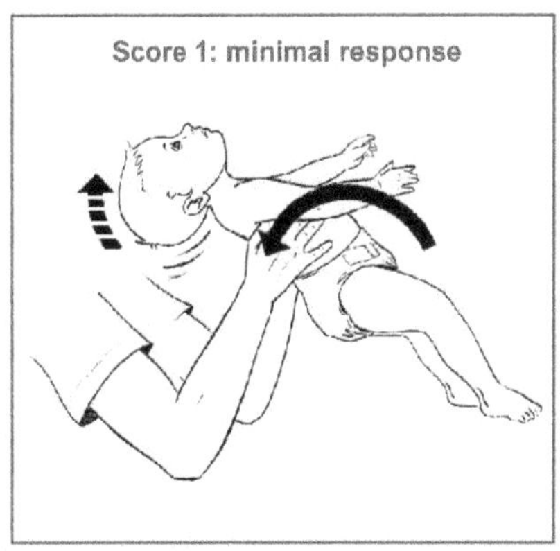

ITEM 4: Head vertical to right side slope

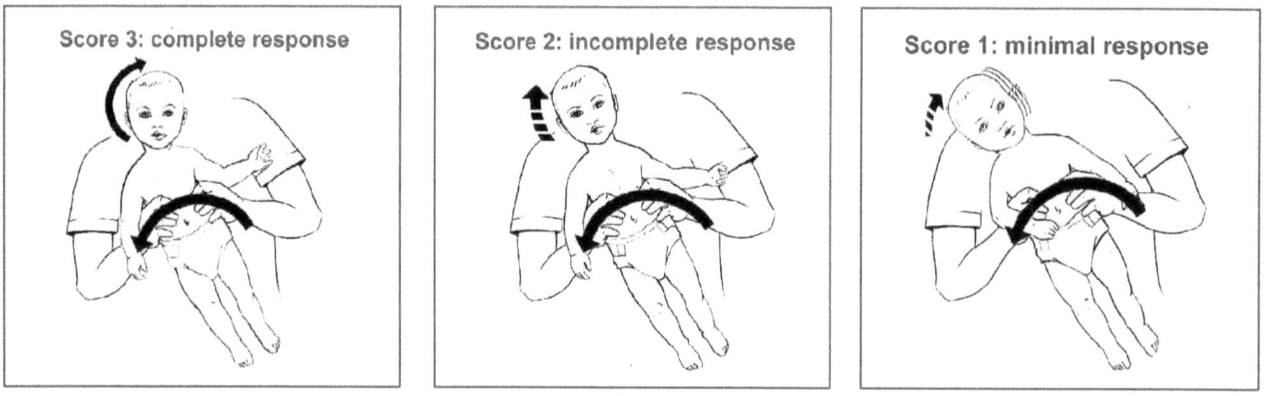

ITEM 5: Head vertical to left side slope

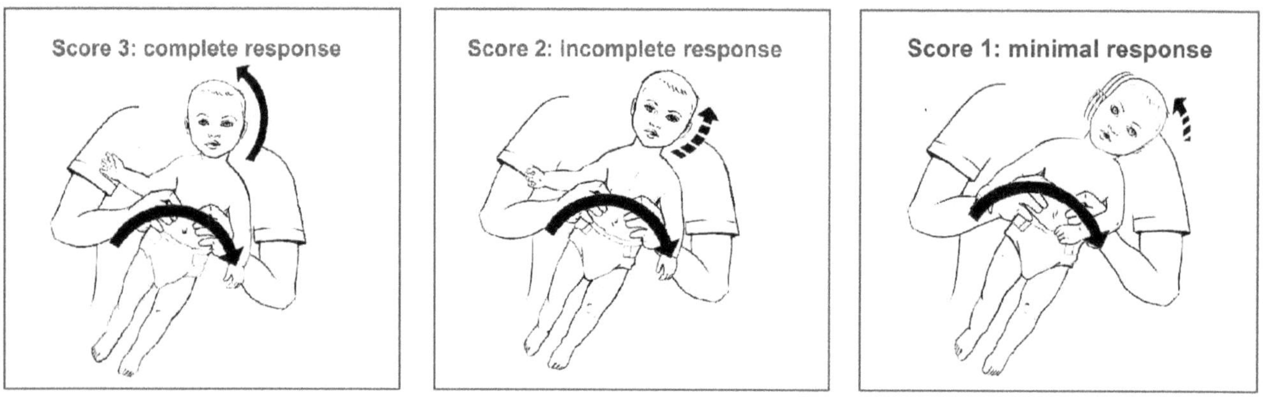

ITEM 6: Neck extension in prone / head off the table

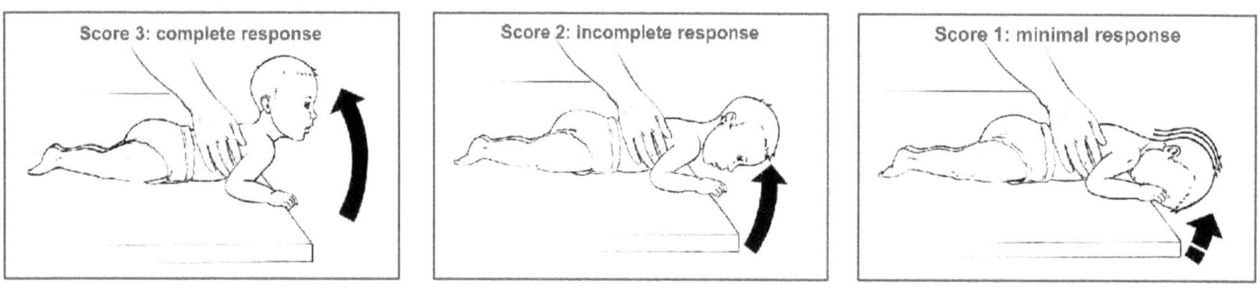

ITEM 7: Neck left side flexion / head off the table

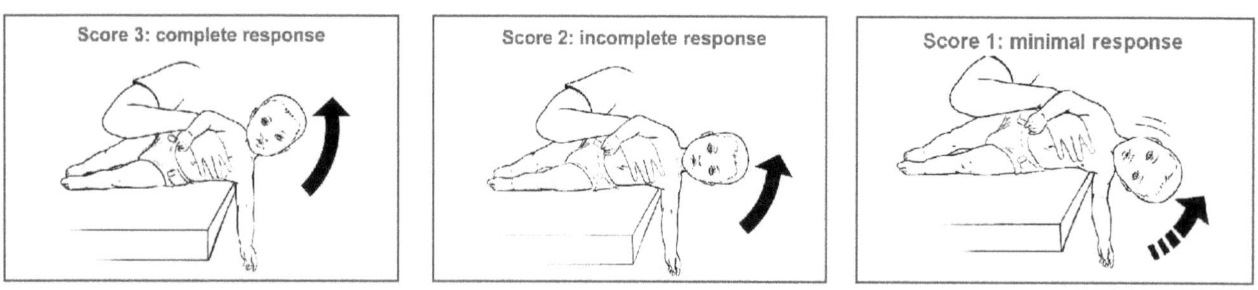

ITEM 8: Neck right side flexion / head off the table

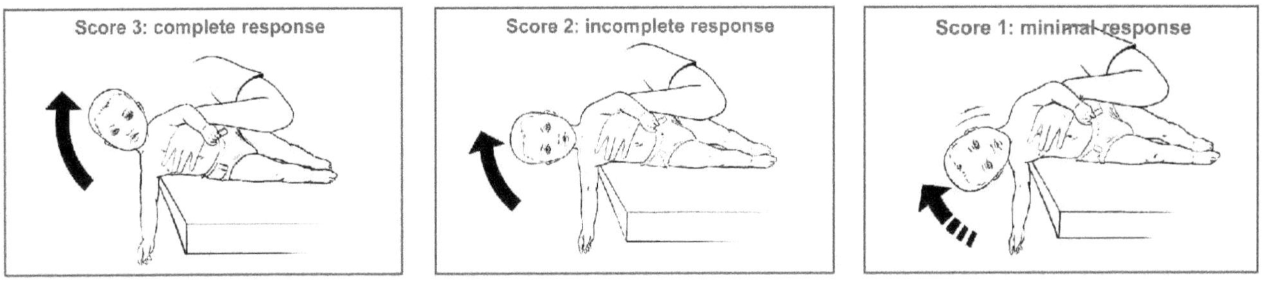

ITEM 9: Neck flexion in supine / head off the table

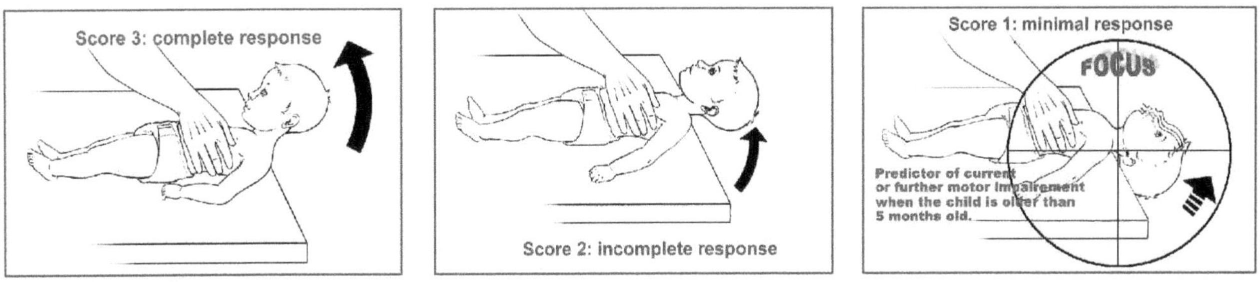

ITEM 10: Arms protective reaction

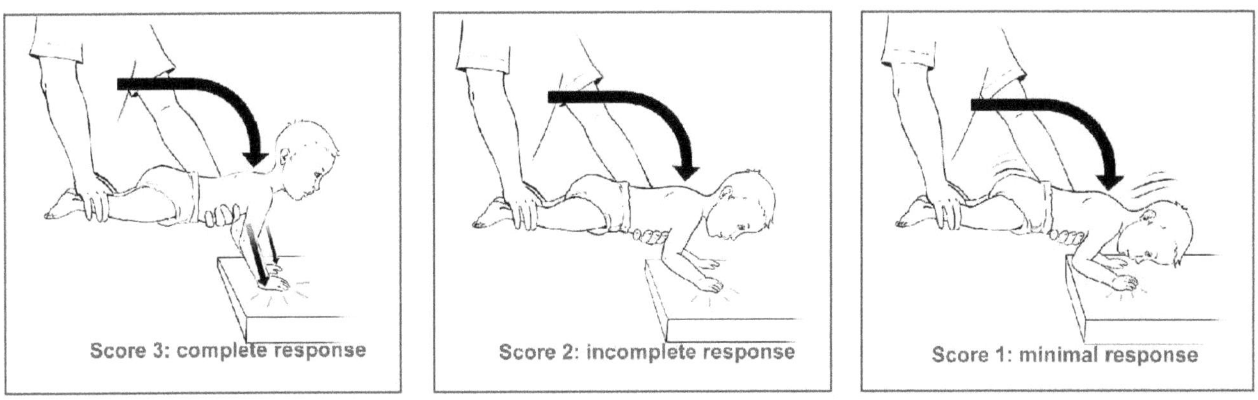

ITEM 11: Left arm sideways protective reaction

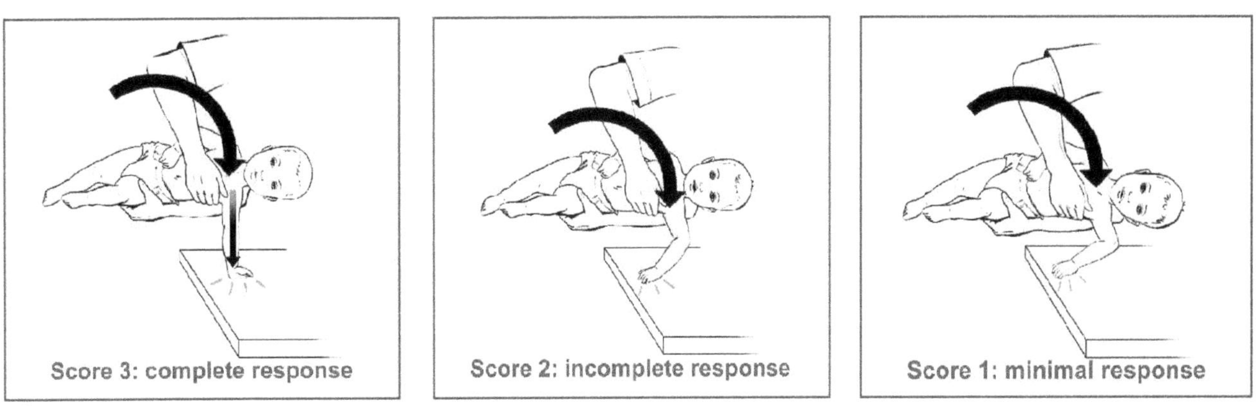

ITEM 12: Right arm sideways protective reaction

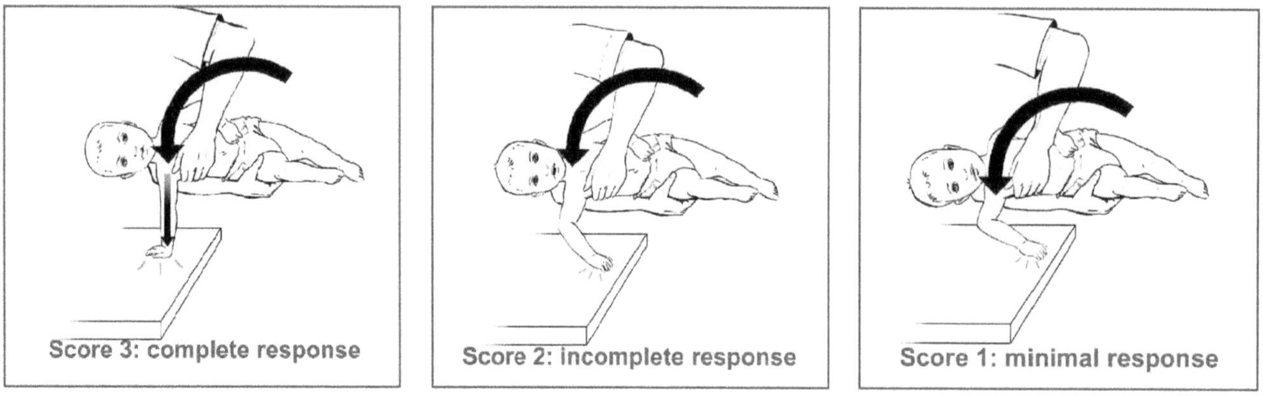

ITEM 13: Rolling from prone to supine position

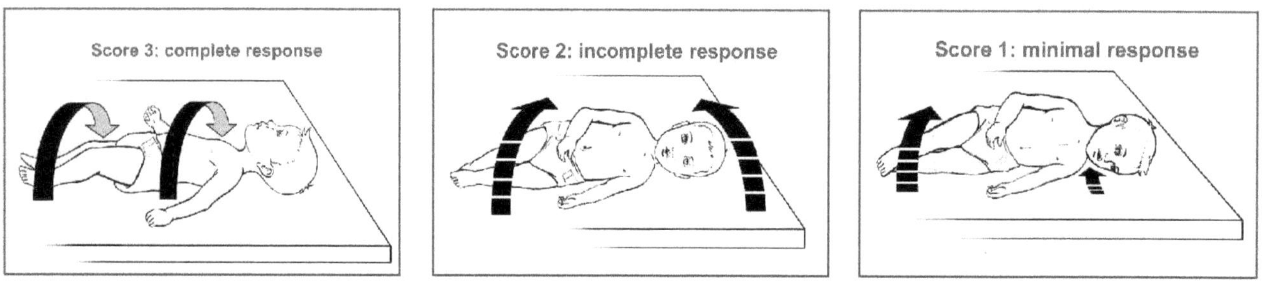

ITEM 14: Rolling from supine to prone position

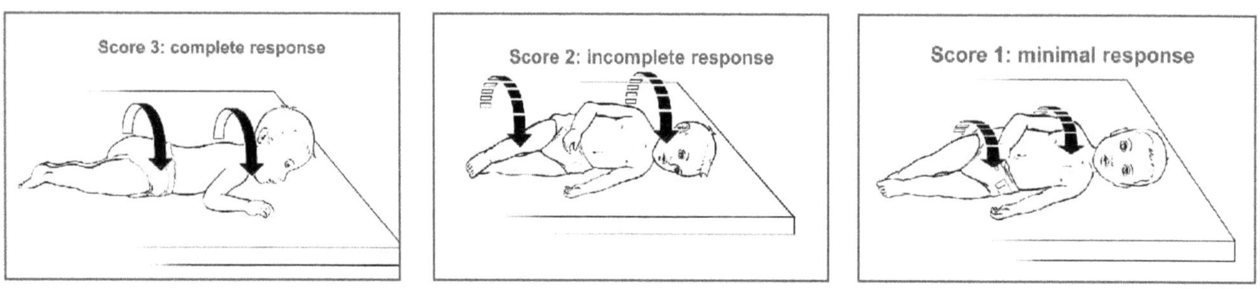

ITEM 15: Self supported sitting posture

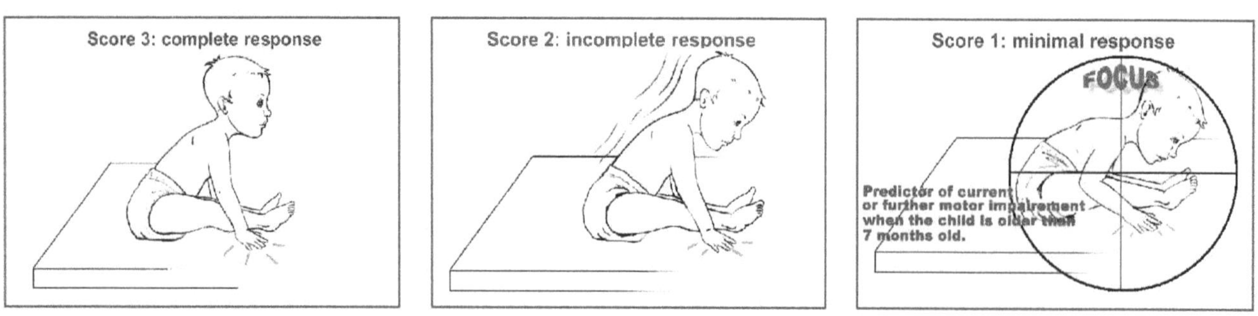

ITEM 16: Unsupported sitting posture

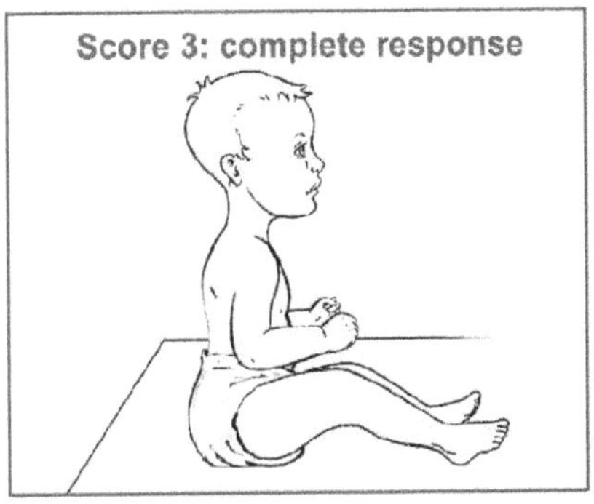

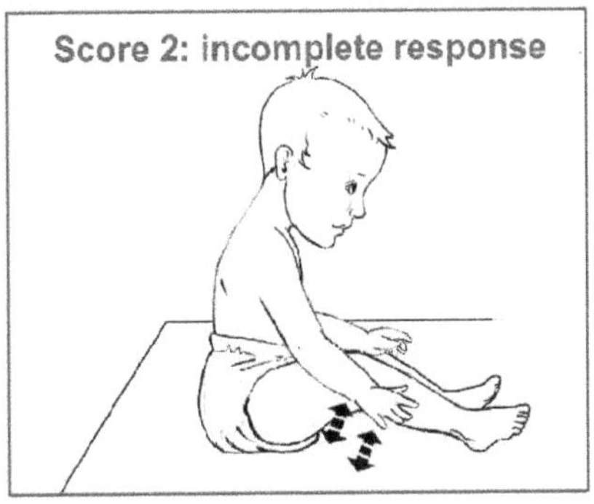

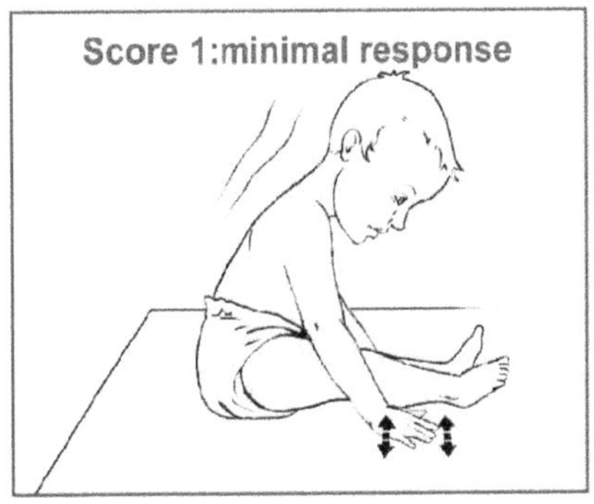

ITEM 17: Trunk vertical to backwards slope

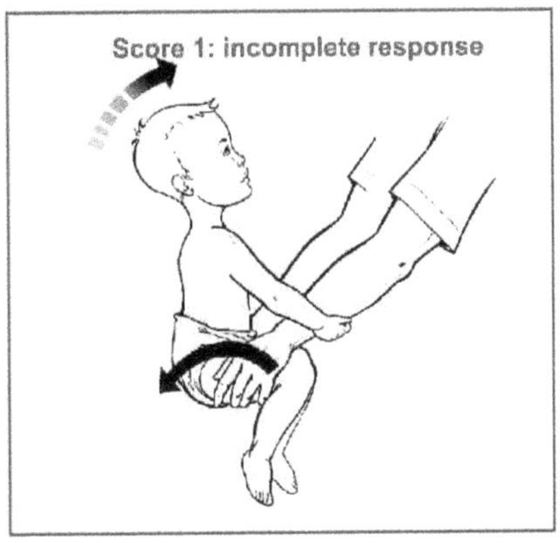

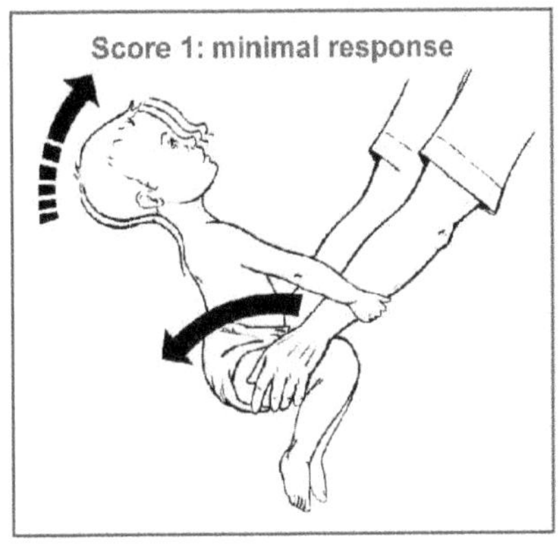

ITEM 18: Trunk vertical to forwards slope

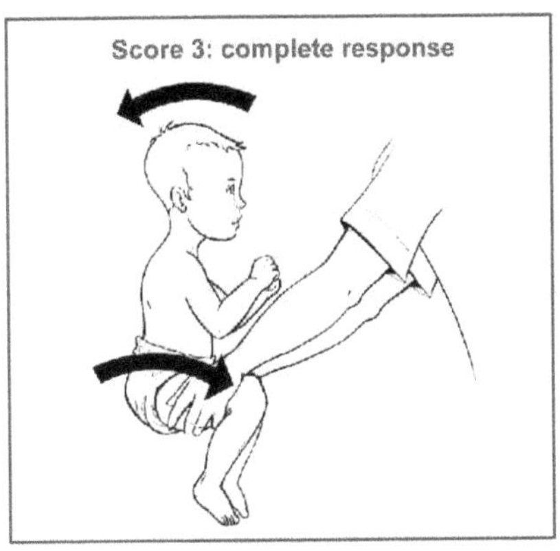

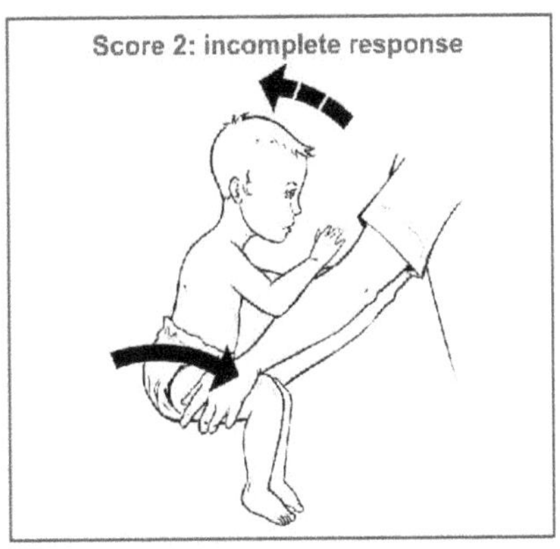

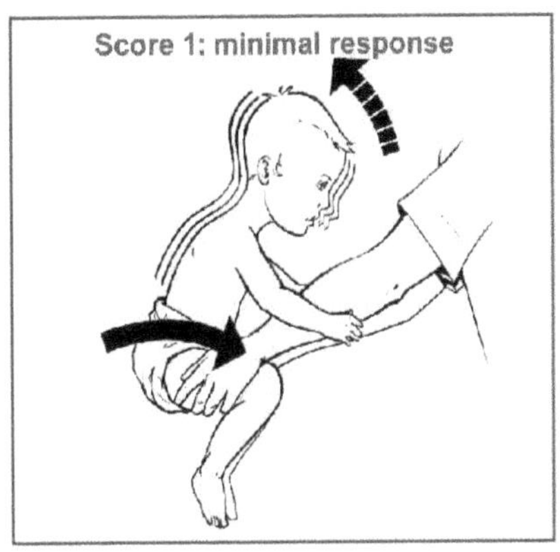

ITEM 19: Trunk vertical to right side slope

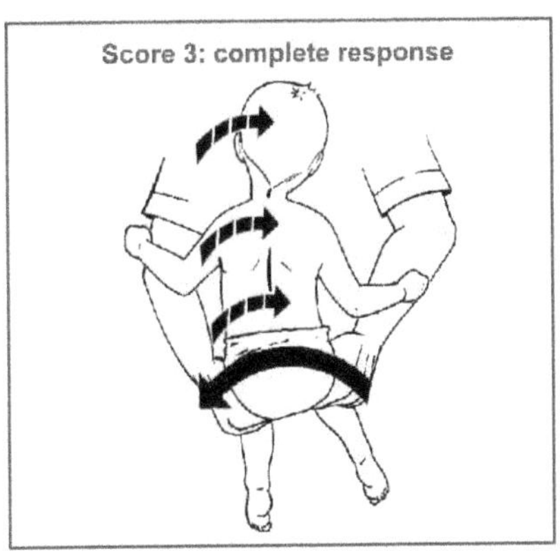

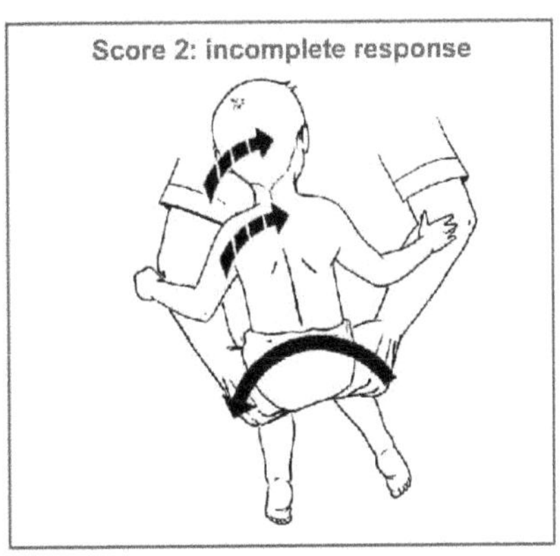

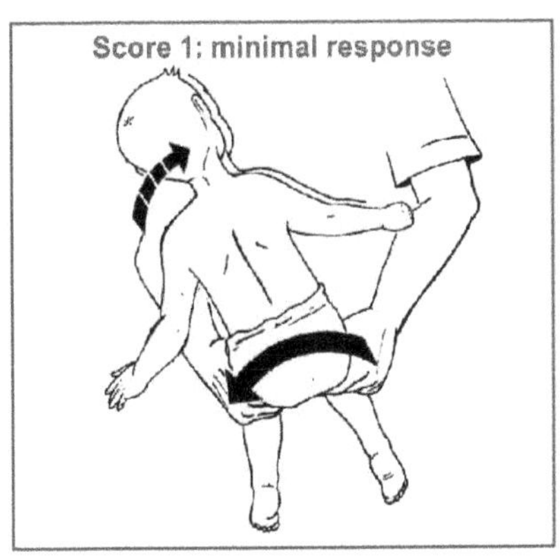

ITEM 20: Trunk vertical to left side slope

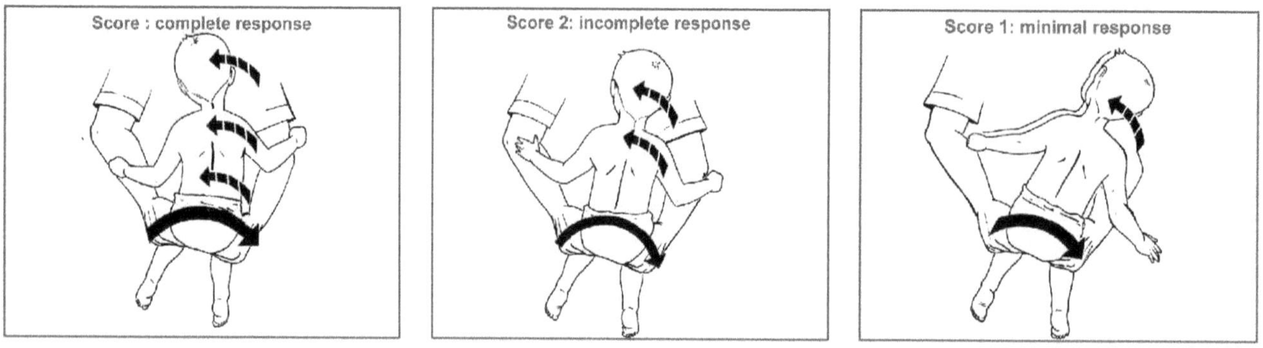

ITEM 21: Trunk extension

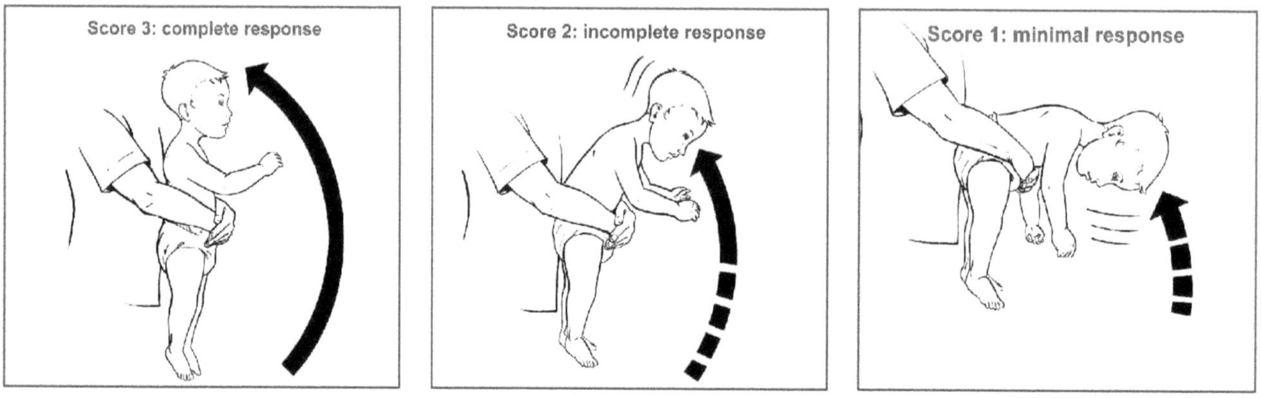

ITEM 22: Weight bearing

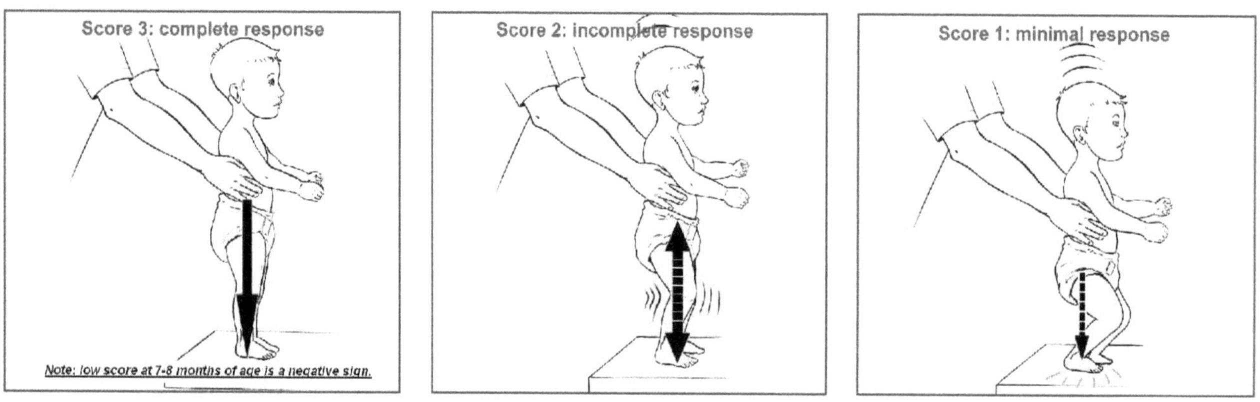

ITEM 23: Independent sitting from prone position

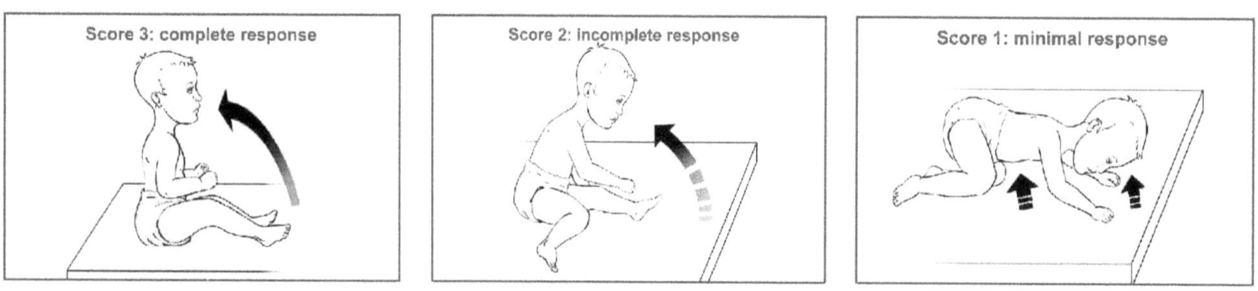

ITEM 24: Independent sitting from supine position

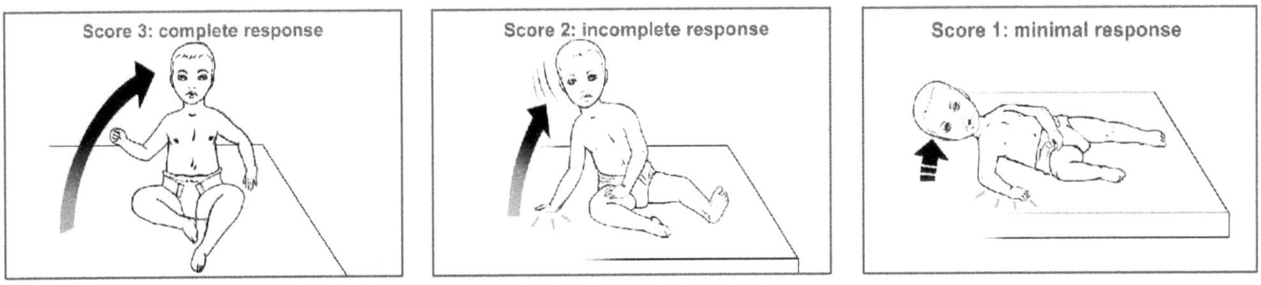

ITEM 25: Quadruped locomotion

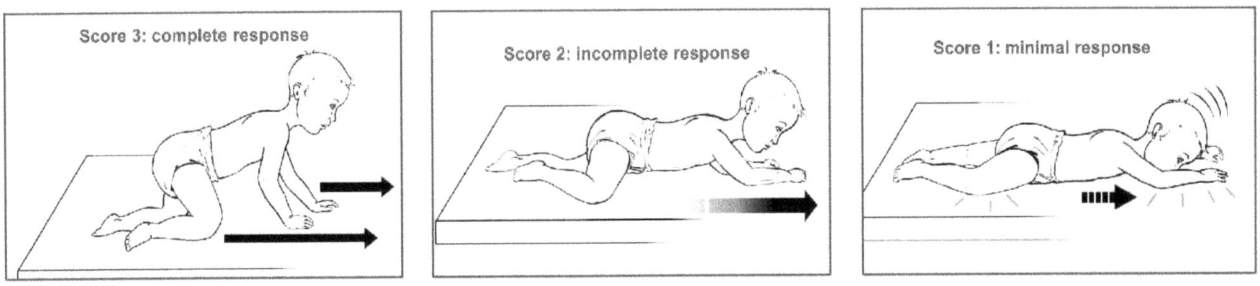

ITEM 26: Leaning / grasping standing posture

ITEM 27: Standing by one horizontal forearm

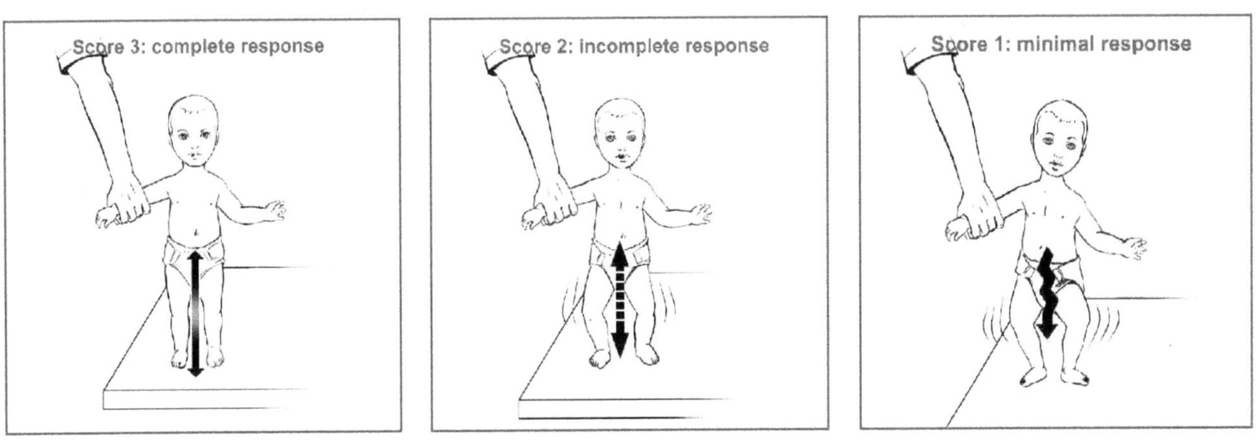

ITEM 28: Standing by hands on wall

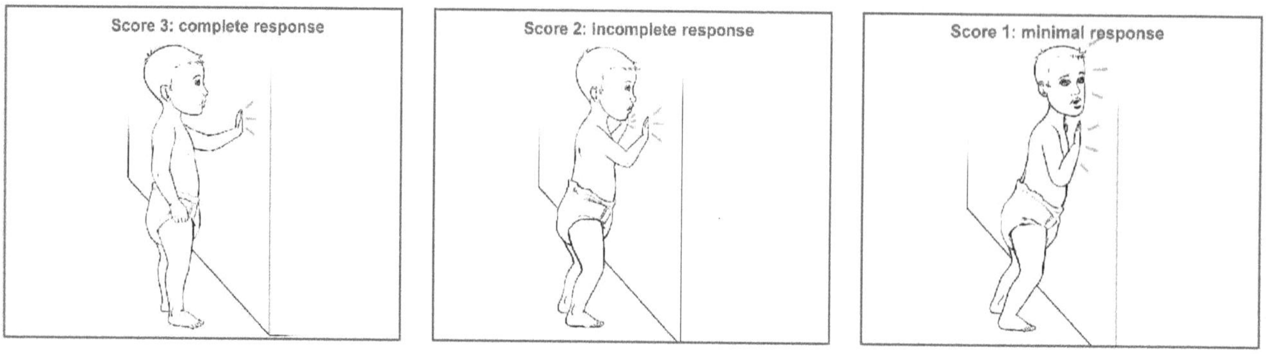

ITEM 29: Kneeling to standing lean / grasp

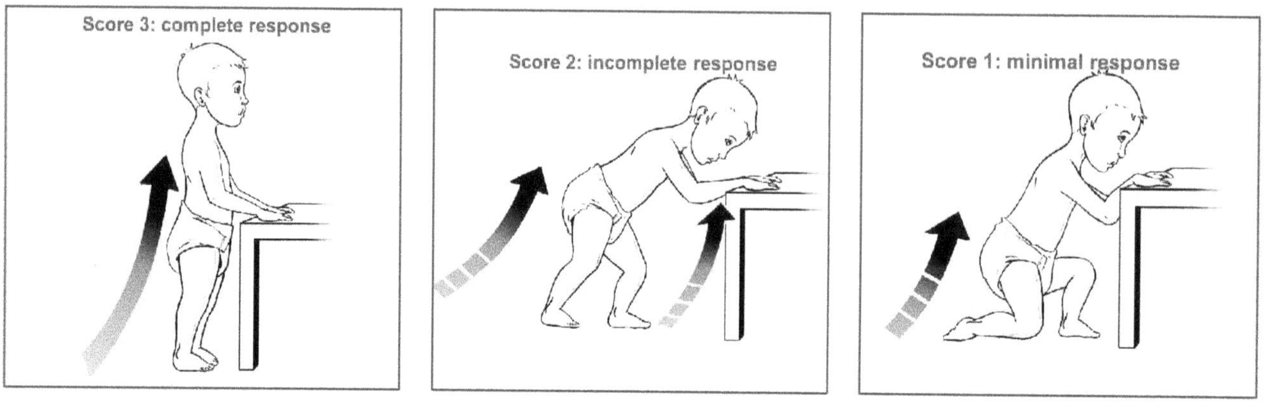

ITEM 30: Kneeling to standing by hands on wall

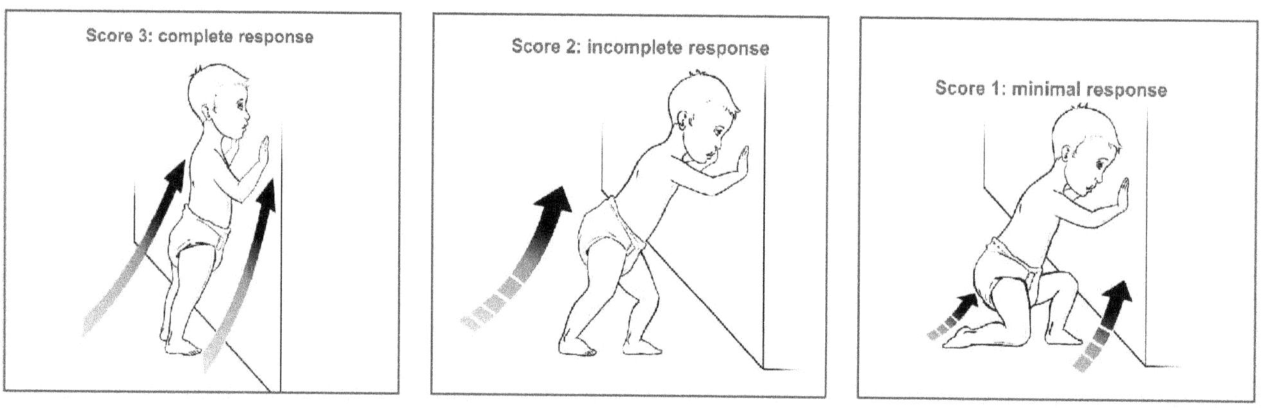

ITEM 31. Leaning / grasping sideways steps

ITEM 32: Steps by one hand / arm horizontal

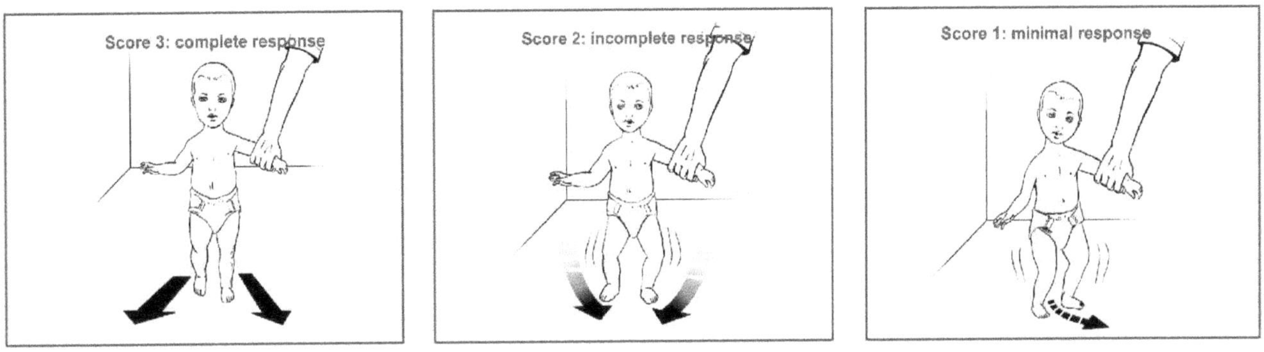

ITEM 33: Steps by hands on wall

ITEM 34: Free standing posture

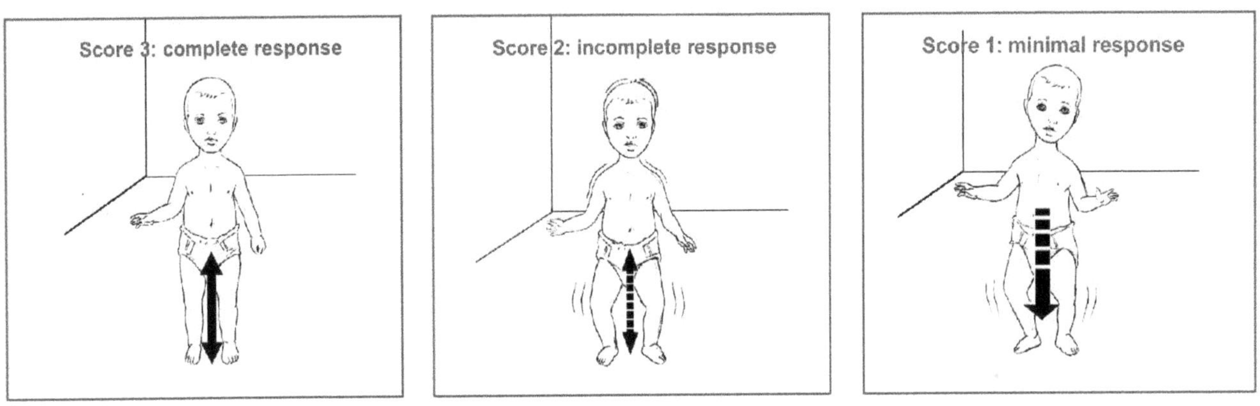

ITEM 35: Free steps

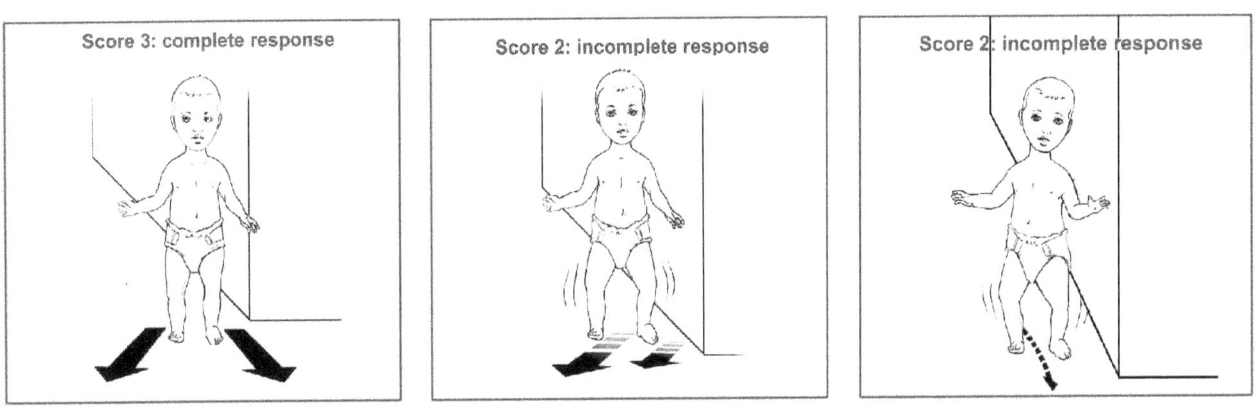

ITEM 36: Free standing from the floor

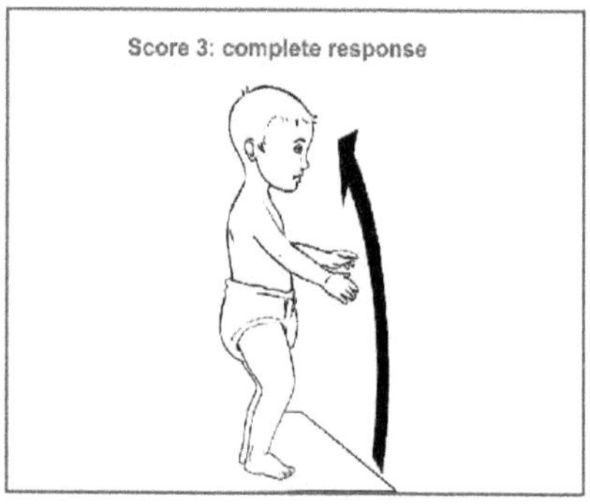
Score 3: complete response

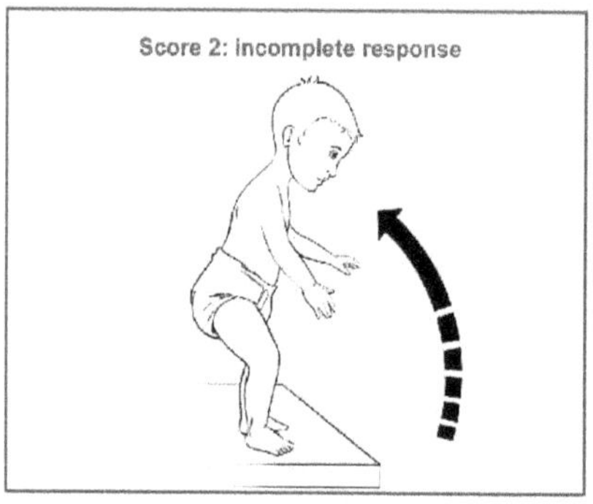
Score 2: incomplete response

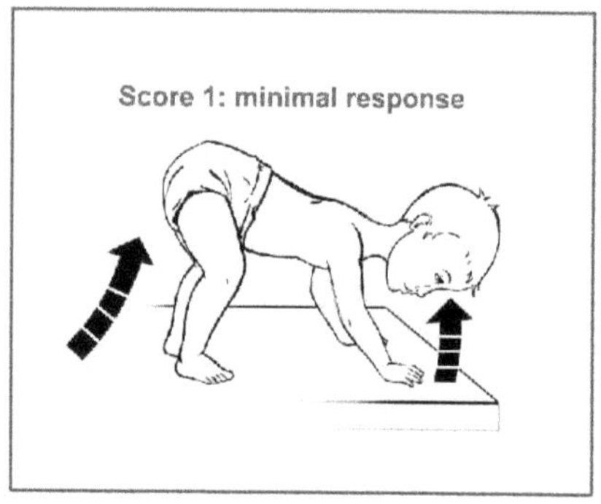
Score 1: minimal response

ITEM 37: Free backwards steps

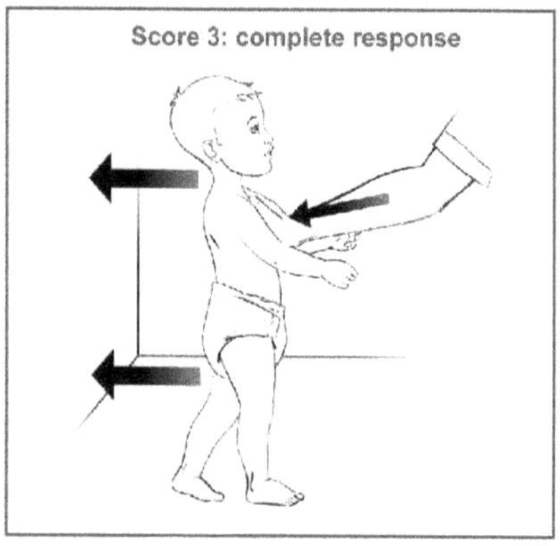
Score 3: complete response

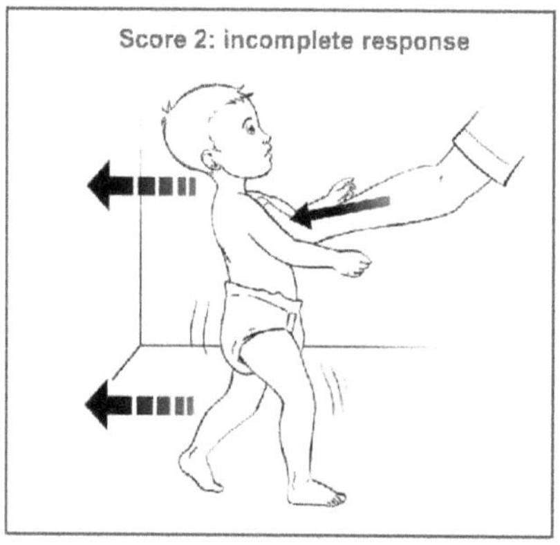
Score 2: incomplete response

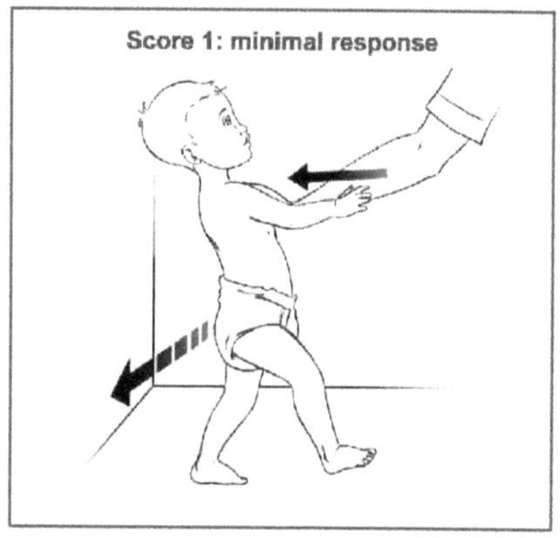
Score 1: minimal response

ITEM 38: Free steps over small obstacle

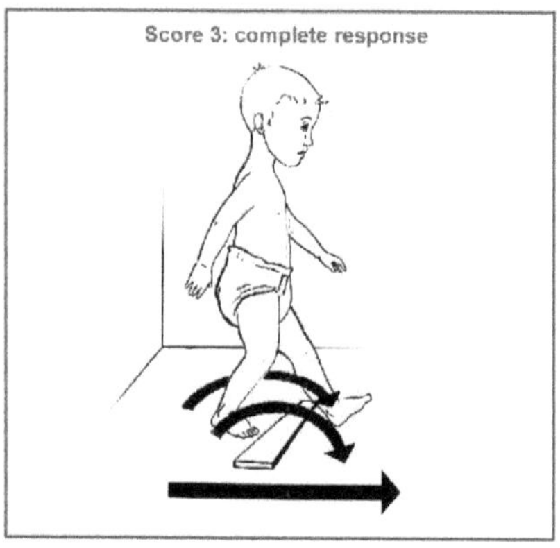
Score 3: complete response

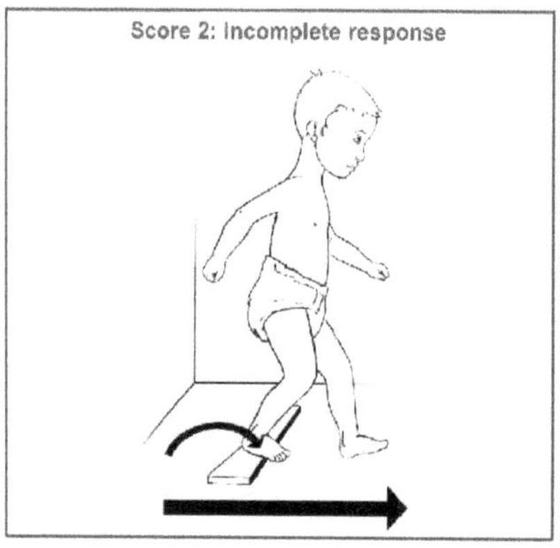
Score 2: incomplete response

Score 1: minimal response

ITEM 39: Free steps out of 10 cm. box

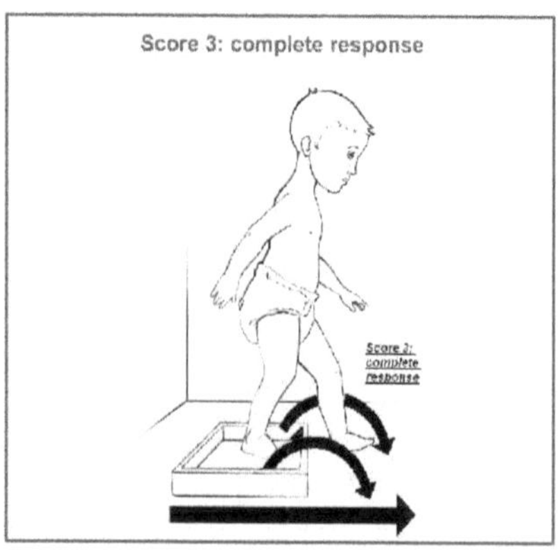

Score 3: complete response

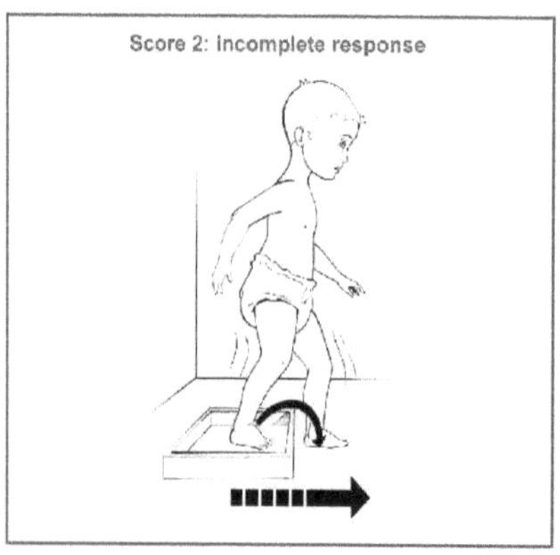
Score 2: incomplete response

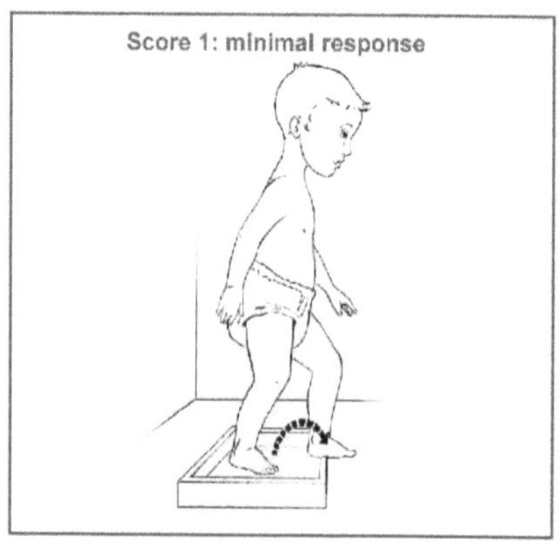
Score 1: minimal response

ITEM 40: Free steps down from 10 cm. box

Score 3: complete response

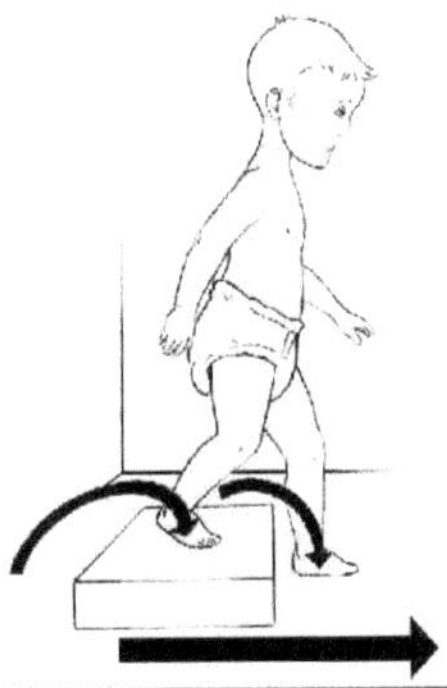

Score 2: incomplete response

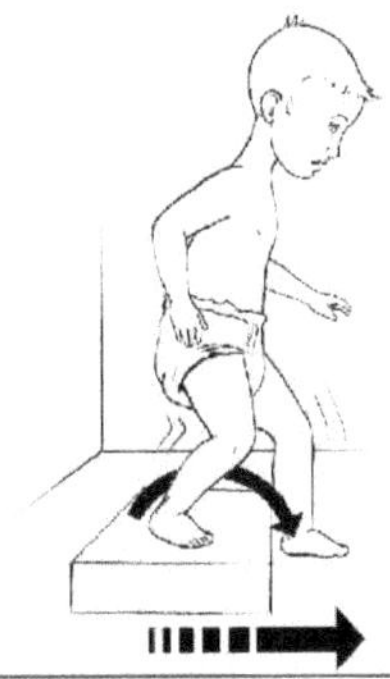

Score 1: minimal response

ITEM 41: Free step up on 10 cm. box

Score 3: complete response

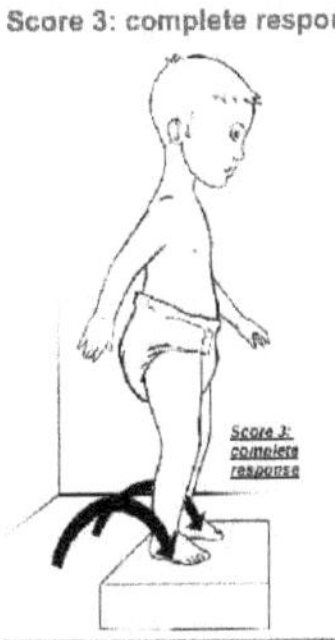

Score 2: incomplete response

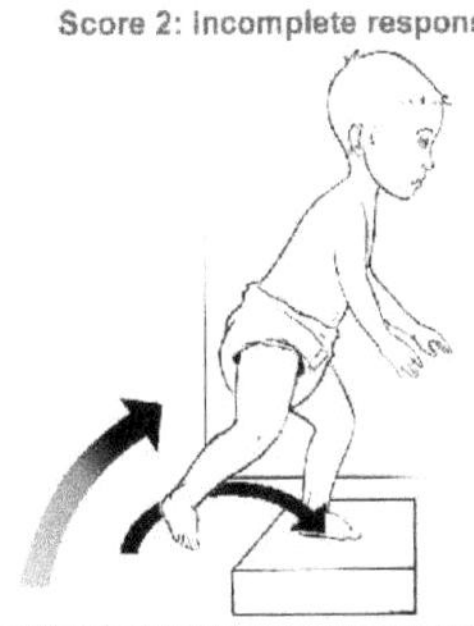

Score 1: minimal response

CME CHARACTERISTICS

1. Provokes the appearance of absent automatic motor functions.

After reviewing the results of the initial assessment, the CME therapist will identify exactly what are the absent motor-postural responses for the chronological age, in order to choose the appropriate exercises for each particular child. The appropriate exercise is the one that induces the child to actively respond with a motor reaction which is not currently under his/her spontaneous control. For instance, for a child 20 months old, only able to take steps using two hands support, the appropriate CME are those providing support below the hips and below the knees. Then, the child can feel the sensation of moving his/her body without hands assistance. In other words, the CME therapist will not reinforce the motor responses that are already under the child's control. If the exercise is not challenging for the child and for the therapist, it is not a good exercise according to CME standards.

2. Child's cooperation and motivation are not requisites in CME.

Children acquire developmental motor control at ***similar chronological ages*** all over the world. The average children gain head control by 4 months of age, sit up by themselves between 7 to 9 months, stand up from kneeling holding on to the crib by 9 to 10 months and walk freely between 12 to 16 months. This undeniable reality indicated to us, that volitional or voluntary participation of the child in his particular motor evolution is secondary to the manifestation of the genetic-environmental driving force that induces the child to control postures and to move according to the automatic neuro-genetic master plan. The motor functions manifest first in a spontaneous manner, and through this experience, the child becomes conscious of the motor functions and gradually becomes able to control it and to use it. Because of this, CME therapy promotes the use of specific exercises intended to "provoke the appearance of the automatic motor reactions independent of the direct cooperation and conscious participation of the child".

In other words, CME therapists do not depend on any requisite to perform the assessment and to apply the therapy.

This assertion doesn't mean that CME therapists would prefer to work with an uncooperative rather than a cooperative child, it only means that we cannot use the argument: ***"the therapy is not working properly today because the child is uncooperative and not motivated"***. By contrast, according to the CME principle, it will always be the therapist's responsibility to find the appropriate ways to make the child respond to the exercises during the therapy session.

3. *Expose the child to the natural influence of the force of gravity with gradual progression to distal support.*

To expose the body segments to the influence of the force of gravity using the most distal support that the child can safely tolerate. After the child becomes able to react positively to any given exercise, a new one, using more distal support should by tried. This progression takes place in direct proportion to the underlying recovery potential of the child's brain. For instance, in the case of a 12 months old child who doesn't control the standing reaction, the CME therapist will provoke the recovery potential using exercises in which the child is exposed to the standing position supported by one thigh and opposite side of the trunk and gradually will be exposed to the standing posture supported by below hips, then by below knees, later on by ankles and finally by soles. The rational behind this point is that the motor delayed child cannot follow the standard laws of cephalous-caudal progression because of the delayed existing condition, as a result the idea of making the child "experience" the motor postures and functions with less support is logical, to make them feel the taste of motor control over the force of gravity.

4. ***Stretching maneuvers are integrated into the CME.***

To maintain and enhance the mobility of the joints and range of motion, especially for hips, knees and ankles, is an important goal in pediatric physical therapy. In the C.M.E. therapy program, the stretching maneuvers are performed into the functional exercises; in this manner the child's brain will receive better proprioceptive information by the combination of 3 components: range of motion, functional antigravity posture and weight bearing. The exercise called ***"Standing by thighs on lap, ABD-ADD"***, is a good example, because it is possible to observe three stretching points: the knee flexors by locking the knee in extension, the hips adductors by opening the legs apart, and the ankles extensor group by applying solid weight bearing downwards, as illustrated in the figure 7 sequence. In others CME are incorporated stretching maneuvers for fingers, wrists, elbows, shoulders, neck, and trunk.

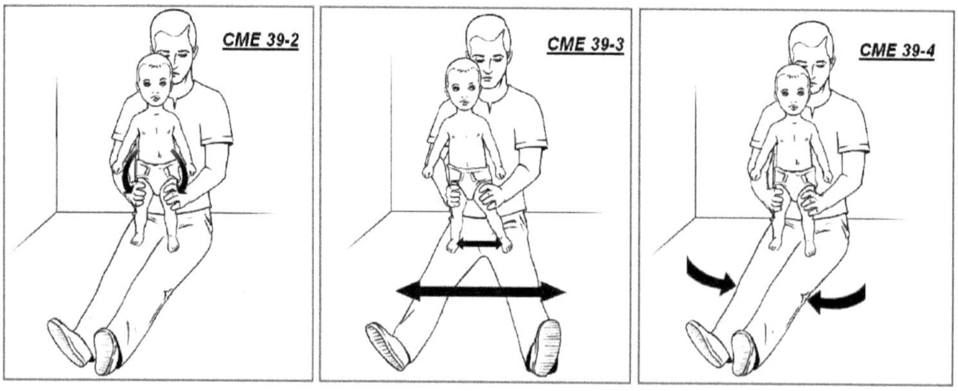

Fig. 7.

5. ***High muscle tone condition in the lower extremities is not an obstacle to stimulate standing position control.***

CME sustains and applies the early stimulation of the standing position, regardless of the high tone condition. For instance, the immediate stimulation of the standing posture in delayed children, older than seven months, with the purpose of developing the control of the vertical trunk which is the basis of any further motor function.

Long ago, it was very common for the parents of spastic CP children to be advised by the physician or/and therapist "do not place the child in standing position because of the risk of increasing the high tone condition", in the belief that standing would increase the high tone response. As a consequence the child was deprived for long periods of time (maybe even years) of experiencing being placed and sustained in the standing posture. Without weight bearing after 7 to 10 months of age many negative reactions can happen, for instance, not to develop the appropriate angle in the femur of the neck creating hip luxation or subluxation, trunk extension can not be achieved with very negative results for postural control, also the shortening of hamstring muscles can occur, and the shortening of the Achilles tendon as well. However on the other hand if, early controlled and constant standing stimulation are used, all these negative consequences can be avoided.

6. A trial period is proposed to demonstrate the short term results of CME.

After the initial motor assessment and with the direct participation of the parent's, 2 or 3 basic absent motor functions are chosen as primary goals to be achieved within the first 8 weeks of application of the CME therapy program. It should be done twice a day, seven days a week. If the goals are achieved the parents are advised to continue with the CME program, otherwise they are encouraged to continue searching for other options of physical therapy.

CME INTERVENTION MODEL

The suggested CME model of intervention is based on two components: the assessment which serves to identify the level of motor maturity and the battery of exercises. CME assessment includes the most significant motor functions of the first sixteen moths of life ranging from basic head control reactions to basic balance reactions in ambulatory state. Also the group of exercises covers all the needs of motor stimulation, beginning from primary head control functions to advanced functions of equilibrium control.

After the evaluation, the motor outcomes will orient the therapist to set up the short terms goals of the CME program to be achieved within the first 8 weeks of treatment. From this moment the therapy program should be applied on a daily basis twice a day for 6 days a week. Afterwards the therapist teaches the parents the home program exercises and determines the amount of exercises and the duration of each session. If the chosen goals are satisfactorily achieved along the initial 8 weeks of treatment, this means that the CME therapy program can help the child to reach high levels of motor progress, Otherwise, parents are encouraged to consult other motor-therapy approaches.

The question regarding how much the therapist needs to work directly with the child versus how much the parents needs to do it, does not have easy answers. This is because everything depends on two factors: a) the motor-postural condition of the child and b) the natural conditions of the parents to learn and apply the exercises in a self-confident manner.

The cost implicated in having a therapist treating the child on a daily basis for eight weeks, obviously is a limiting factor for the families. Nevertheless CME therapy was created in the spirit of simplicity and effectiveness in order to allow parents or any other adult person in the family to become capable of doing the exercise home program. An unavoidable prerequisite to apply the CME home program is to receive direct theoretic and practical training from a genuinely certified CME therapist, and those are the ones in possession of a CME I, CME II or CME III registered diploma signed by Ramon Cuevas, creator of CME therapy.

The main role of the CME practitioner in addition to applying directly the exercises, is to introduce the necessary new exercises into the program in direct relation with the child's responses. The CME wide variety of different exercises and the perseverance in the intervention are essential clues for success, both susceptible to being controlled; nevertheless the most important element is the child's underlying brain potential to achieve motor control.

Finally, to understand the meaning of CME it is necessary to acknowledge the notion of ***"proto-reaction"*** which is the combination between the word "proto": first in time, first in importance, primary, and the word reaction: an action in response to some influence or force. Consequently proto-reaction means the most important action in the child's motoric evolution. The CME theoretical framework sustains the assumption that the ***"proto-reaction"*** is the weight bearing in vertical posture which will evolve to functional standing position within the first 8 months of life. This action is the most significant response to the gravity force influence, and can be observed from 13 weeks of age in normal children.

CONTRAINDICATIONS

There are a few contra indications to using CME therapy:

1) Any diagnosis of degenerative nature affecting the neuro-muscular tissue (progressive diseases).

2) Diagnosis of Osteogenesis Imperfecta.

3) Uncontrolled seizures.

4) Child younger than 3 months of age, except when the therapist in charge is CME II or CME III diplomat.

5) To use this manual as a therapy guide, in disregard of the DISCLAIMER, by any Physical or Occupational Therapist, not in possession of a CME diploma level II or level III, awarded by the Cuevas Medek Exercise Continuing Education Program, and signed by Ramon Cuevas.

6) To use the inserted information to treat a child without the advice of a CME certified therapist.

CME SUB-INDEX

EXERCISE Page

EXERCISE **Page**

EXERCISE	Page

EXERCISE Page

FORMAT DESCRIPTION

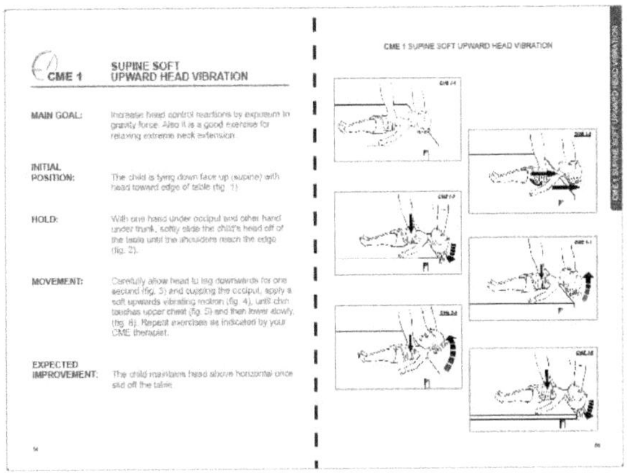

The 99 CME are presented under the same format, with the description placed in the left page and the illustrations in the right one. First is the number assigned to the exercise plus a short descriptive name, subsequently follow 5 titles:

Exercise Configuration: *Describes the proper use of materials in some exercises.*

MAIN GOAL: Describes the basic purpose of the exercise from the developmental point of view.

INITIAL POSITION: Describes the body position of the child, prior to receiving the support of the therapist.

HOLD: Indicates how the hands of the therapist should be placed on the infant's body.

MOVEMENT: Describes the dynamic action of the therapist on the child body in order to provoke the automatic-functional reaction of the infant.

EXPECTED IMPROVEMENT: Describes the child's final optimal response to any given exercise.

CME 1 SUPINE SOFT UPWARD HEAD VIBRATION

MAIN GOAL: Increase head control reactions by exposure to gravity force. Also it is a good exercise for relaxing extreme neck extension.

INITIAL POSITION: The child is lying down face up (supine) with head toward edge of table (fig. 1).

HOLD: With one hand under occiput and other hand under trunk, softly slide the child's head off of the table until the shoulders reach the edge (fig. 2).

MOVEMENT: Carefully allow head to lag downwards for one second (fig. 3) and cupping the occiput, apply a soft upwards vibrating motion (fig. 4), until chin touches upper chest (fig. 5) and then lower slowly, (fig. 6). Repeat exercises as indicated by your CME therapist.

EXPECTED IMPROVEMENT: The child maintains head above horizontal once slid off the table.

CME 1 SUPINE SOFT UPWARD HEAD VIBRATION

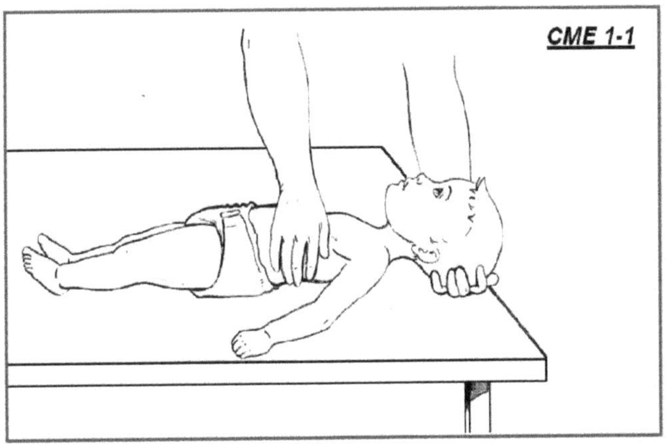

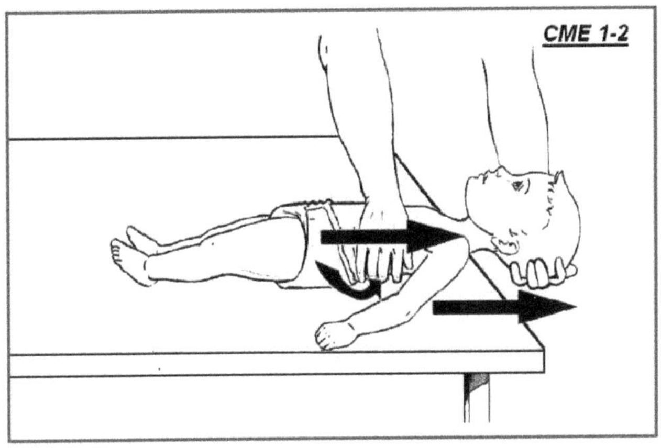

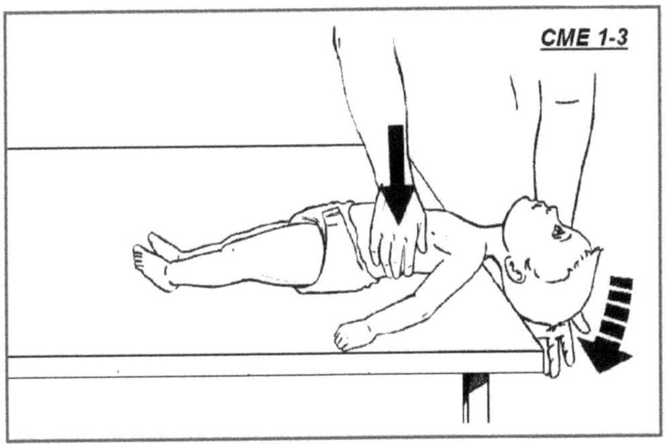

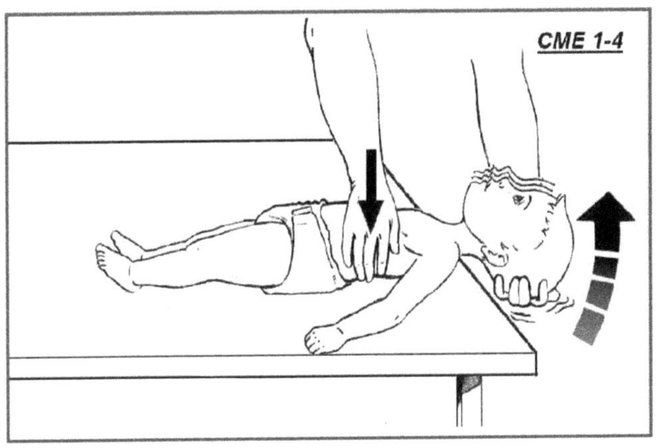

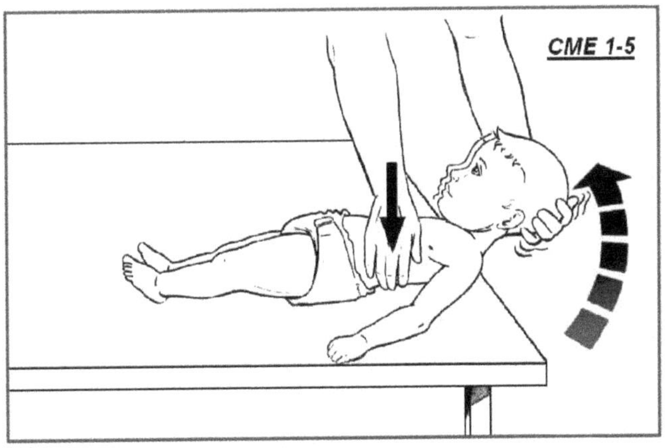

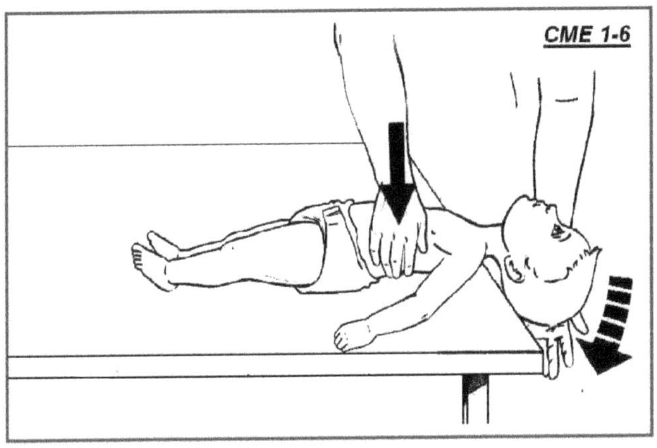

CME 2 LATERAL SOFT UPWARD HEAD VIBRATION

MAIN GOAL: Increase head control reaction by exposure to gravity force. Also is a very good exercise for functional stretching of SCM muscles.

INITIAL POSITION: The child is lying down in a lateral position (fig.1).

HOLD: Standing behind the child grab with your right hand the right arm over the trunk, place left hand under the left side of the face and slide the child until head and arm are off the edge, (figs. 1 & 2).

MOVEMENT: Release the head (fig. 3) and then apply soft upward vibrating movement (fig. 4), until the ear almost touches the shoulder as indicated in figure 5, then go back slowly to the initial position with no vibration and release the head at the end of the range (fig. 6). Follow your CME practitioner´s indications.

EXPECTED IMPROVEMENT: The head stays aligned horizontally or higher after sliding off the table.

CME 2 LATERAL SOFT UPWARD HEAD VIBRATION

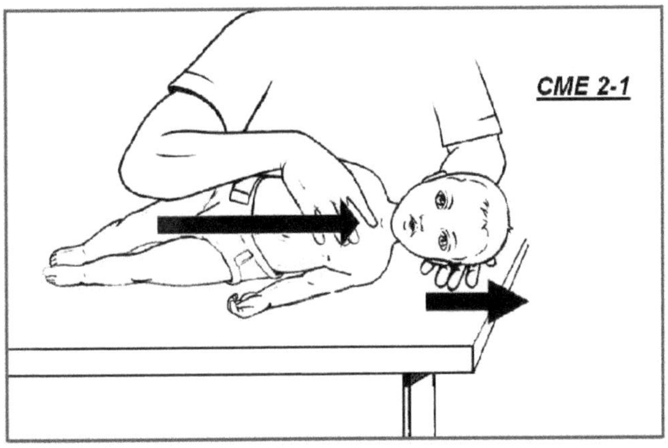

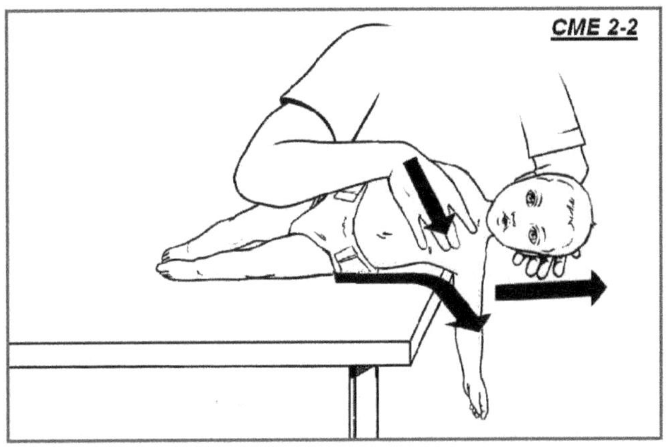

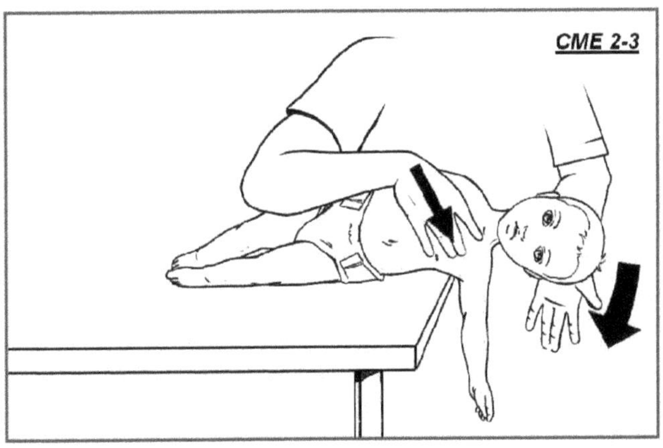

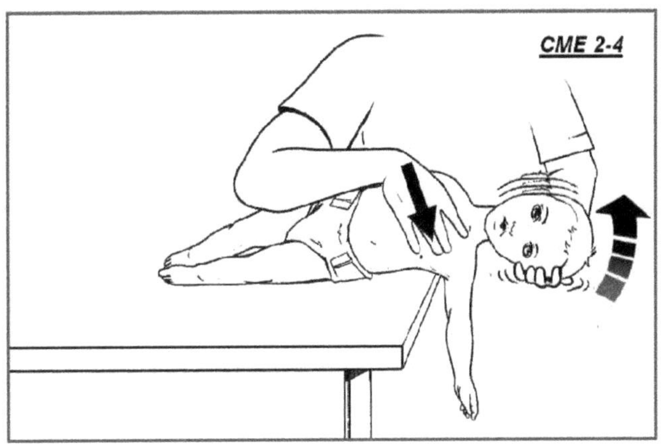

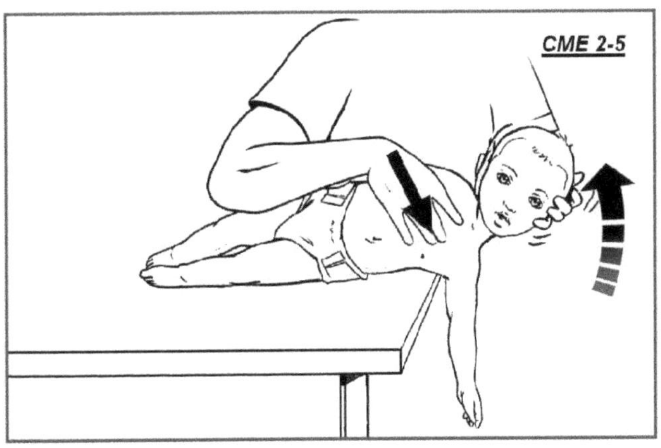

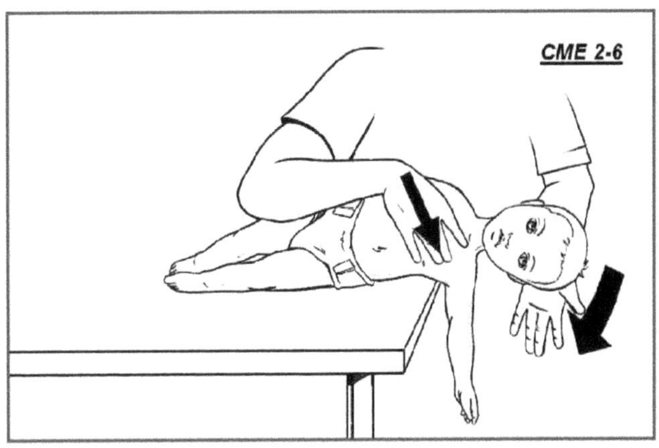

CME 3 PRONE SOFT UPWARD HEAD VIBRATIONS

MAIN GOAL: Induce head control reaction by exposure to gravity force.

INITIAL POSITION: The child is lying down on his stomach or prone position with head toward the edge of the table (fig.1).

HOLD: Place right hand under his chest and with the left hand by forehead, softly slide the child out of table as indicated in figure 2.

MOVEMENT: Release the head for a moment (fig. 3), then apply on his forehead a soft upward vibrating movement (fig. 4), until his head reaches 45 degrees over the table as indicated in figure 5. Come back slowly to the initial position without vibrations and release the head at the end of the range (fig 6). Stop for one second and repeat the movement according to indications.

EXPECTED IMPROVEMENT: The child holds up the head aligned with the table or higher.

CME 3 PRONE SOFT UPWARD HEAD VIBRATIONS

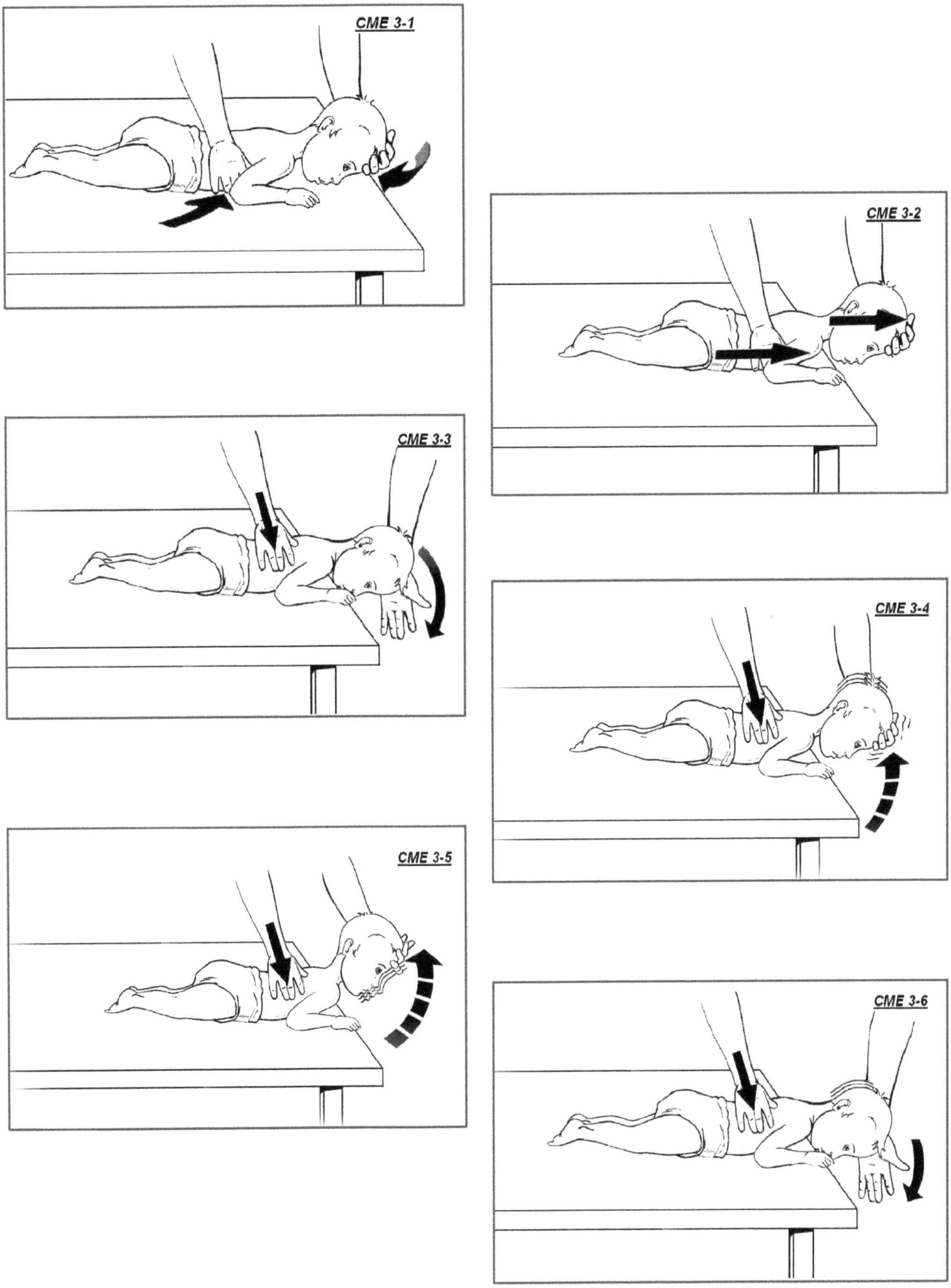

CME 4 VERTICAL HEAD BY CHIN AND NAPE

MAIN GOAL: To provoke vertical stability of the head.

INITIAL POSITION: Lying down face up with legs bent on the edge of the table (fig.1).

HOLD: Place left hand under the child's chin with the thumb on the left cheek and index and middle finger on the right cheek, then cup the nape with her right hand, (figure 2).

MOVEMENT: Carefully move the child up to sitting position supporting the trunk between your forearms as indicated in figures 3 and 4. Now quickly release each hand with alternate pattern, (chin-nape chin-nape) by counting of ten (fig. 5). Place down the child to the supine position. Repeat the exercise according your CME therapist´s indications.

EXPECTED IMPROVEMENT: The child keep´s the head vertically aligned whenever the head is free.

CME 4 VERTICAL HEAD BY CHIN AND NAPE

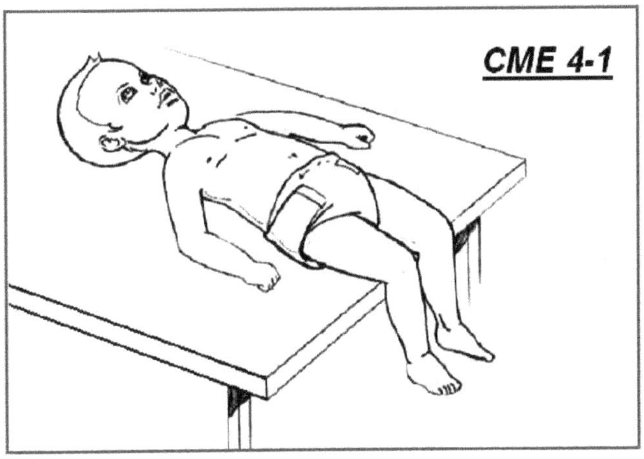

CME 4-1

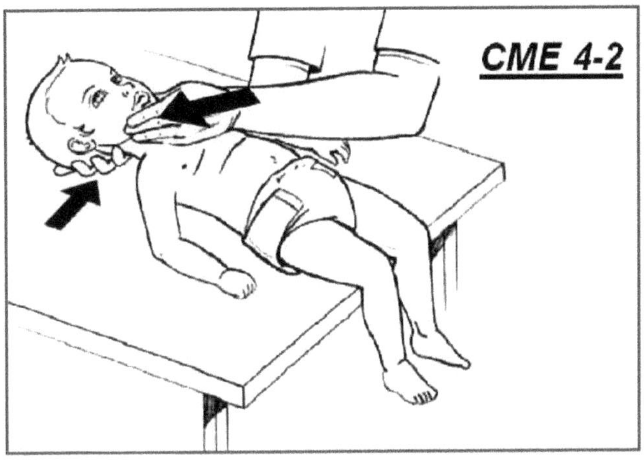

CME 4-2

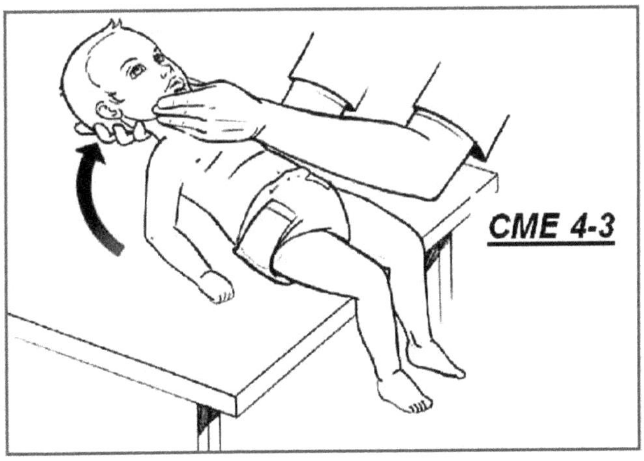

CME 4-3

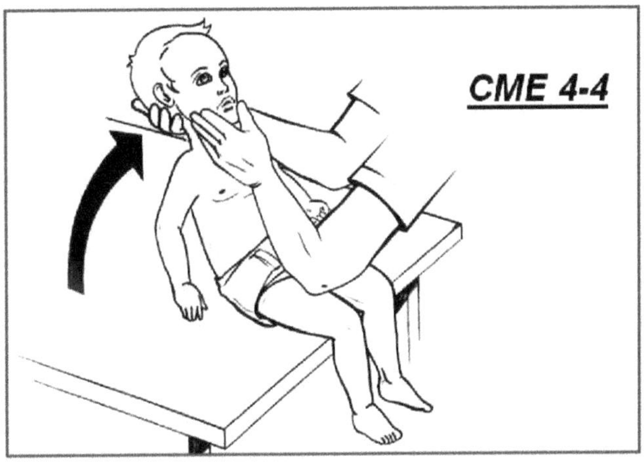

CME 4-4

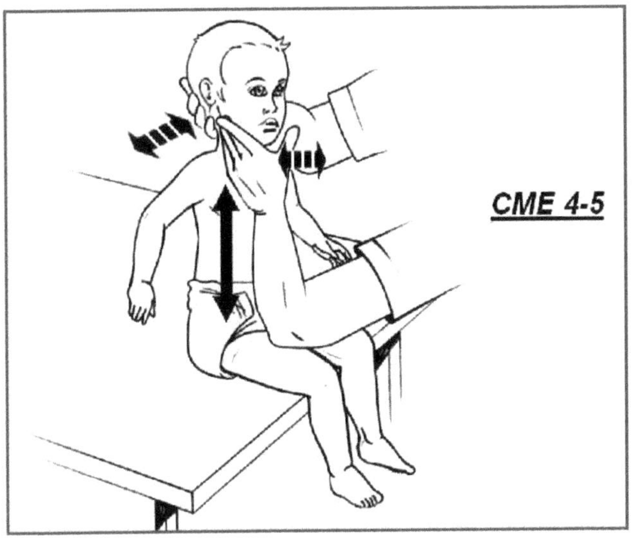

CME 4-5

CME 5 SERIES OF CURVES

MAIN GOAL: To provoke head-trunk alignment.

INITIAL POSITION: Child is lying down in supine position with head close to the top of the table with legs perpendicular to you (fig. 1).

HOLD: Firmly bring arms and mid-trunk close together (fig. 1).

MOVEMENT: Pivoting the child's head and upper trunk to the left side and the hips and legs to the other side with a pivot range of 45 degrees on each curve (fig. 2), while sliding the whole body with a smooth "S" type of motion until the hips reach the bottom of the table (figs. 3,4 & 5).

EXPECTED IMPROVEMENT: The child keeps the head aligned with trunk all along the descending curves and also can turn his face against the direction of the movement.

CME 5 SERIES OF CURVES

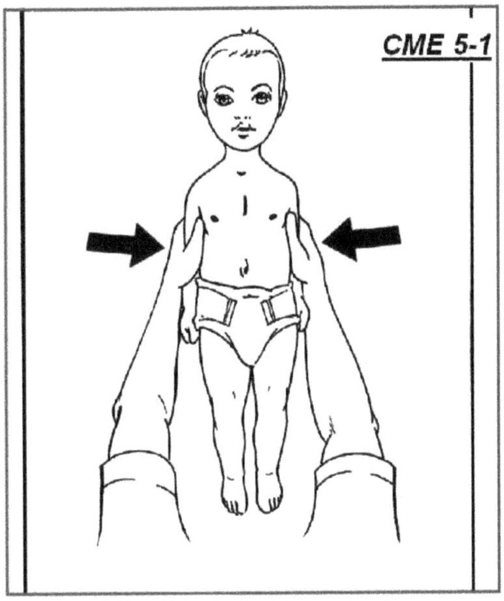

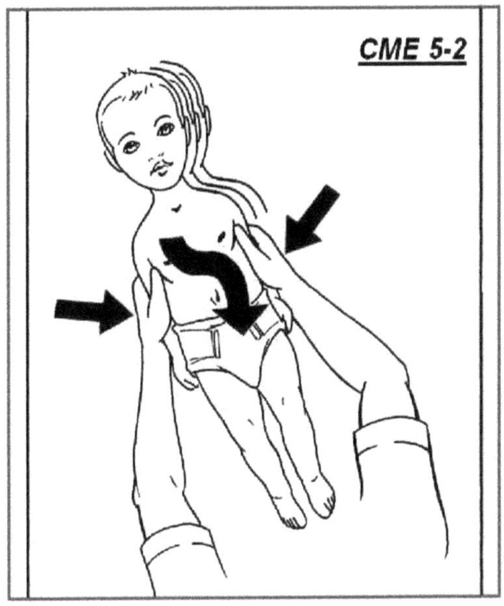

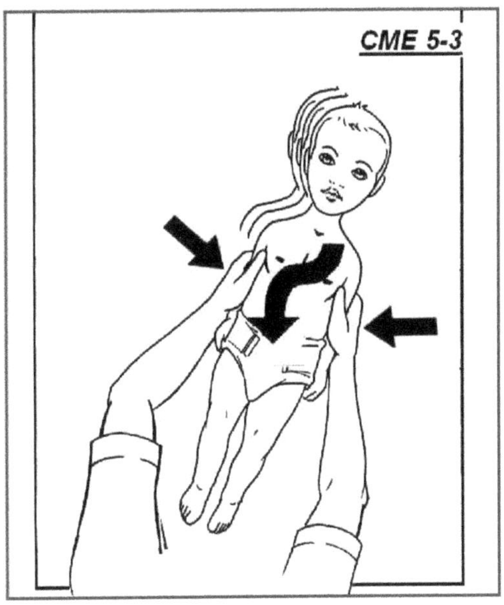

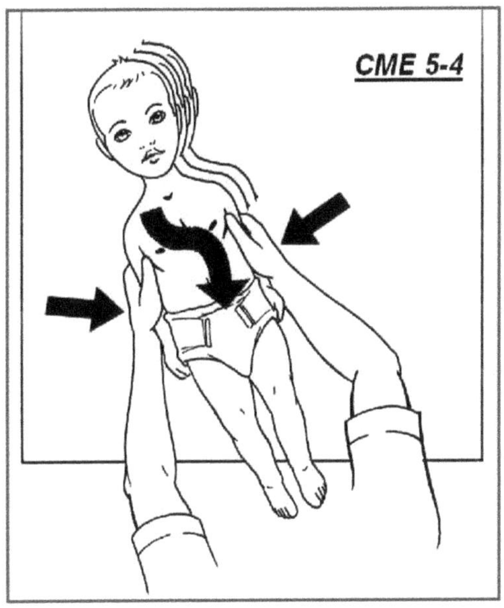

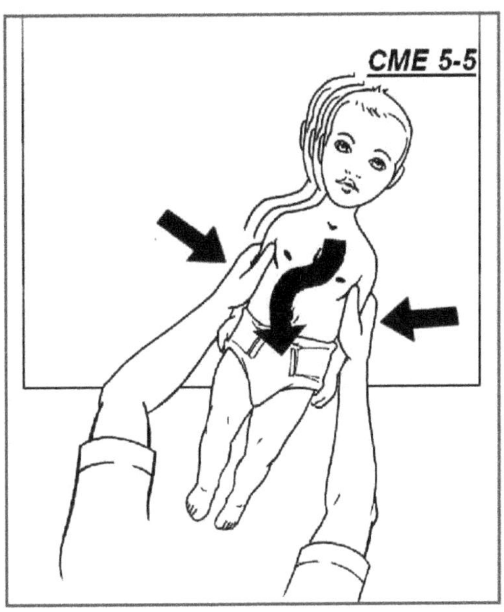

CME 6 COMPASS MOTION

MAIN GOAL: Increase head and trunk stability.

INITIAL POSITION: The child is lying down with his face up with legs on the table.

HOLD: Firmly support around arms and trunk as indicated in figure 1.

MOVEMENT: While holding firmly, apply on the child's trunk a compass-type movement towards his left side keeping the pivot steady under the knuckles (fig. 2). Trunk and head slide to the left, and hips and legs move to the right until a complete 90 degrees range (fig. 3). Reverse the direction of the pivoting motion until you reach the initial position (figs. 4, 5 and 6). Stop for a few seconds at the center and then repeat the movement toward the other side. Keep the pivot centered under knuckles at all time. Repeat as indicated by your CME therapist.

EXPECTED IMPROVEMENT: The child will be able to keep the head aligned firmly with the trunk at all times.

CME 6 COMPASS MOTION

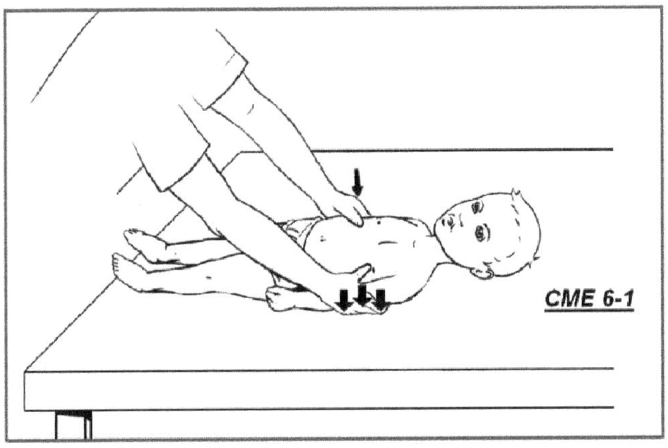

CME 6-1

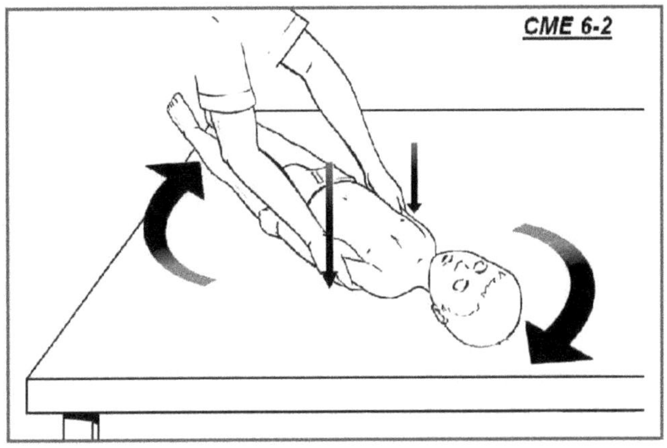

CME 6-2

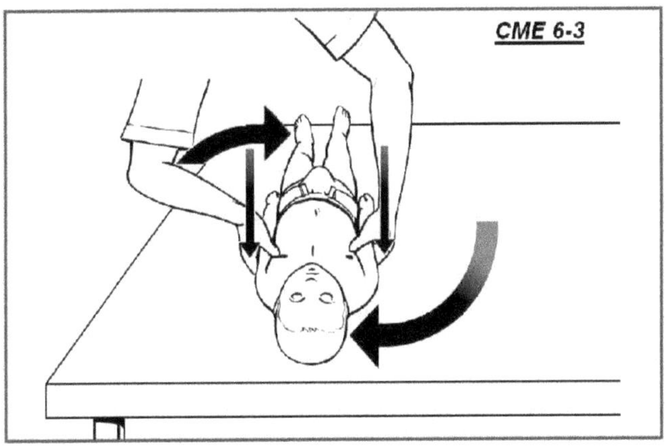

CME 6-3

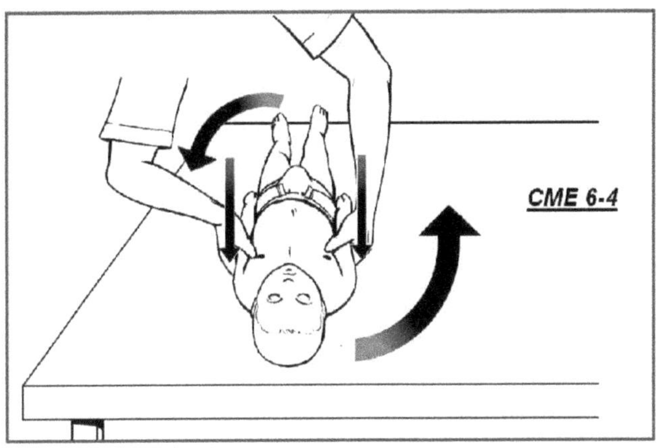

CME 6-4

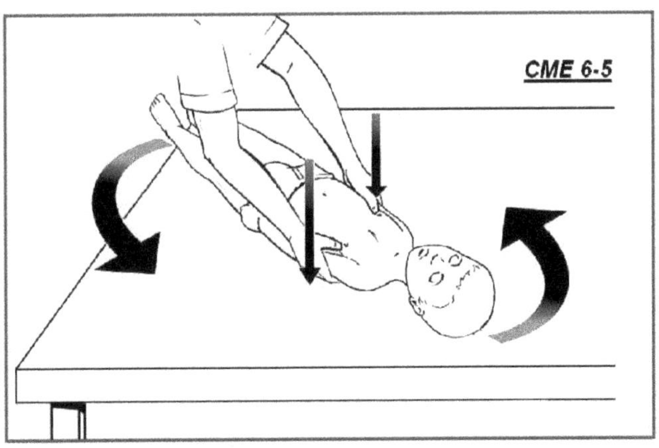

CME 6-5

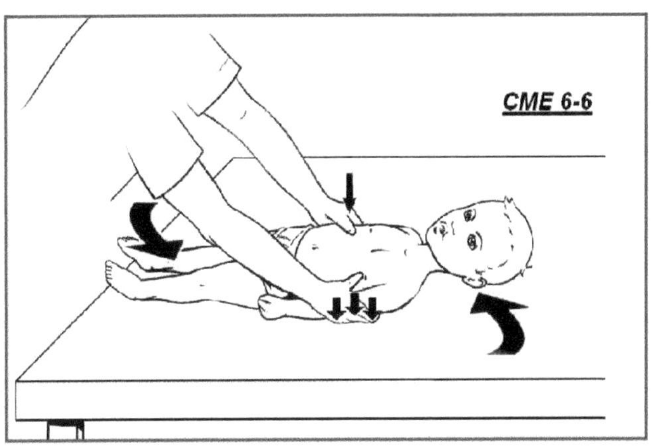

CME 6-6

CME 7 COMPASS MOTION 180 DEGREES

MAIN GOAL: Increase head and trunk stability.

INITIAL POSITION: The child is lying down face up with legs on the table.

HOLD: Firmly support arms and trunk as indicated in figure 1.

MOVEMENT: While holding firmly, apply a compass-type movement towards his left side keeping the pivot steady under the knuckles (fig. 2). Trunk and head slide to the left, and hips and legs moves to the right until complete 180 degrees has been reached (fig. 3, 4 and 5).

Continue on next page

CME 7 COMPASS MOTION 180 DEGREES

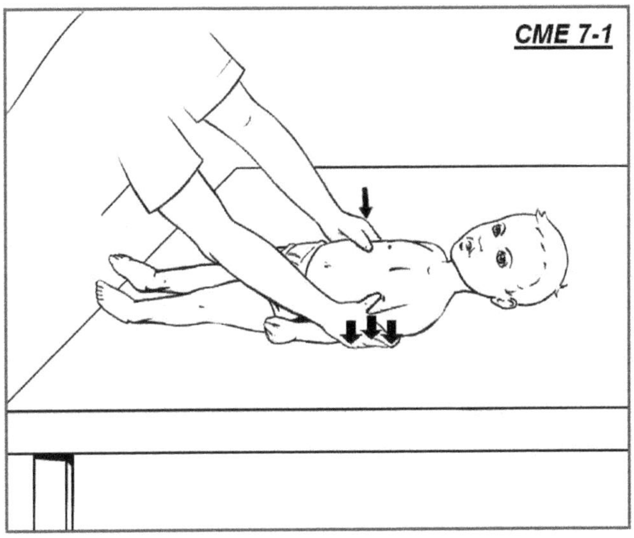
CME 7-1

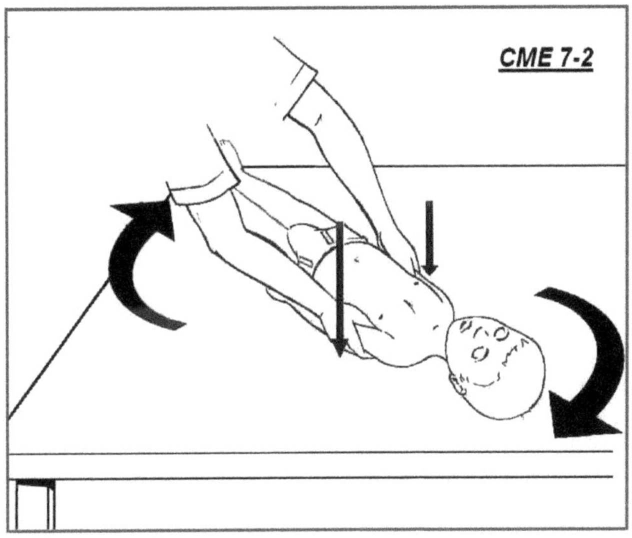
CME 7-2

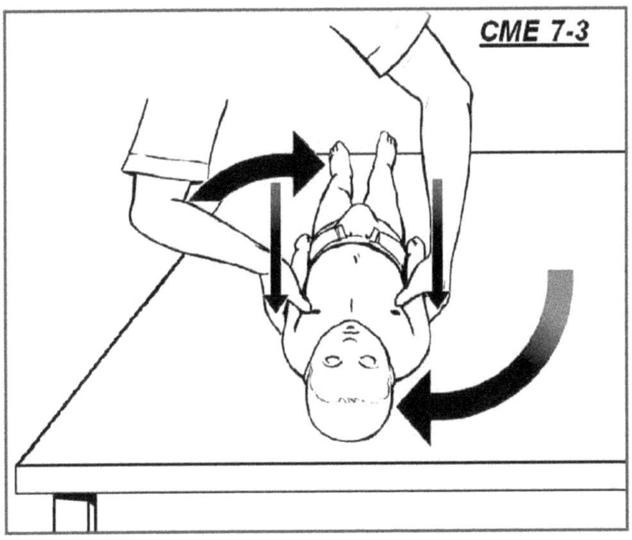
CME 7-3

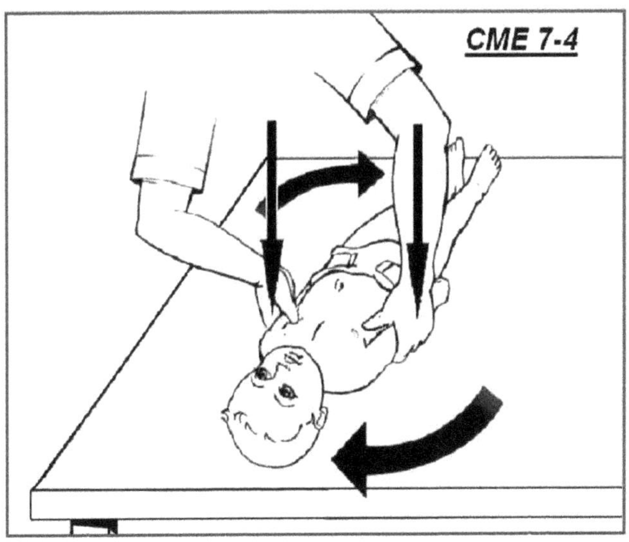
CME 7-4

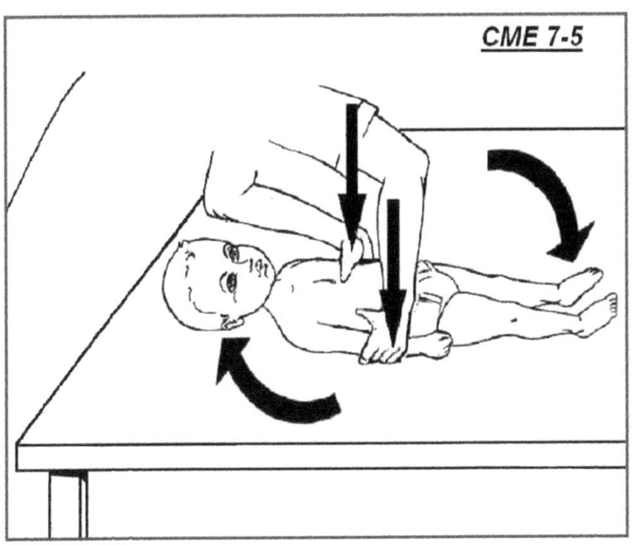
CME 7-5

CME 7 COMPASS MOTION 180 DEGREES

From this position now apply a spinning movement in reverse direction (toward the right side of the child) using the same pivoting motion, until you reach the initial position (figs. 6, 7, 8, 9 and 10). Stop for a few seconds at the center and then repeat the movement toward the other side. Keep the pivot centered under knuckles at all time. Repeat as indicated by your CME therapist.

EXPECTED IMPROVEMENT: The child can hold the head firmly aligned with the trunk at all times, furthermore, she/he can turn her/his face against the direction of spinning.

CME 7 COMPASS MOTION 180 DEGREES

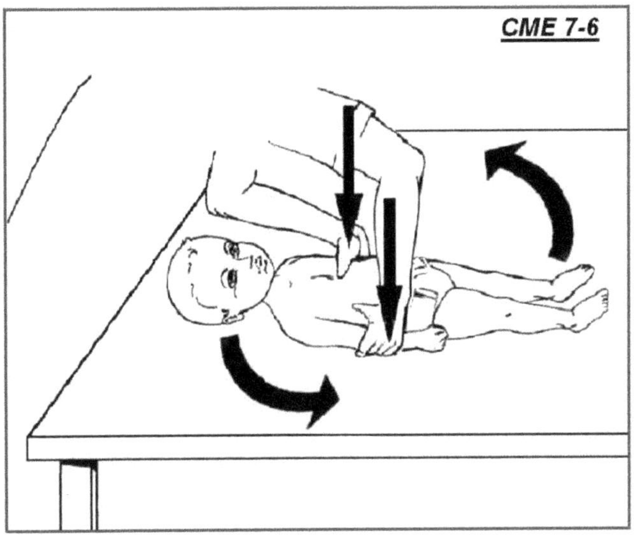
CME 7-6

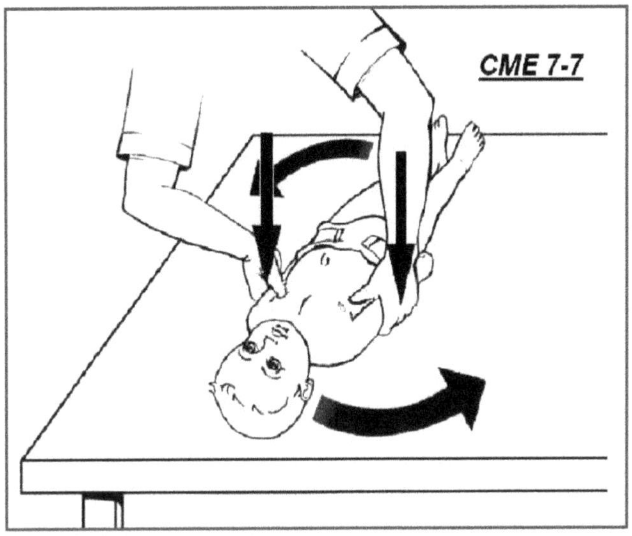
CME 7-7

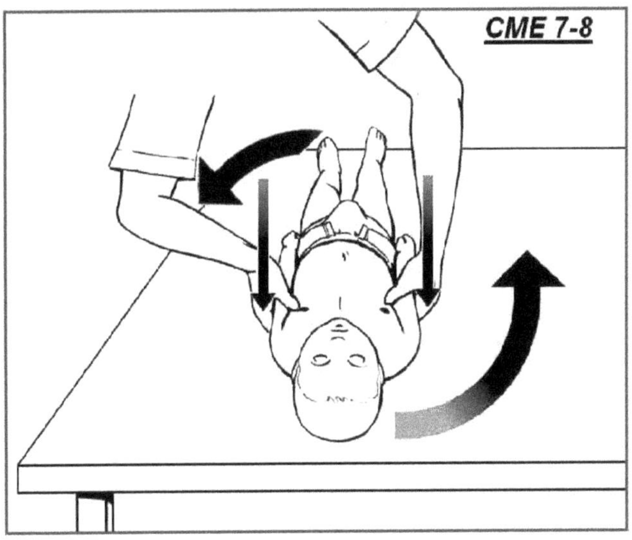
CME 7-8

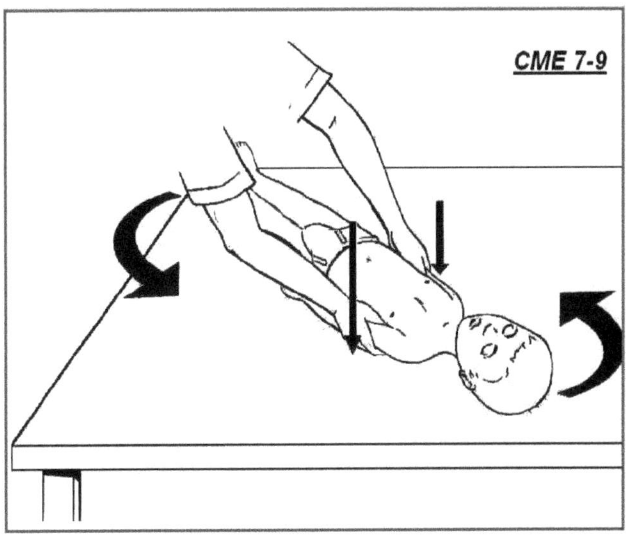
CME 7-9

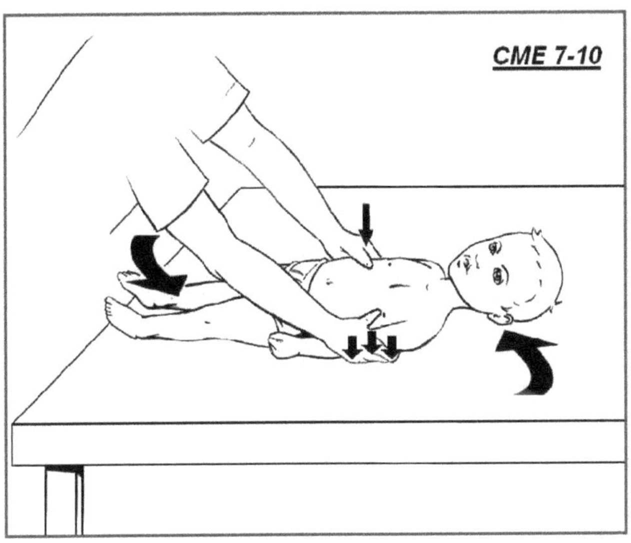
CME 7-10

CME 8 FREE HEAD OFF OF THE TABLE

MAIN GOAL: Increase head-trunk dynamic stability.

INITIAL POSITION: The child is placed in supine position (lying down face up) with head towards the edge of the table (fig.1).

HOLD: Firmly support arms and trunk, above elbows, as indicated in figure 1.

MOVEMENT: Turn the child over his/her left side (fig. 2) to firmly slide the head out of the table (fig. 3), hold the lateral position for the count of 5, then pivot the child to face up position (fig. 4), holding for the count of 5, softly turn the child over his/her right side and stay for 5 counts (fig. 5). Finally move the child inside the table (fig. 6) and pivot 90 degrees to face up position as indicated in the figure 7. Repeat the exercise as indicated by your CME therapist.

EXPECTED IMPROVEMENT: The child should keep the head firmly aligned with the trunk at all times.

CME 8 FREE HEAD OFF OF THE TABLE

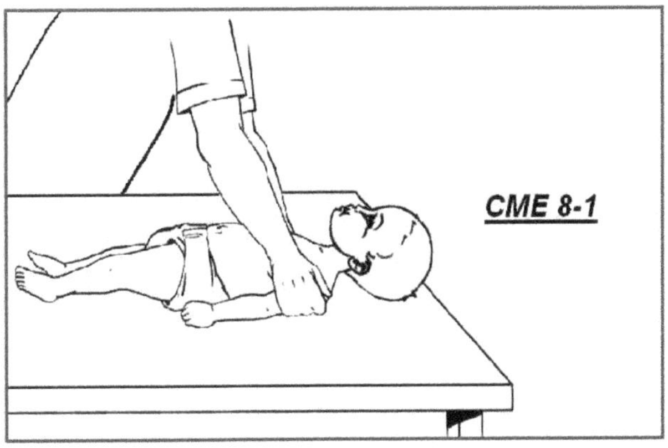

CME 8-1

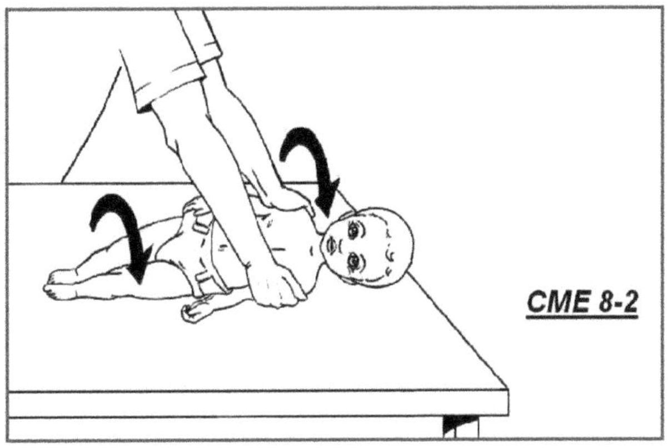

CME 8-2

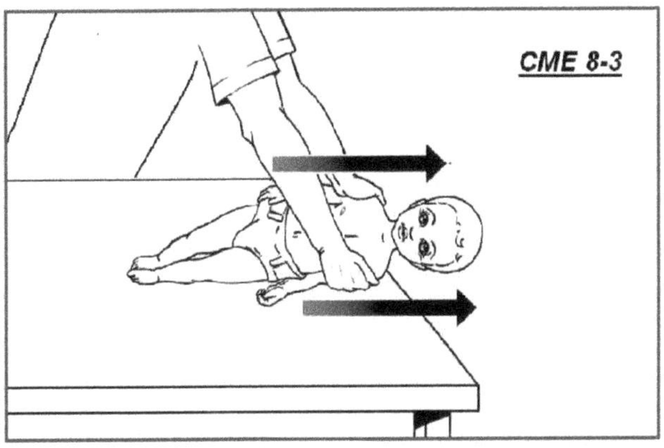

CME 8-3

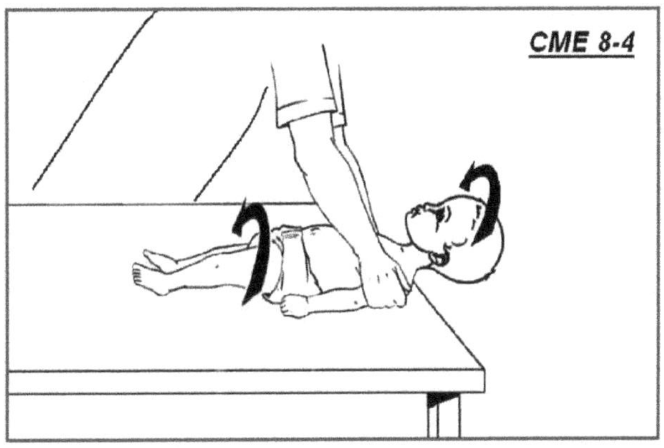

CME 8-4

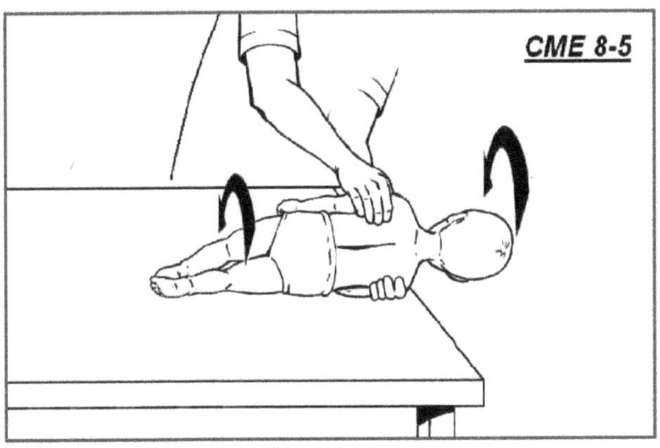

CME 8-5

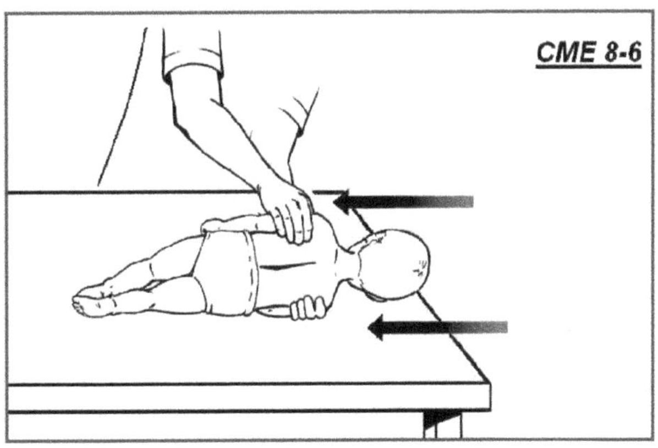

CME 8-6

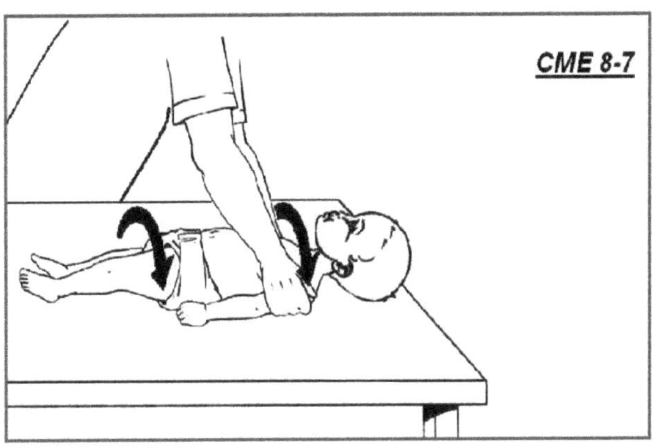

CME 8-7

CME 9 HORIZONTAL 180 DEGREES AERIAL GYRATION

MAIN GOAL: Increase head-trunk stability.

INITIAL POSITION: The child is lying in prone position (face down on the table.

HOLD: Hold firmly at the mid trunk with both hands.

MOVEMENT: Softly lift the child vertically in front of you as indicated in figure 1, then, incline his/her body to a full horizontal position (fig. 2). Turn the child to a lateral position by always pivoting in front of your face without losing the horizontal position (fig. 3), and continue the gyration until to arrive to full supine position (fig. 4), keeping horizontal alignment until the child's head is in contact with your right shoulder (fig. 5). (At all times during the gyratory movement the shoulders and hips of the child should stay horizontally aligned).

Continue on next page

CME 9 HORIZONTAL 180 DEGREES AERIAL GYRATION

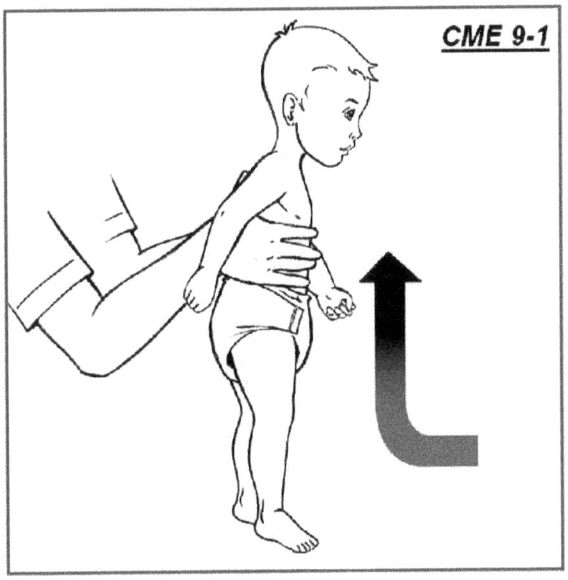

CME 9-1

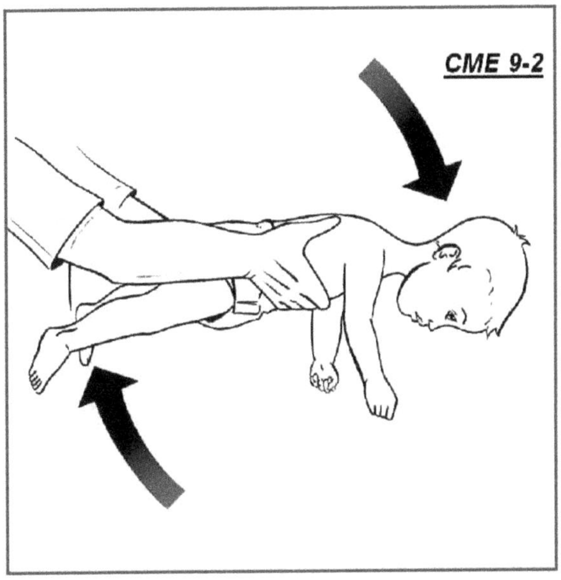

CME 9-2

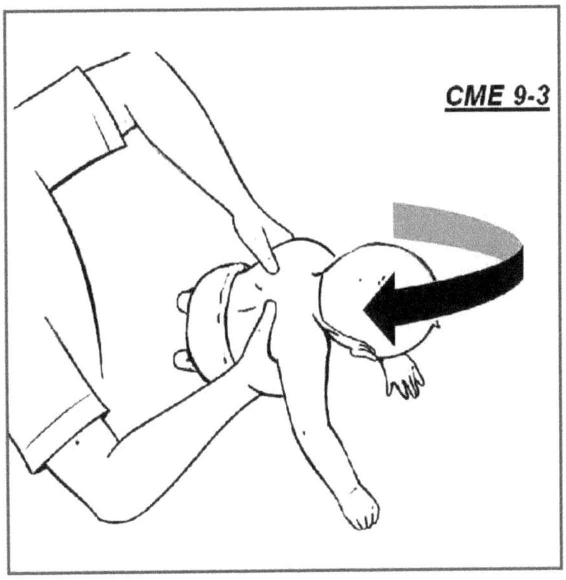

CME 9-3

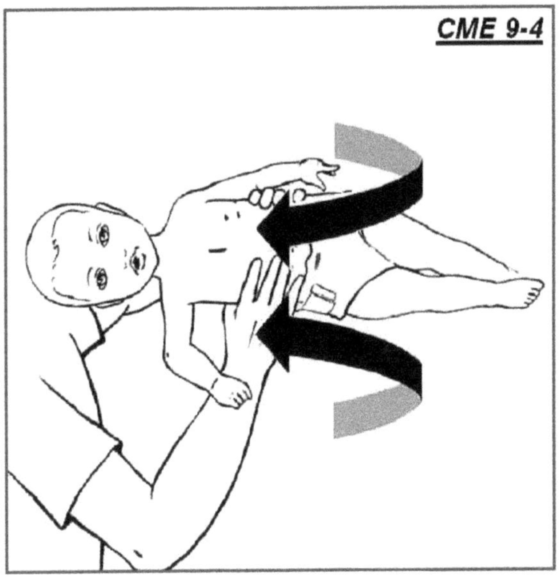

CME 9-4

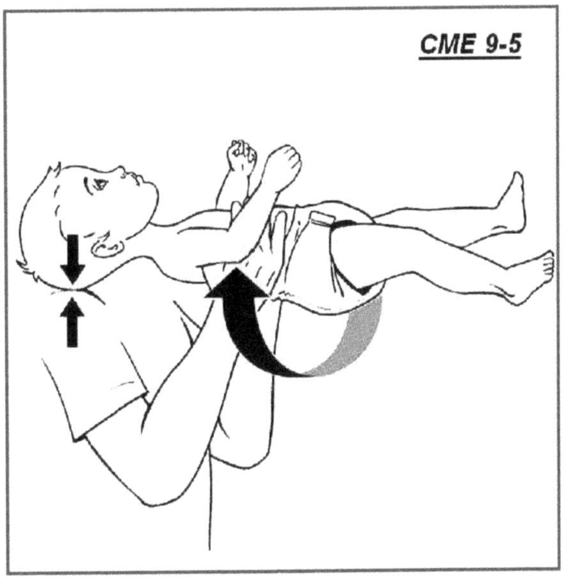

CME 9-5

CME 9 HORIZONTAL 180 DEGREES AERIAL GYRATION

From this horizontal position (fig, 6), turn the child to lateral position on your right hand (fig. 7), and softly reverse the direction of the gyratory movement until you reach initial position as illustrated in the figure sequence of 8 and 9. Place the child vertical (fig. 1), and start the same movement again toward your left shoulder. Repeat the exercise according your CME practitioner´sinstructions.

EXPECTED IMPROVEMENT: The child will be able to keep the head horizontally or higher all along the gyratory movement.

CME 9-6

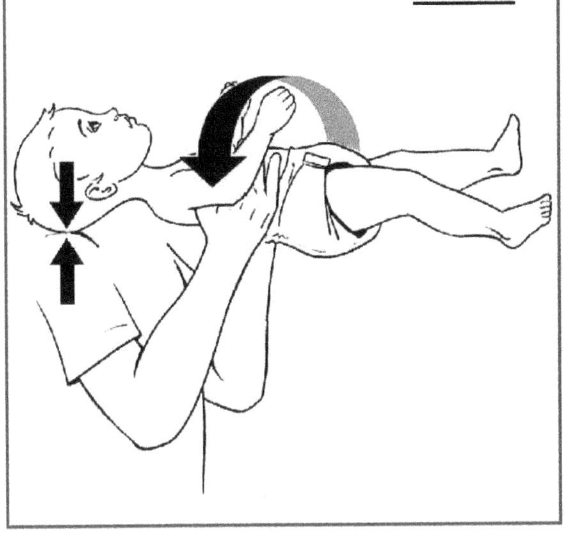

CME 9-7

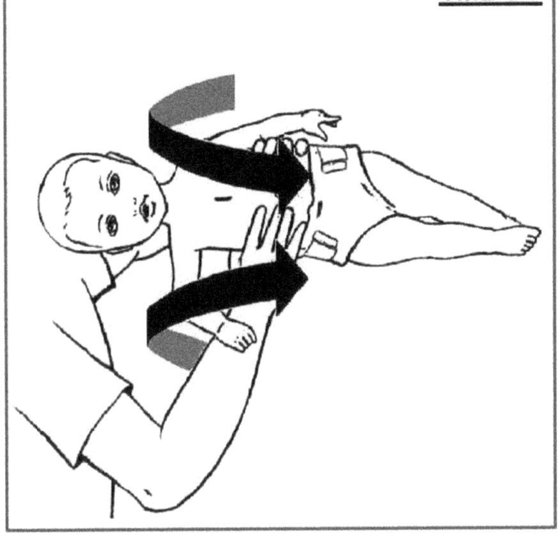

CME 9-8

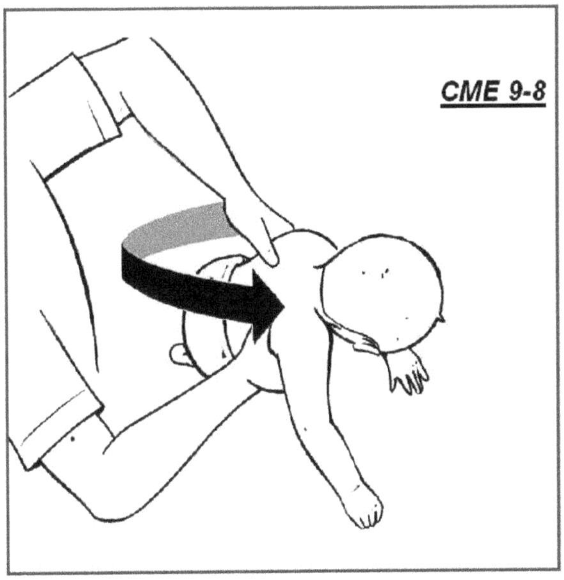

CME 9-9

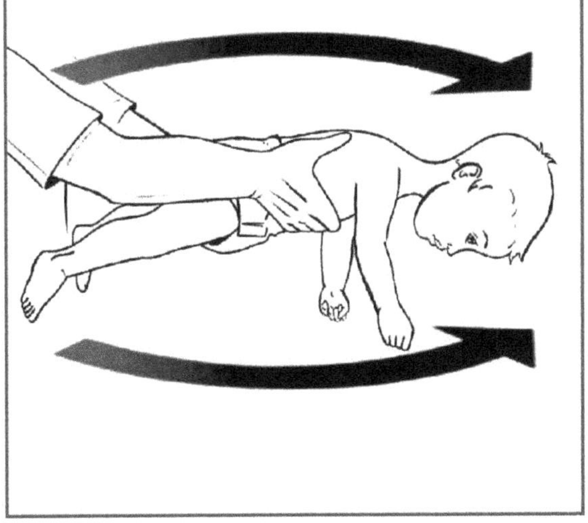

CME 10 AERIAL SQUATTING

MAIN GOAL: Increase head-trunk stability in vertical position.

INITIAL POSITION: The child is in supine position crosswise on the table.

HOLD: With the left hand firmly grab both legs together by mid lower legs, bend the knees and firmly place feet on the table. With the right hand softly and firmly grab by the neck and nape, (fig. 1). DO NOT DIG NAILS ON THE SOFT TISSUE OF THE NECK.

MOVEMENT: Push the child gently with your right hand to a sitting position as illustrated in figure 2, then, lift her/his body upwards to an aerial squatting position (fig. 3). Quickly move your right hand down to the buttocks and place your elbows on the table as indicated in figure 4. Hold the position with maximum head vertical alignment for the count of 10. Place the child back on the table. Repeat the exercise as indicated by your CME therapist.

EXPECTED IMPROVEMENT: The child should keep the head aligned with the trunk at all times

CME 10 AERIAL SQUATTING

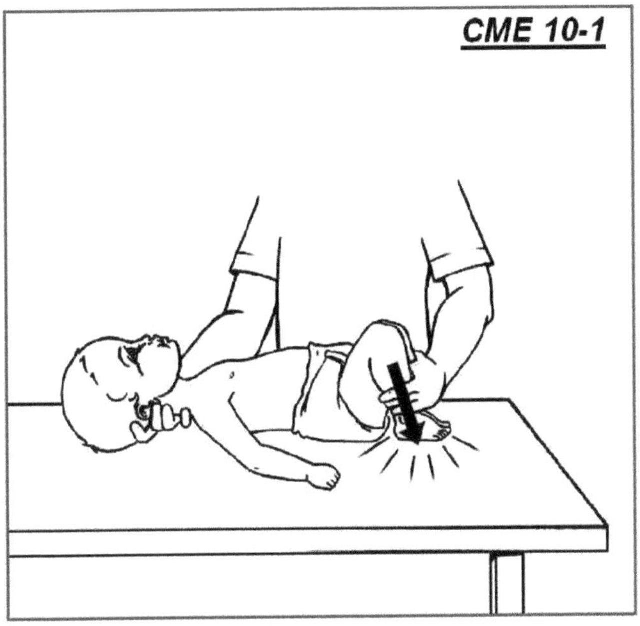

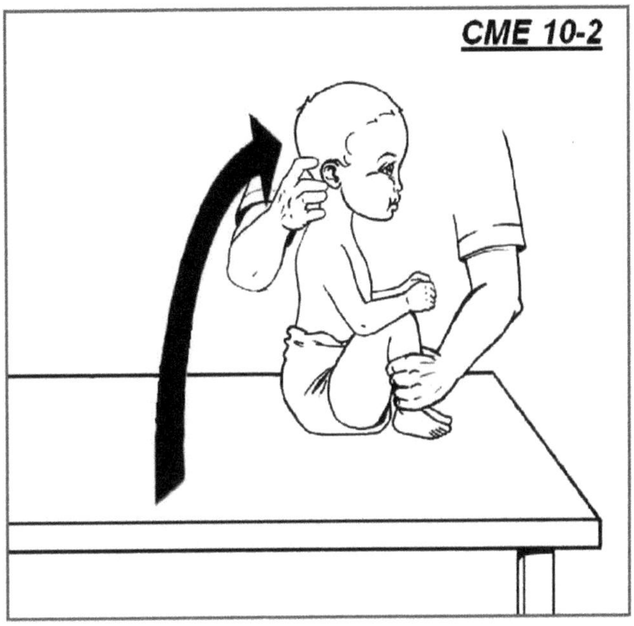

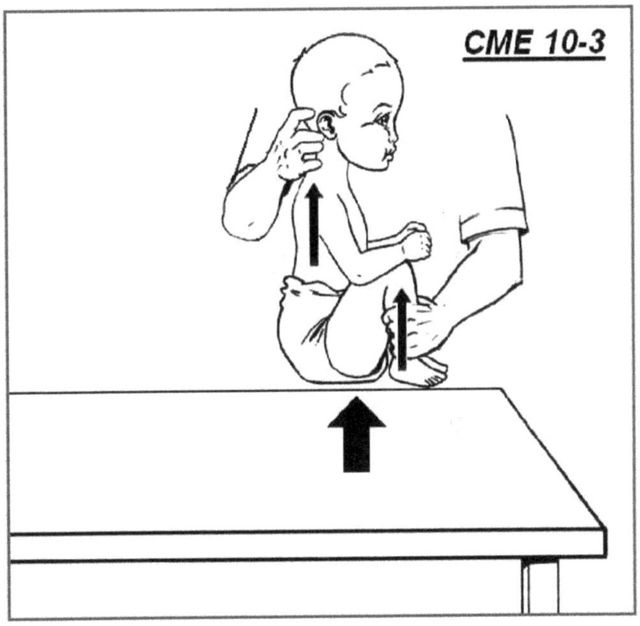

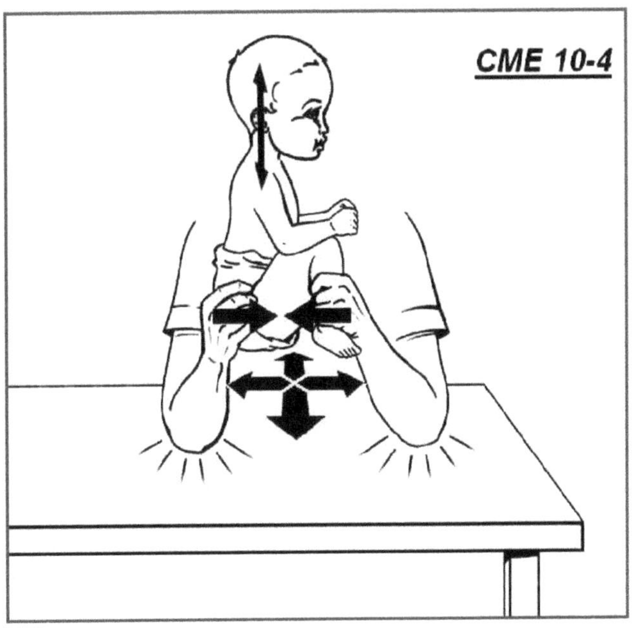

CME 11 AERIAL CHAIR

MAIN GOAL: Increase head-trunk vertical alignment.

INITIAL POSITION: The child is lying on table in prone position (face down) as illustrated in the figure 1.

HOLD: Under the stomach with both forearms to pick up the child to a vertical position supported on your chest as shown in figure 2. Move your right hand under the right hip with palm up and thumb over thigh, bending hip 90 degrees. Do the same on the left hip. Each hand very close to the hip joint, in order to keep the child in full contact with your trunk, as indicated in figure 3.

MOVEMENT: Move the child away from your trunk, and with rapid and accurate movements keep the child's head aligned vertically to the count of 10 as indicated in figure 4. Repeat the exercise as indicated by your CME practitioner.

EXPECTED IMPROVEMENT: The child should keep the head aligned with the trunk at all times.

CME 11 AERIAL CHAIR

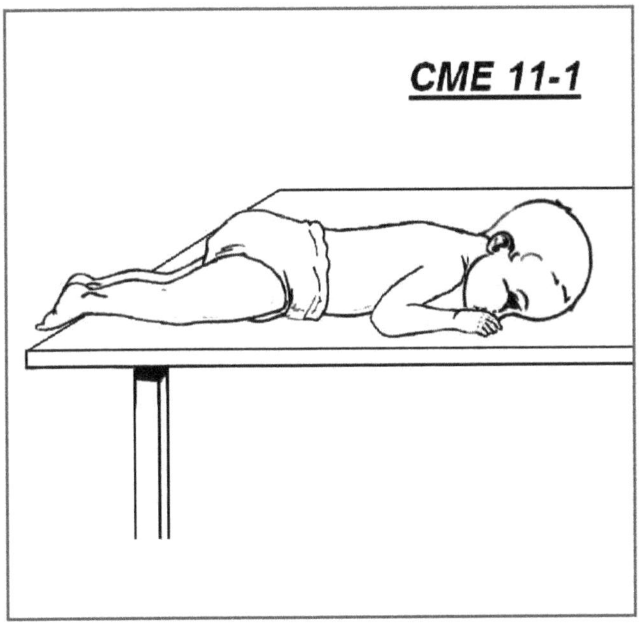

CME 11-1

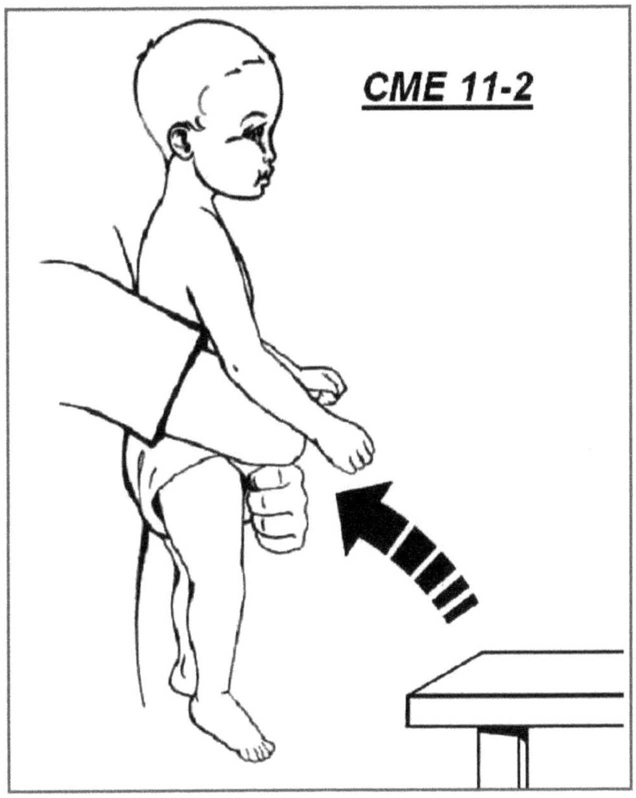

CME 11-2

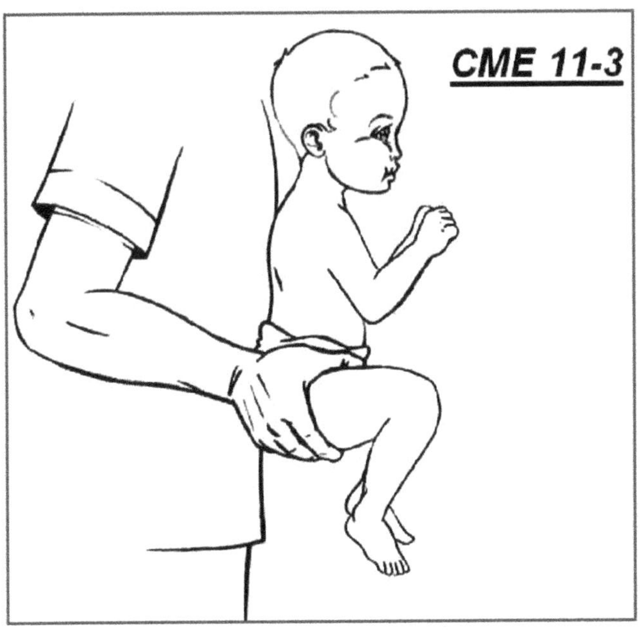

CME 11-3

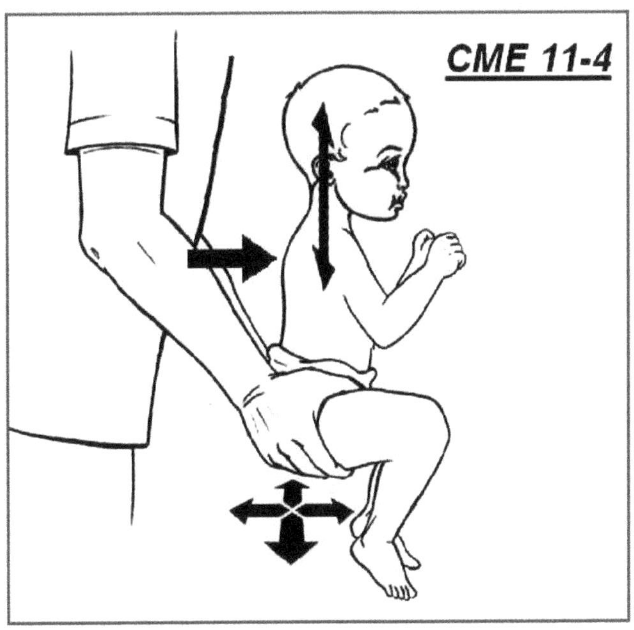

CME 11-4

CME 12 KNEELING BY CHEST

MAIN GOAL: To provoke trunk-thighs alignment reaction in vertical position.

INITIAL POSITION: Child is lying down on his/her stomach or prone position, with both feet on the edge of table.

HOLD: Put your right hand on the center of the child's chest, and put your left hand on both legs together below the knees, approaching your left forearm towards the child's back as indicated in figures1 and 2.

MOVEMENT: Pivoting on child's knees, lift his/her trunk softly until making contact on your inclined left forearm (fig. 2), and proceed upwards until the child reaches the vertical kneeling position (figs. 3, 4 and 5). Maintain his/her trunk in full vertical alignment to the count of thirty. Repeat the exercise as indicated by your CME therapist.

EXPECTED IMPROVEMENT: The child will keep trunk and thighs aligned at all times without pushing to trunk flexion.

CME 12 KNEELING BY CHEST

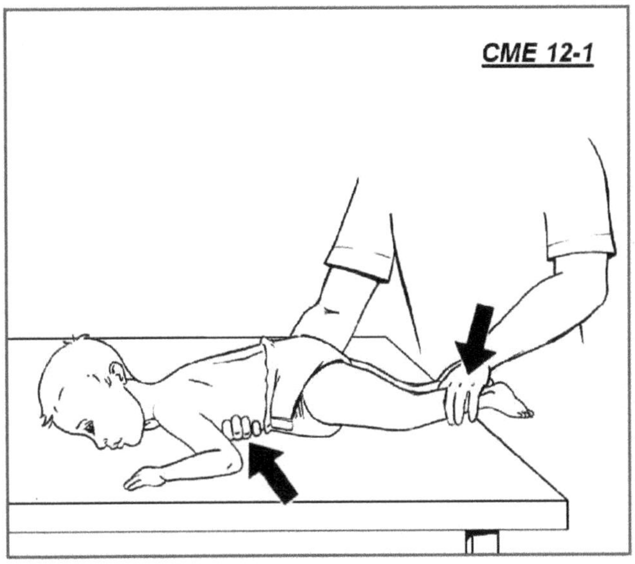

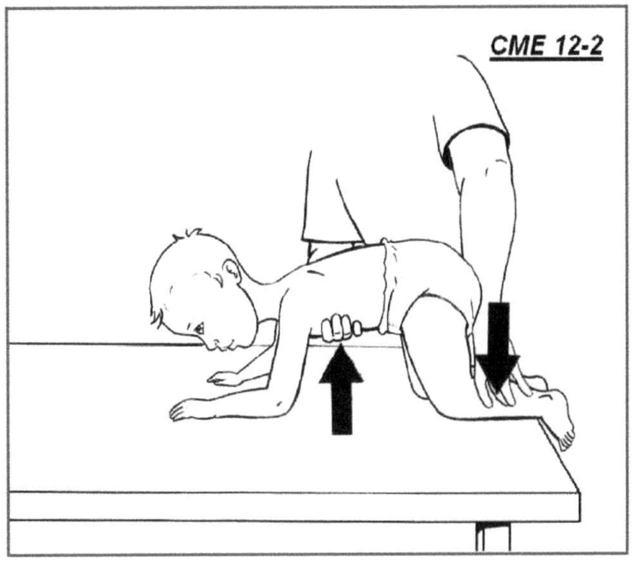

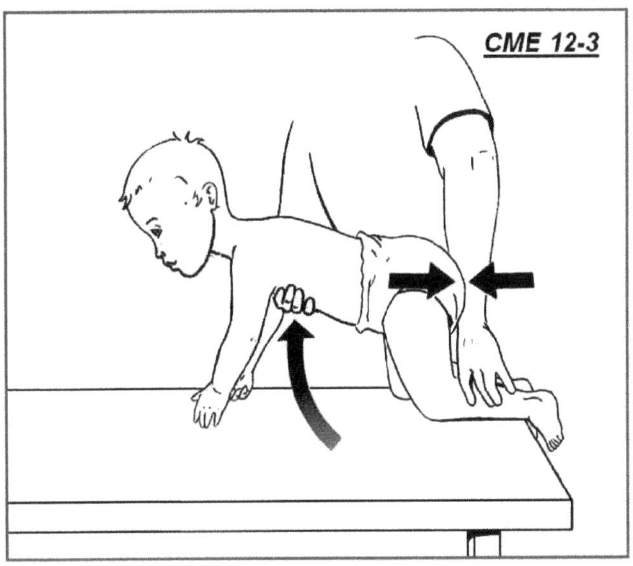

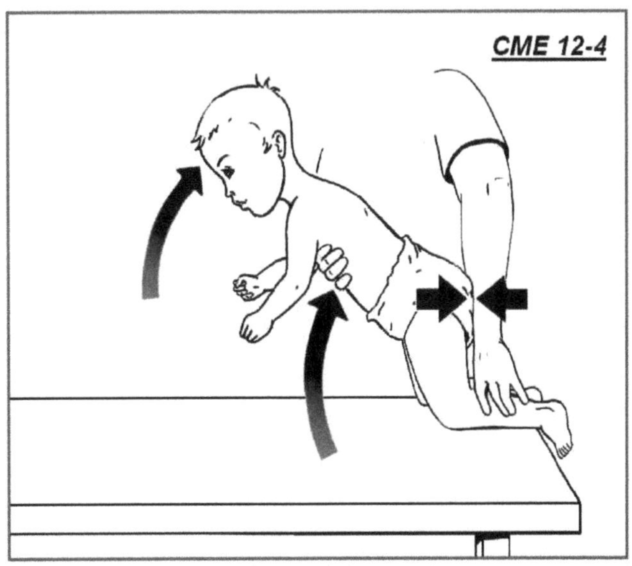

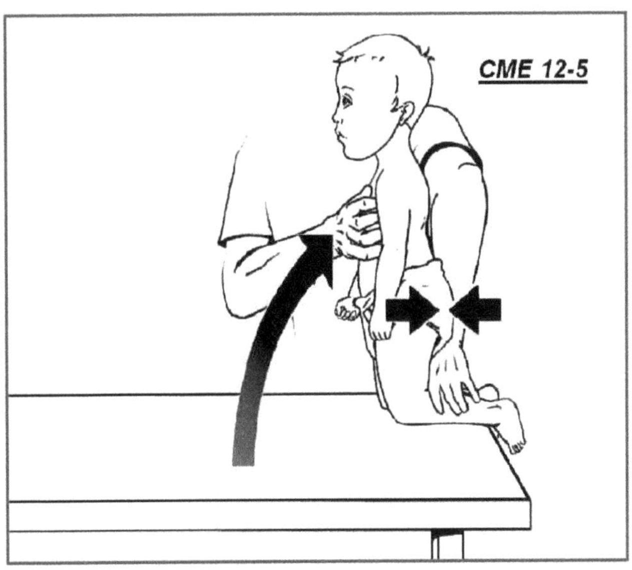

CME 13 LOTUS POSITION

MAIN GOAL: To provoke trunk vertical stability and basic balance reactions.

INITIAL POSITION: The child lies down supine (face up) on the table.

HOLD: With your left hand palm up, grab the right ankle and then cross left ankle on top keeping both knees bent (fig. 1). With the right hand hold under neck and right shoulder right (fig. 2). Do not apply unnecessary pressure on neck with your nails or/and fingertips.

MOVEMENT: Roll the child completely over her/his right side (fig. 2) and slowly push the trunk upwards by pivoting on buttocks over your left forearm (fig. 3), until reaching upright sitting position (fig. 4). Now stay steady for 10, then tapping on the shoulder by ten strokes (fig.5), and finally let him/her free to the count of 10 (fig. 6). Follow instructions of your CME practitioner.

EXPECTED IMPROVEMENT: The child keeps the vertical stability while the trunk is free.

CME 13 LOTUS POSITION

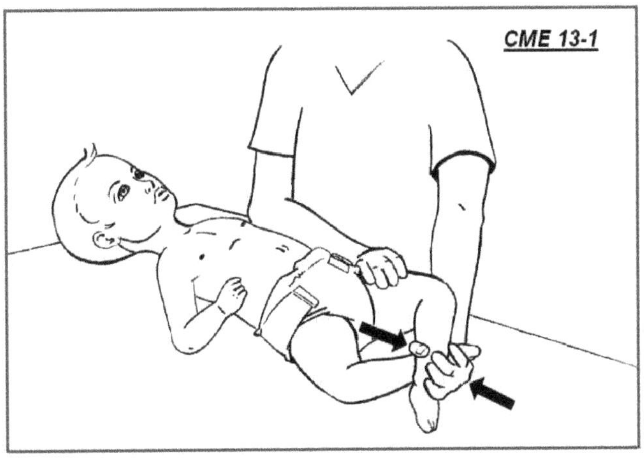

CME 13-1

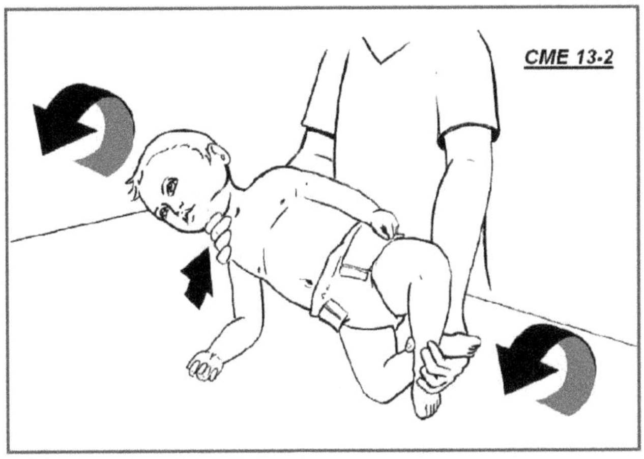

CME 13-2

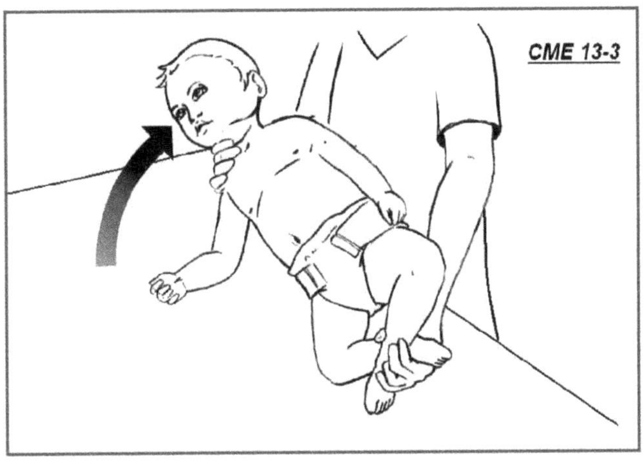

CME 13-3

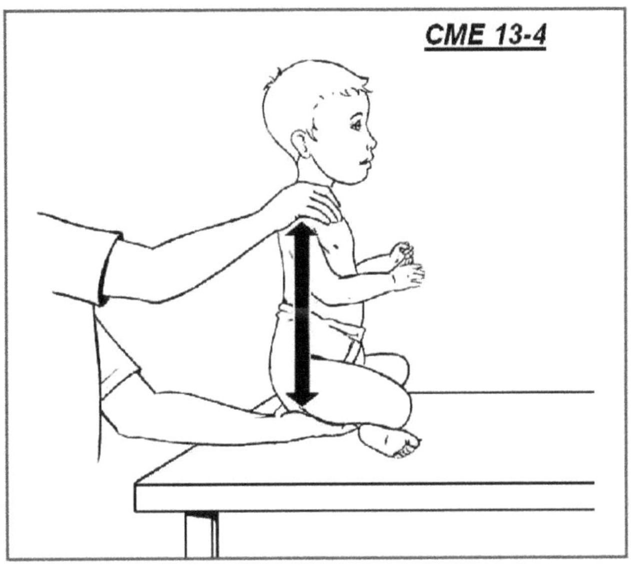

CME 13-4

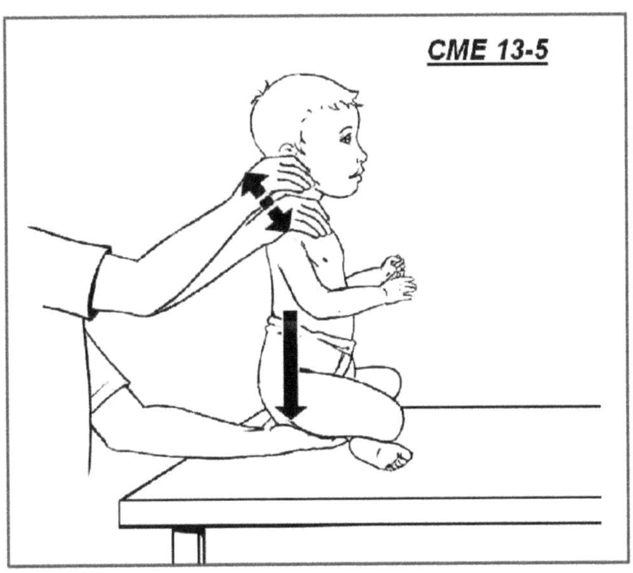

CME 13-5

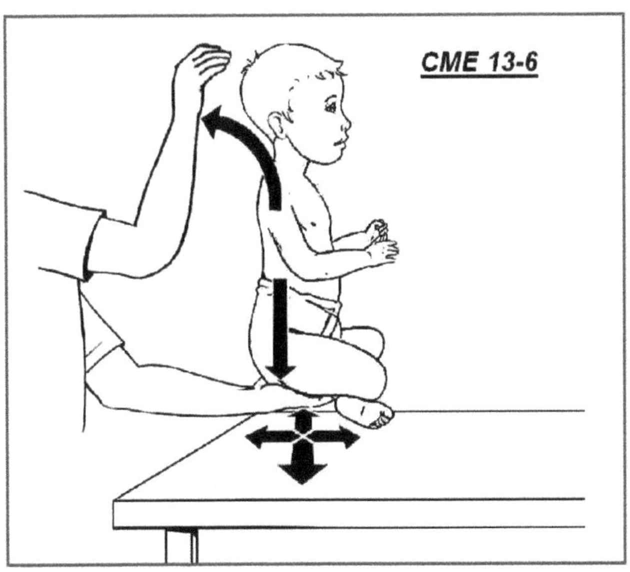

CME 13-6

CME 13 LOTUS POSITION

CME 14 KNEELING BY MID TRUNK, TILT AND RELEASE

MAIN GOAL: To provoke head-trunk-hips alignment.

INITIAL POSITION: Child is lying down on his/her stomach or in a prone position, with both feet off of edge of the table.

HOLD: With your right hand at the center of the child's trunk, and with your left hand holding both legs together below the knees (fig. 1).

MOVEMENT ONE: Place the child in a crawling position (fig. 2), push the trunk softly backwards until making contact with your inclined left forearm (fig. 2), and proceed until the child reaches the kneeling position (fig. 3). With the child's trunk firmly supported between hand and forearm, apply lateral inclination to the right side keeping the trunk aligned vertically and in full contact with your left forearm (fig. 4), then reverse the movement inclining the trunk to the right side (fig. 5). Keep this rhythm until five slanting movements to each side have been completed.

MOVEMENT TWO: Once the first movement is completed, stop at the center and rapidly move the hand on and off from the trunk in a series of ten movements (fig. 6). Follow the indications of your CME therapist.

EXPECTED IMPROVEMENT: The child maintains the trunk straight up all along the exercise.

CME 14 KNEELING BY MID TRUNK, TILT AND RELEASE

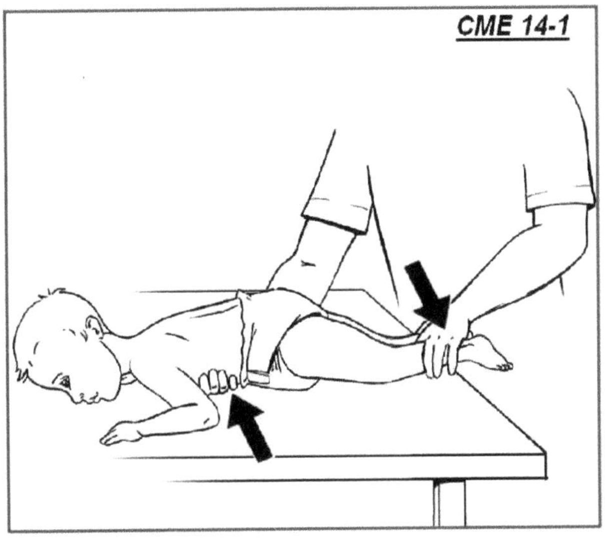

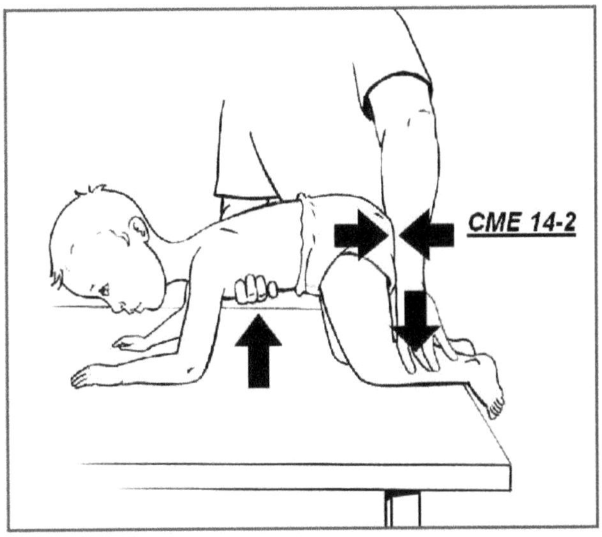

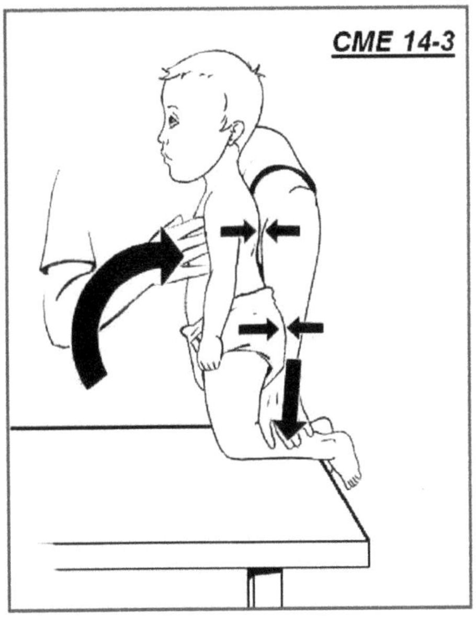

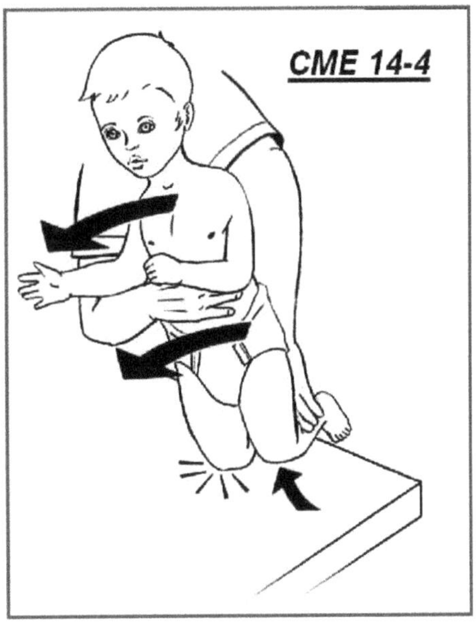

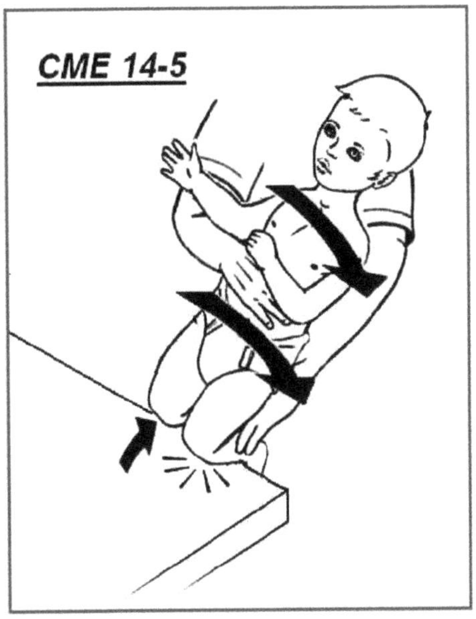

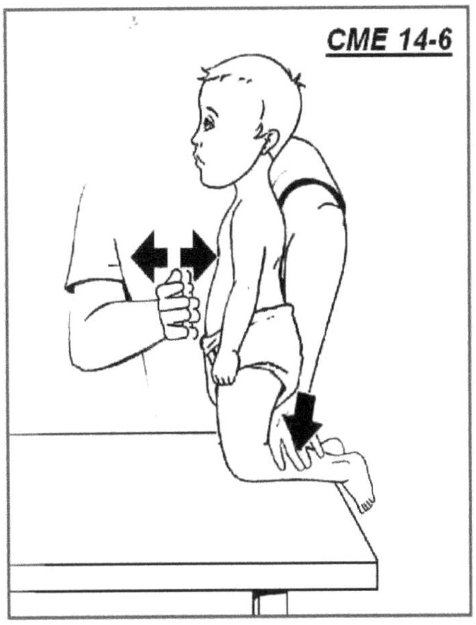

CME 15 SUPINE TO SITTING BY MID TRUNK

MAIN GOAL: To instigate the trunk upright response.

INITIAL POSITION: Child is lying down on his back or prone position, with both knees bent at edge of the table, perpendicular to you (fig.1)

HOLD: Firmly and gently put both hands on the mid trunk as indicated in figure 1.

MOVEMENT: Turn the child completely to the left side until the shoulders become perpendicular to the table as illustrated in figure 2. Now pivot the child over the left buttock, with the child moving slowly upward until reaching the sideways sitting position as indicated in figure 3. The child reaches the sitting position with his/her trunk facing toward the right side. Then, slowly begin to spin her/his trunk towards the right side as shown in figure 5.

Continue on next page

CME 15 SUPINE TO SITTING BY MID TRUNK

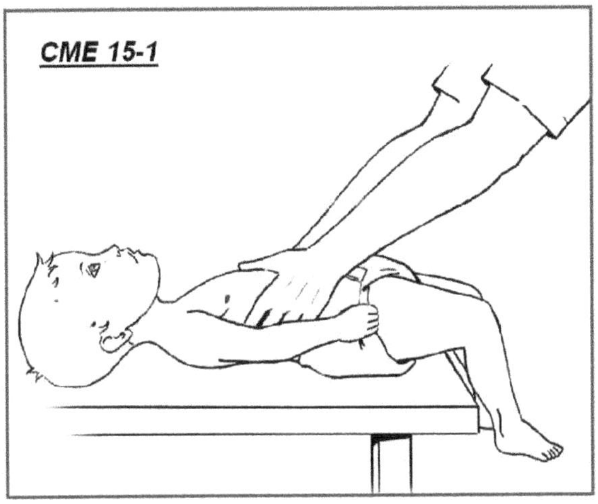
CME 15-1

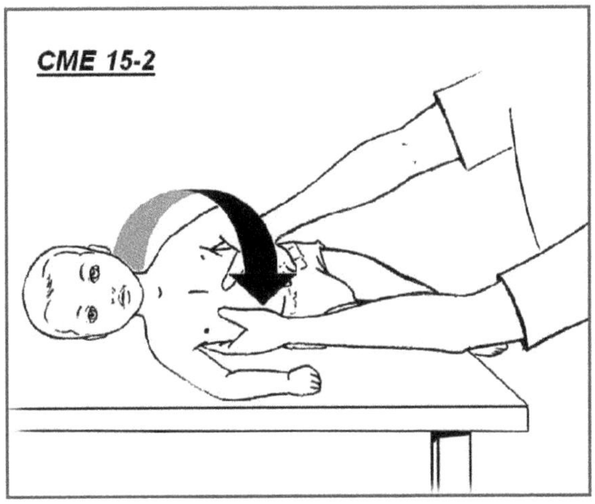
CME 15-2

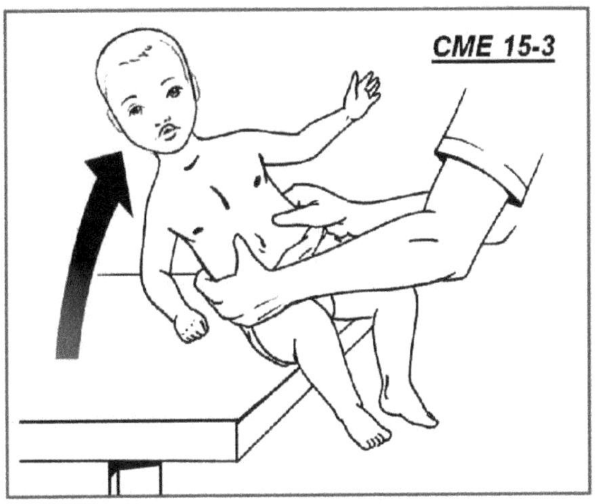
CME 15-3

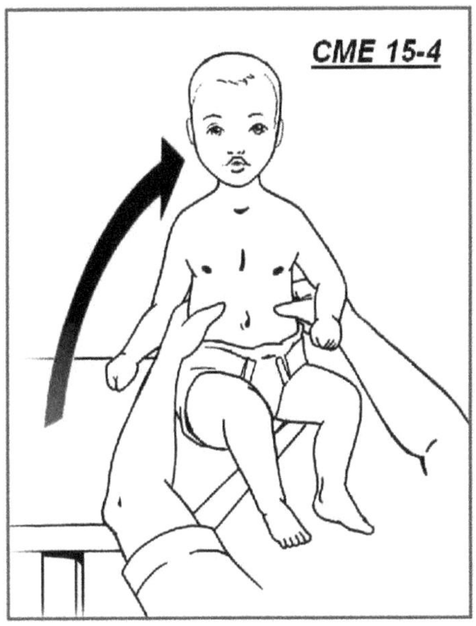
CME 15-4

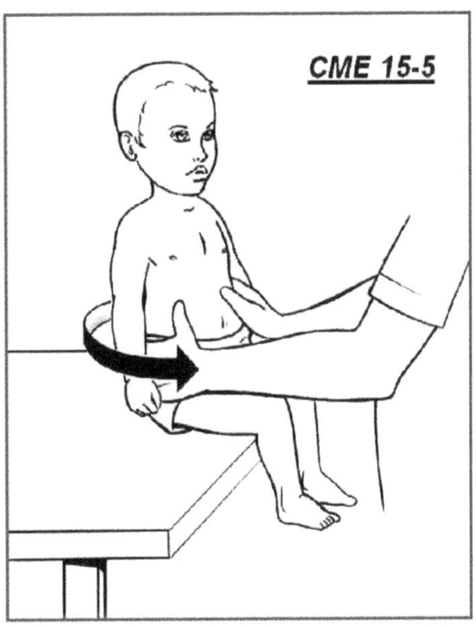
CME 15-5

CME 15 SUPINE TO SITTING BY MID TRUNK

And figure 6, taking care to maintain the trunk totally vertical during the entire spinning movement. Now the child is in full lateral position and you pivot him/her over the left buttock slowly downwards and sideways as illustrated in figures 7 and 8, finishing the movement by turning the child face up (supine position) (fig. 9).

EXPECTED IMPROVEMENT: The child actively anticipates the direction of the movement going up, and also assists the movement by pushing up on the table with his hand.

CME 15 SUPINE TO SITTING BY MID TRUNK

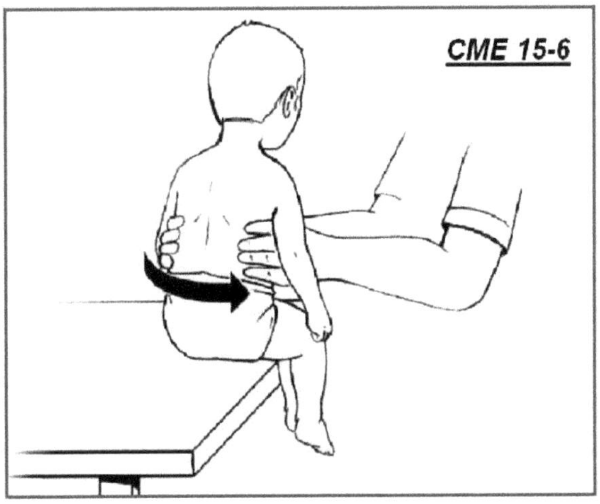

CME 15-6

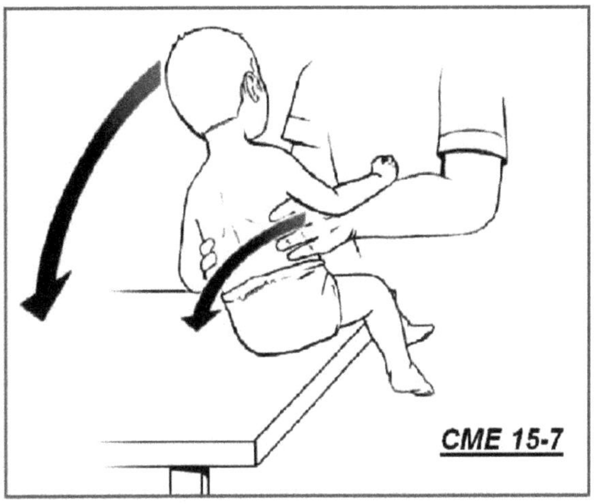

CME 15-7

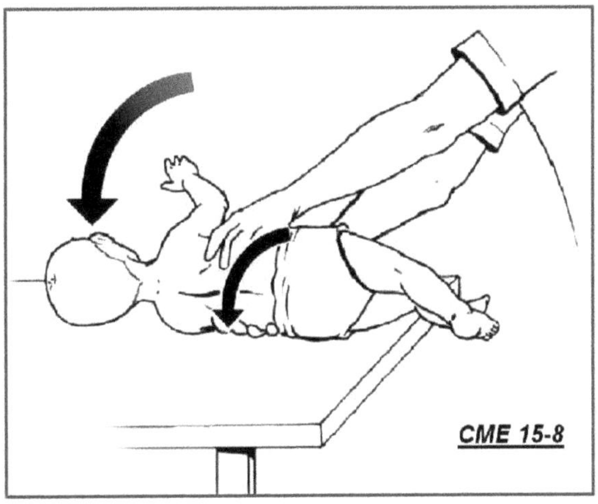

CME 15-8

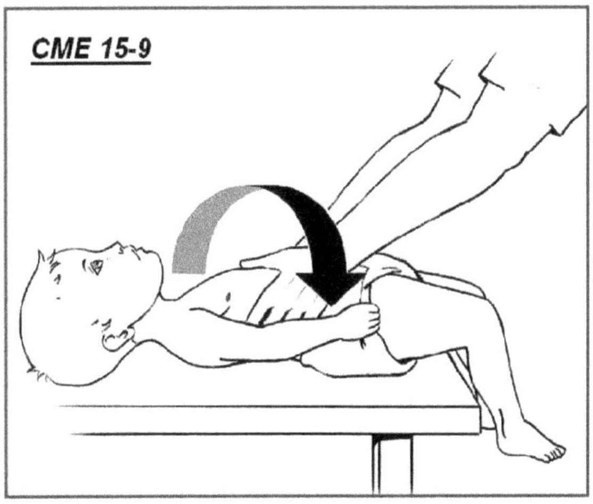

CME 15-9

CME 16 SUPINE TO SITTING BY PELVIS

MAIN GOAL: To provoke the upright trunk response with hand pushing reaction.

INITIAL POSITION: The child is lying on his/her back (prone position), with both knees bent at edge of table, parallel to you.

HOLD: Firmly with both hands at the top of his/her hips as indicated in figure 1. (Do not dig into the abdomen with your thumbs).

MOVEMENT: Turn the child completely sideways to his/her right side until the shoulders are perpendicular to the table (fig. 2). Then drive the trunk slowly upwards (fig. 3), until reaching the upright sitting position (fig. 4). Place right hand on the neck (fig. 5), and lower the child back to the table to supine position as illustrated in figure 6. Repeat the exercise as indicated by your CME therapist.

EXPECTED IMPROVEMENT: The child anticipates the reaction and moves the head and trunk towards the sitting position easily.

CME 16 SUPINE TO SITTING BY PELVIS

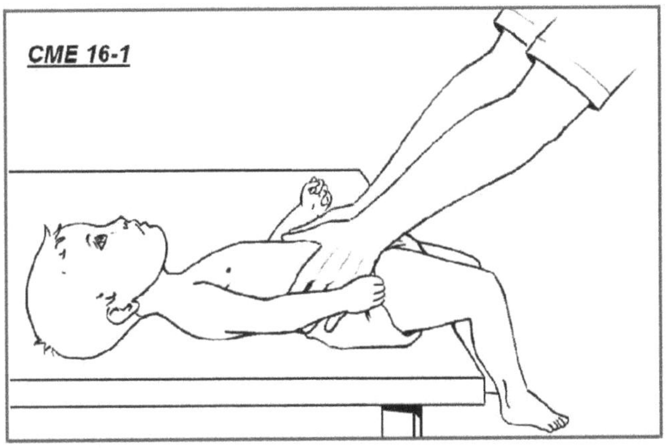

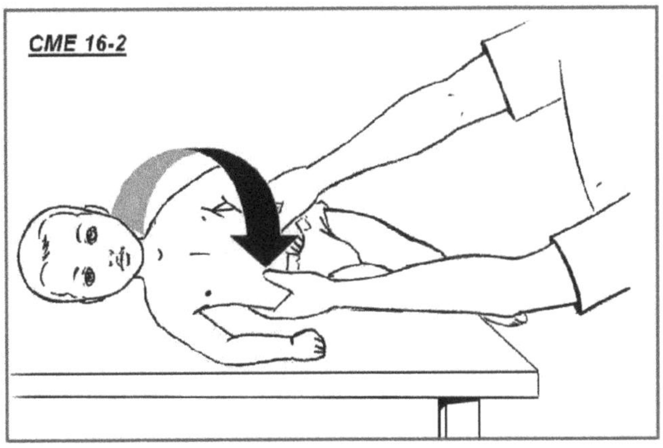

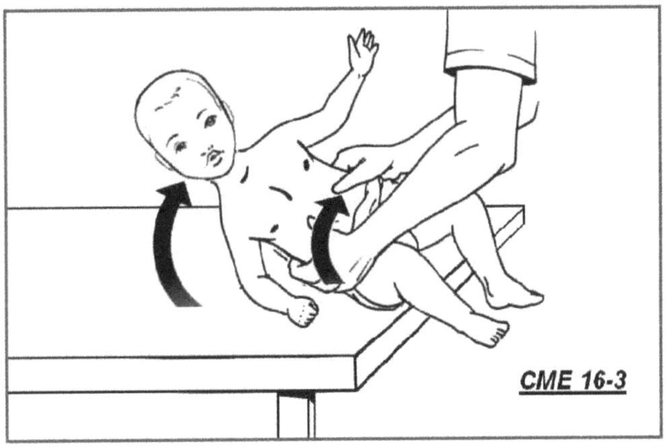

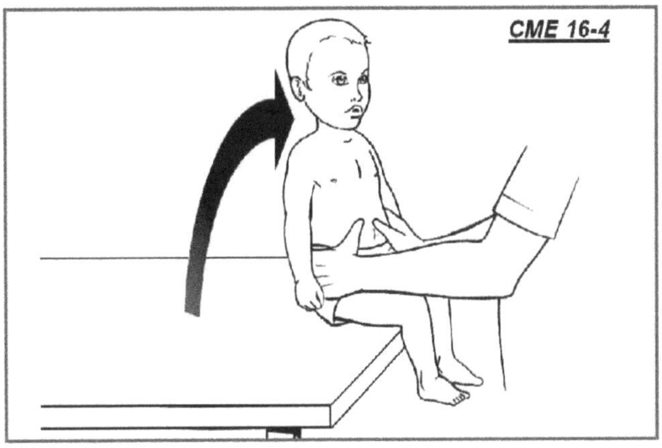

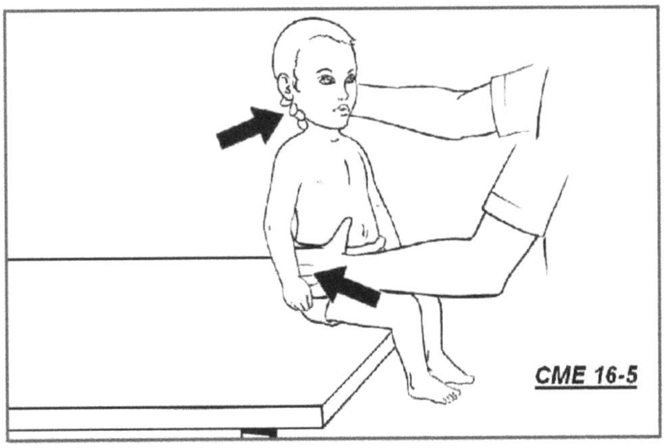

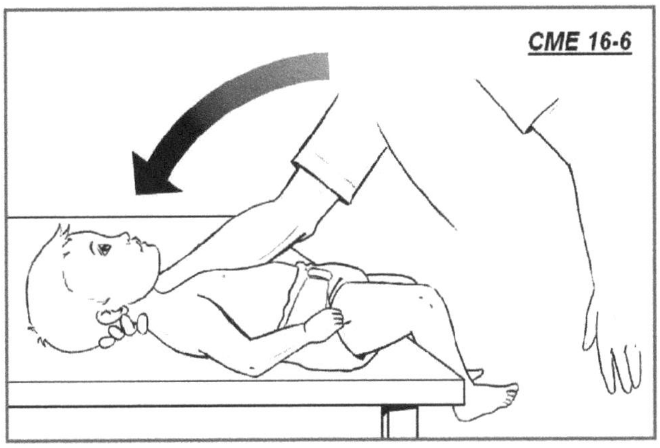

CME 17 SUPINE TO SITTING BY SANDWICH

MAIN GOAL: To provoke the trunk's automatic upright response by a static beginning.

INITIAL POSITION: Child is lying down in a supine position, with both feet off the edge of the table, perpendicular to you.

HOLD: Tuck right hand under the mid point of the back, and the left on top of the mid trunk exactly as a mirror image of the right hand (fig.1).

MOVEMENT: Without spreading your fingers, firmly compress with both hands the center of the trunk and turn the child completely sideways to the left side until the shoulders are perpendicular to the surface of the table (fig. 2). Firmly contain the lateral position with both hands waiting for the child's spontaneous upright sitting reaction as shown in figure 3. As soon as the child begins to move to a sitting position, you should assist the response with both hands until he/she reaches the sitting position as indicated in figure 4. Follow the instructions of your CME therapist.

EXPECTED IMPROVEMENT: The child moves rapidly by him/herself to a sitting posture.

CME 17 SUPINE TO SITTING BY SANDWICH

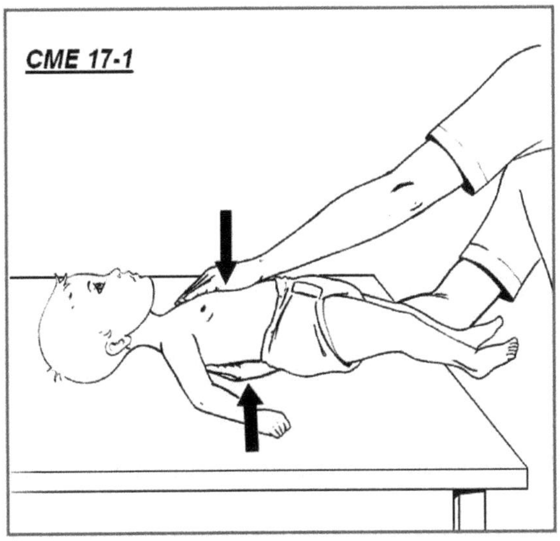

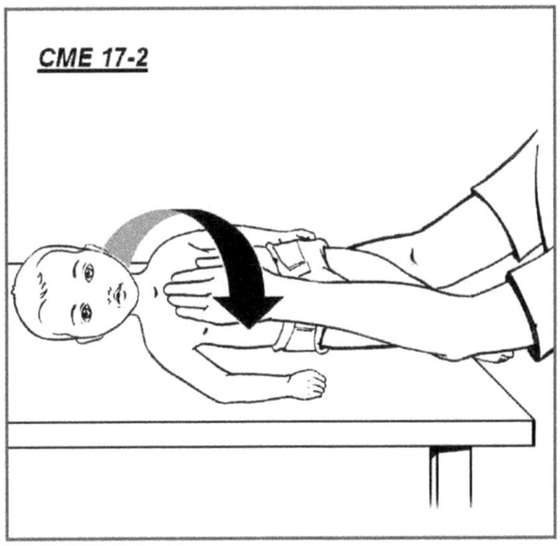

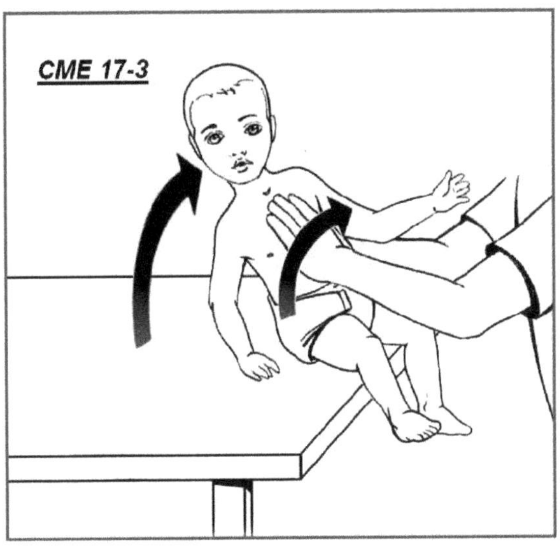

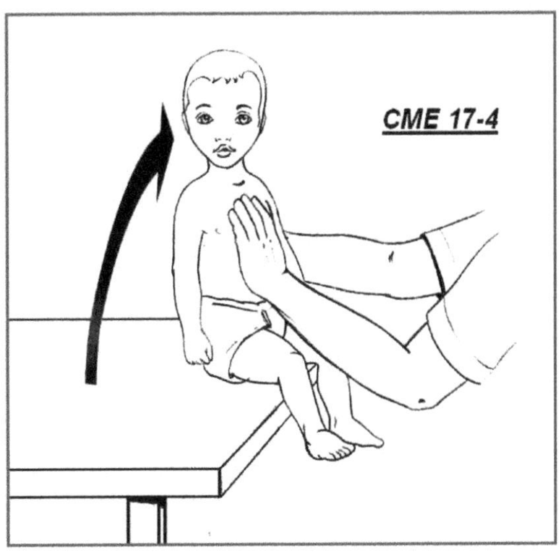

CME 18 SUPINE TO SITTING BY LOW SANDWICH

MAIN GOAL: To provoke automatic trunk upright response by static beginning.

INITIAL POSITION: Child is lying down in supine position, with both feet off the edge of the table, perpendicular to you.

HOLD: Tuck right hand under the lower back (base of palm over sacrum bone), and your left hand on top of the lower abdomen (belly button at the center of your hand) exactly as a mirror image of the right hand (fig.1).

MOVEMENT: Without spreading your fingers, firmly compress with both hands to the center of the trunk and turn the child completely sideways, pivoting over his/her right side until the shoulders reach the vertical position as illustrated in figure 2. Waiting now for the child to become active and spontaneous upright sitting reaction as illustrated in figure 3. As soon as the child moves to a sitting up position, you should assist the response (fig. 4), until the child reaches the final sitting position (fig. 5). Follow your PT instructions.

EXPECTED IMPROVEMENT: The child sits up quickly as soon as he/she is placed sideways.

CME 18 SUPINE TO SITTING BY LOW SANDWICH

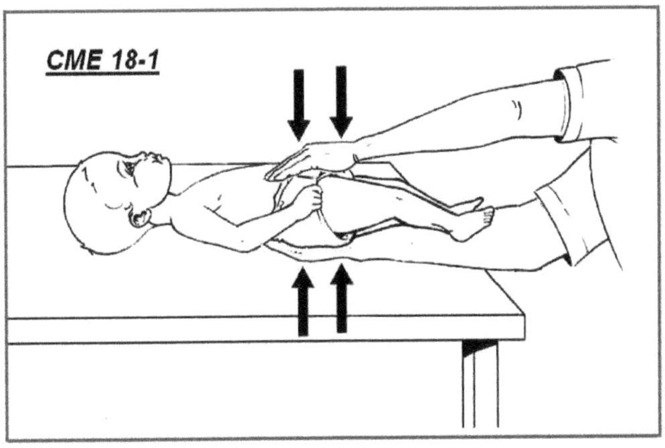

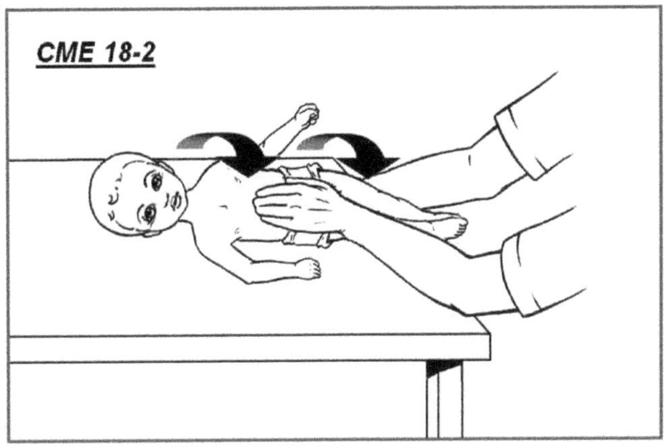

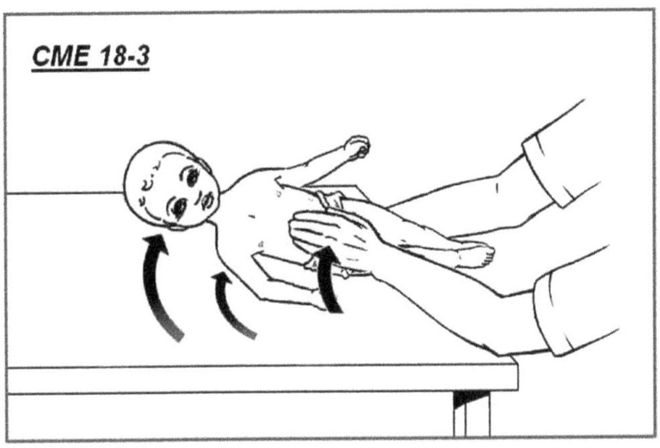

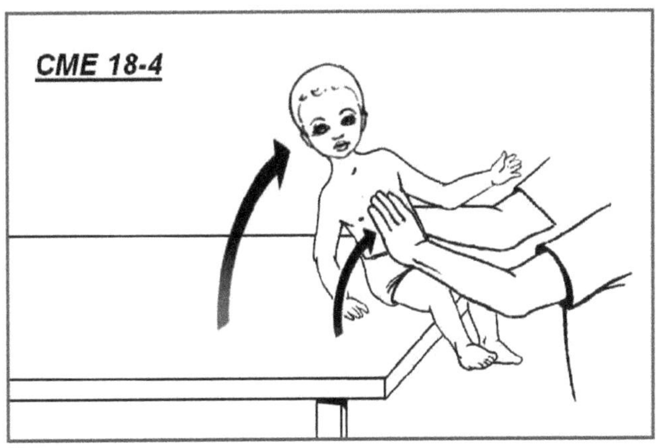

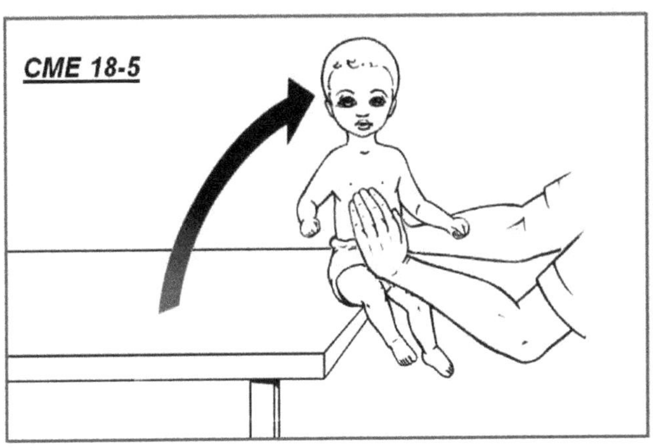

CME 19 SUPINE TO SITTING BY ABOVE KNEES

MAIN GOAL: To provoke automatic trunk upright response.

INITIAL POSITION: Child is lying down in supine position, with both knees bent at edge of the table, perpendicular to you.

HOLD: With both hands by thighs above the knees as illustrated in the figure 1.

MOVEMENT: With both hands turn the child completely on his/her side, pivoting over his/her right side until the shoulders reach the vertical position as illustrated in figure 2, and with a non stop pace, carefully pull diagonally downwards with both thighs in order to provoke the automatic upright sitting reaction as illustrated in figure 3. As soon as the child moves to a sitting up position, you should assist the response (fig. 4), until the child reaches the final sitting position (fig. 5). Put one hand on nape (fig. 6) and place the child back on the table in a supine position (fig. 7). Follow the instructions indicated by your CME therapist.

EXPECTED IMPROVEMENT: The child sits up quickly as soon as he/she is placed sideways.

CME 19 SUPINE TO SITTING BY ABOVE KNEES

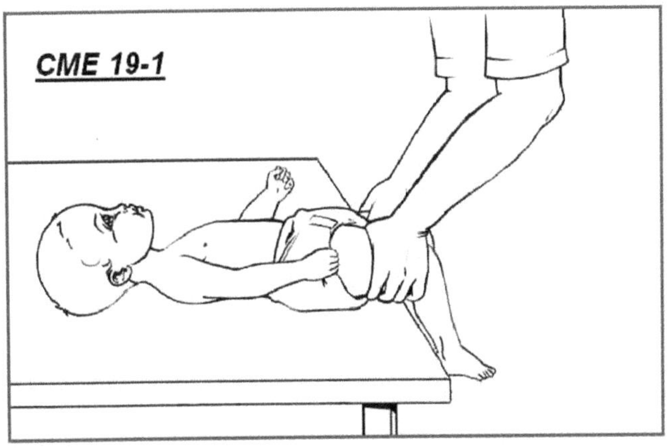

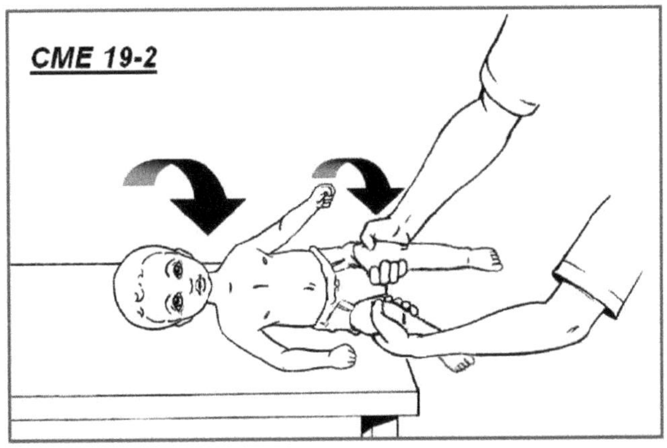

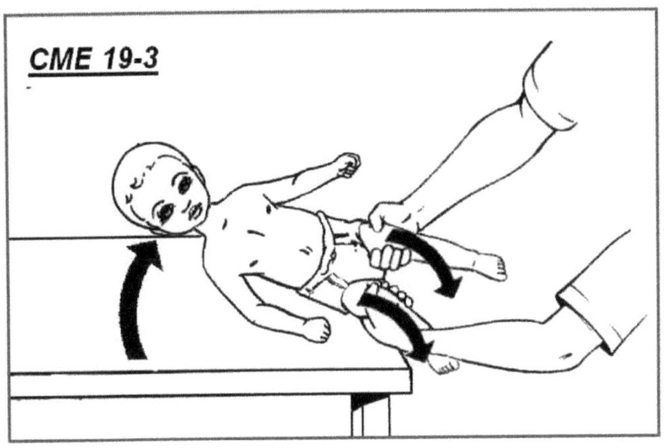

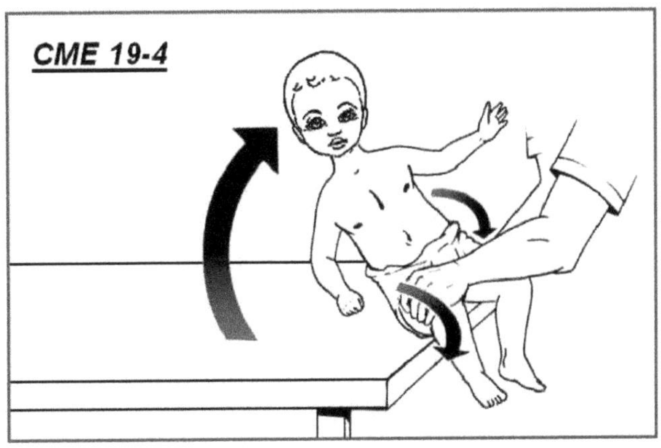

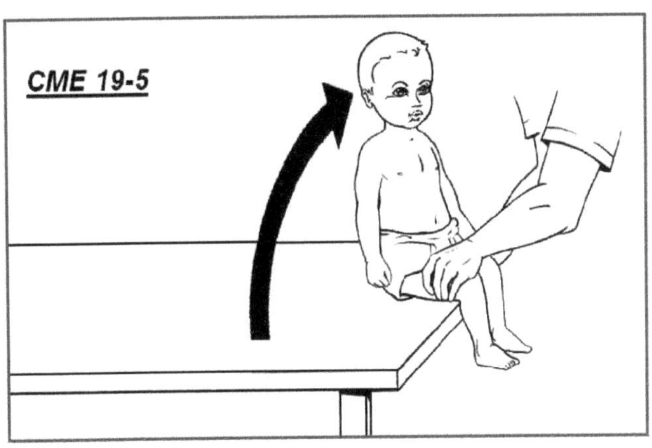

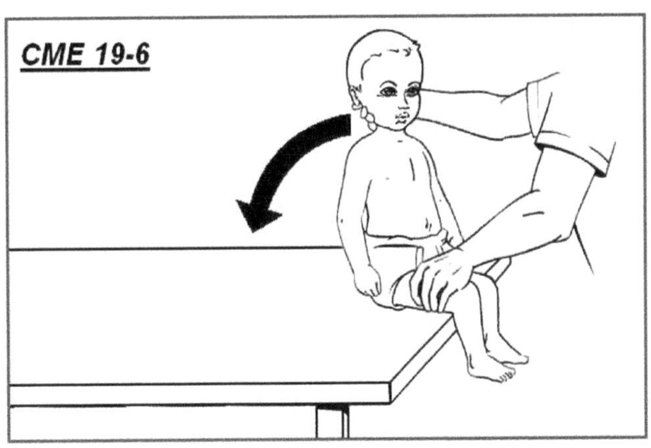

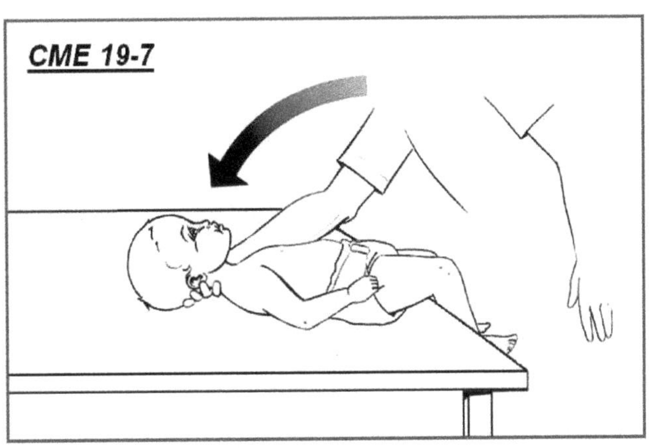

CME 20 SUPINE TO SITTING FROM BEHIND

MAIN GOAL: To provoke sitting up reaction counteracting the strong hyper extension of trunk.

INITIAL POSITION: Lying face up on the table crosswise to you with head toward the left side, (fig. 1).

HOLD: With left hand under his/her lower back grasp the trunk above the left hip (side of pelvis), and with your right hand above the right hip. Both thumbs extended towards the lower back, (fig. 1).

MOVEMENT: Turn the child completely over his/her left side as indicated in figure 2. Then with both hands pivot the trunk upwards over the left buttock taking special care in inducing functional hip flexion as illustrated in figure 3, until reaching full vertical sitting posture as shown in figure 4. Repeat the exercise according to the instructions of your CME practitioner.

EXPECTED IMPROVEMENT: The child sit up quickly without fighting the trunk extension.

CME 20 SUPINE TO SITTING FROM BEHIND

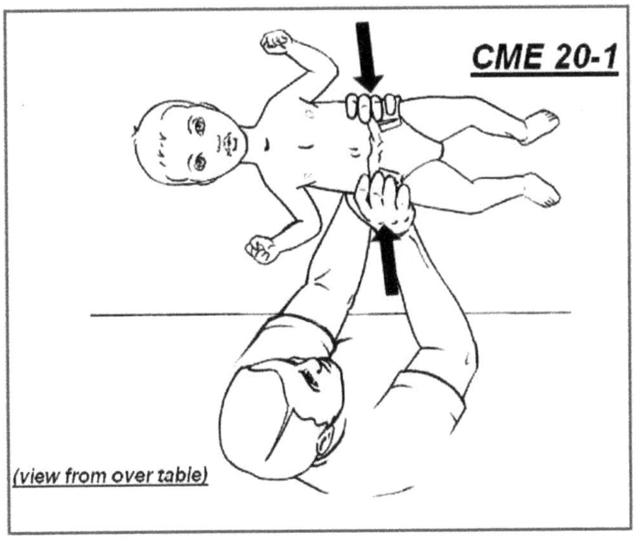

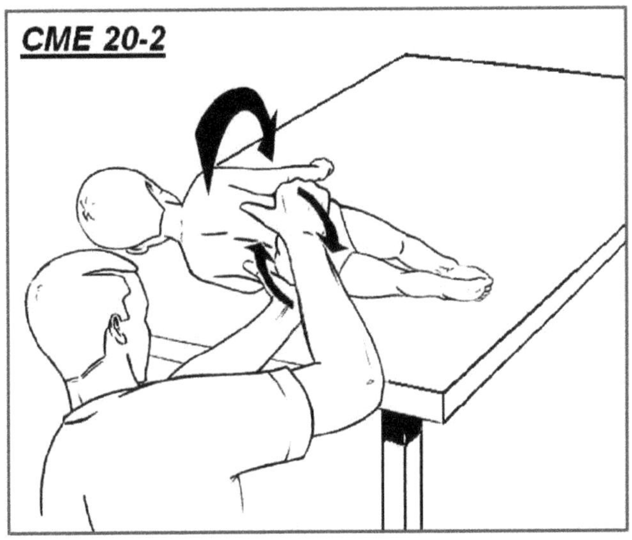

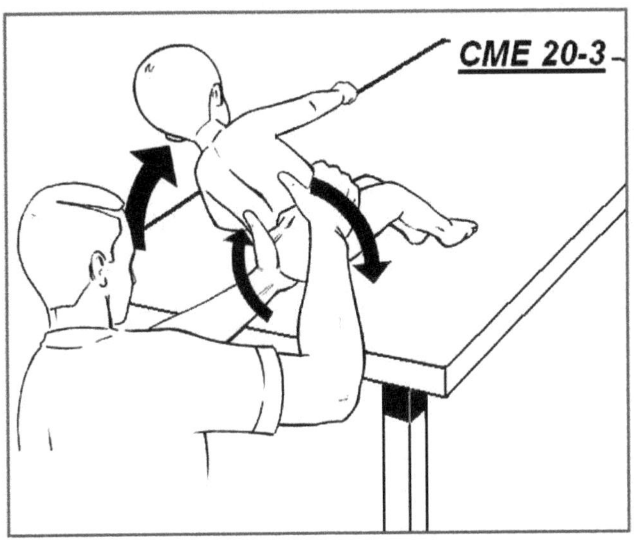

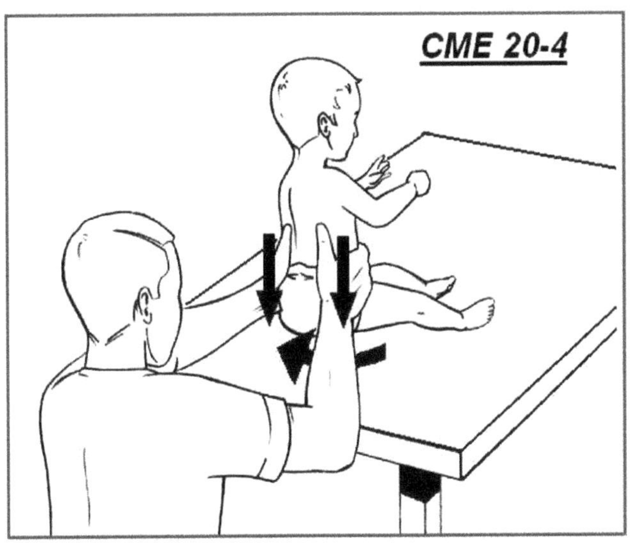

CME 21 SUPINE TO SITTING BY SHOULDER AND OPPOSITE LEG

MAIN GOAL: To provoke lateral sitting up reaction.

INITIAL POSITION: Child is lying on his/her back (supine position) crosswise to you with head to the left side (fig.1).

HOLD: With left hand on right shoulder as follows: little and ring finger under armpit, middle and pointer finger are placed straight behind the shoulder over scapula bone (details in circle), and thumb over the shoulder. With your left hand hold the left leg below the knee and bend the right leg 90 degrees (fig.1).

MOVEMENT: Moving both hands simultaneously, pivot her/his trunk totally sideways towards you, (fig. 2), placing the right foot on the table (fig. 3). Stop moving and wait for the count of 10 in order to provoke the active response of the child to drive the trunk to upright sitting position pushing also with the right hand on the table, (figs. 4 and 5).

EXPECTED IMPROVEMENT: The child moves with no difficulty the head and trunk towards an upright sitting position and pushes up with his/her hand on the table.

CME 21 SUPINE TO SITTING BY SHOULDER AND OPPOSITE LEG

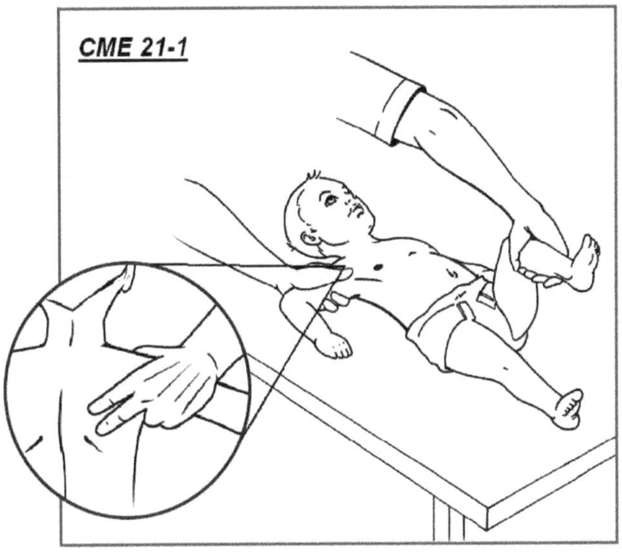

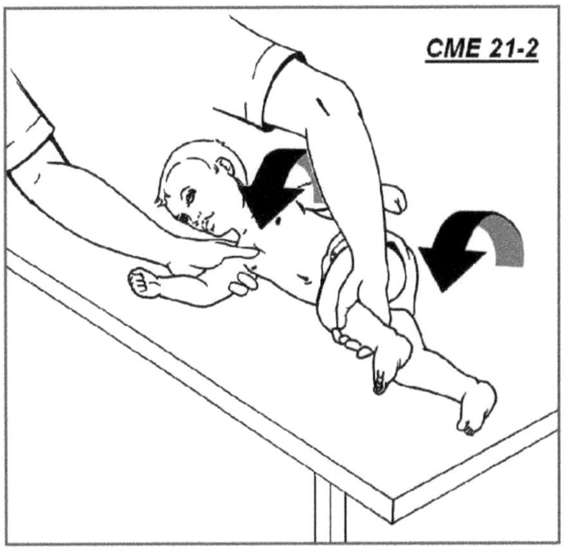

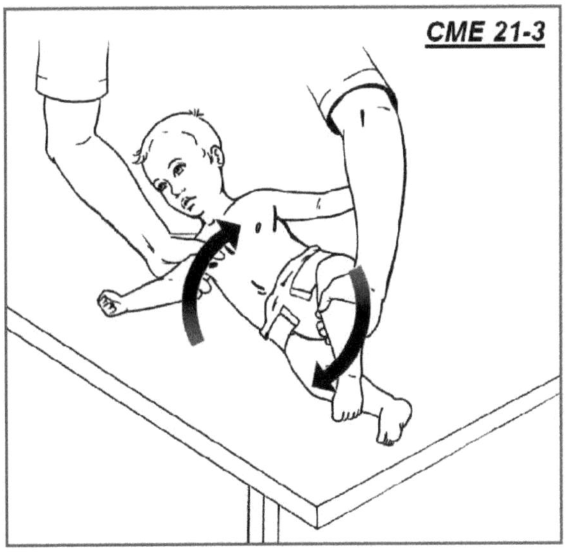

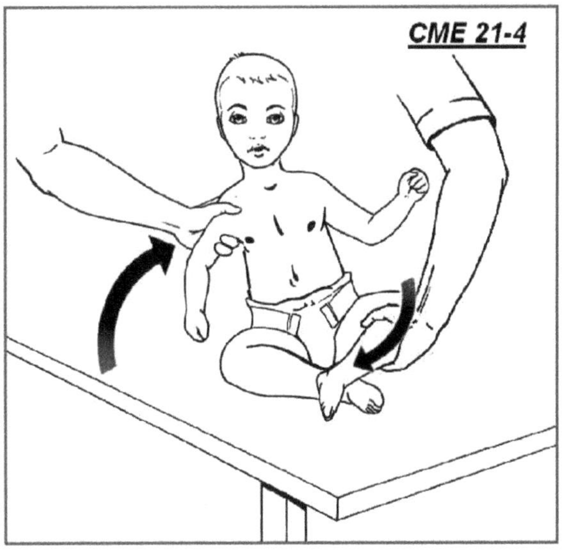

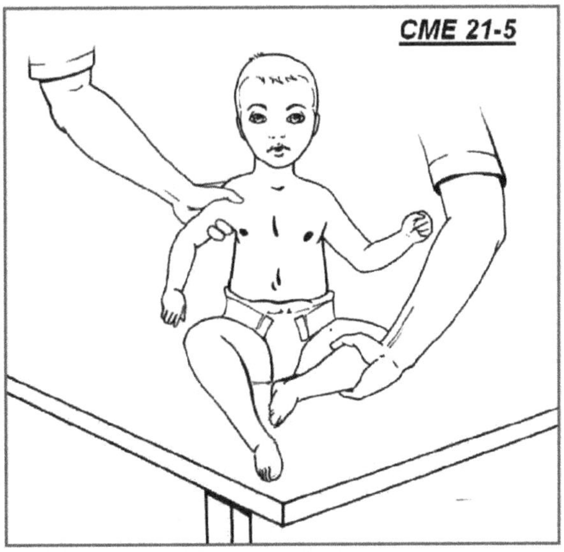

CME 22 TRUNK EXTENSION BY ABDOMEN

MAIN GOAL: To provoke active trunk vertical extension.

INITIAL POSITION: Child is lying in prone position or face down on the table, placed perpendicular to you as shown in the figure 1.

HOLD: Tuck both forearms under the abdomen and compress buttocks and legs against your lower abdomen as indicated in the figure 2.

MOVEMENT: Slowly slide the child out of the table as illustrated in the figure 3, then, apply strong pressure with vibrations against his/her abdomen with the right forearm (fig. 4), until the child actively moves up to a full trunk vertical extension as shown in figure 5. Repeat the exercise as recommended by your CME therapist.

EXPECTED IMPROVEMENT: The child reaches a vertical trunk position by quick and easy extension.

CME 22 TRUNK EXTENSION BY ABDOMEN

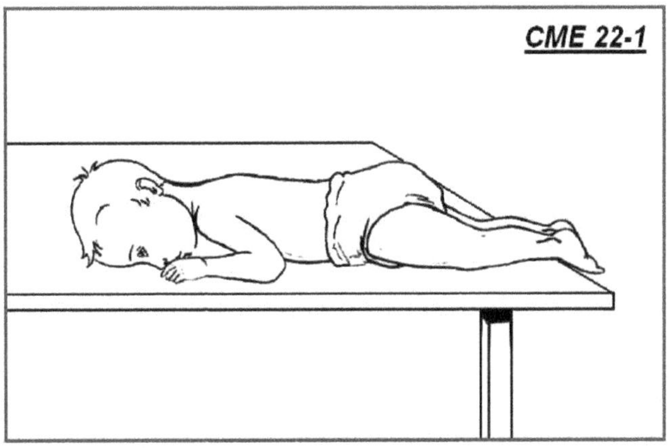

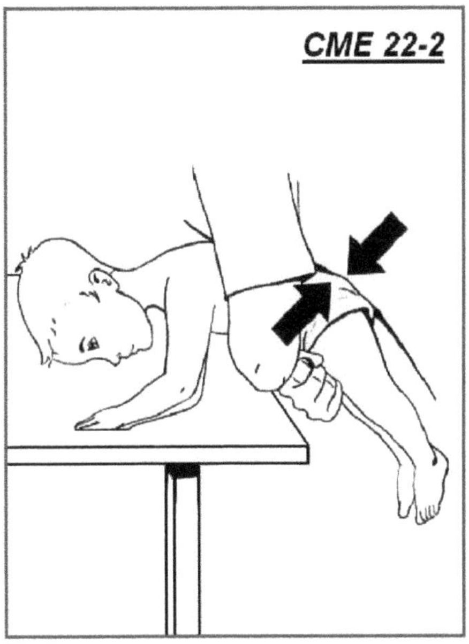

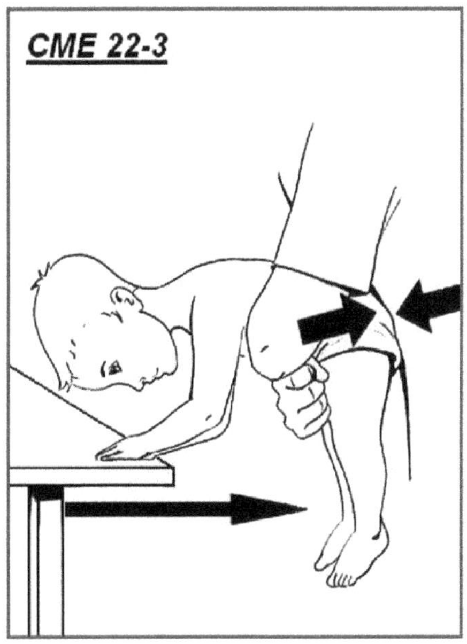

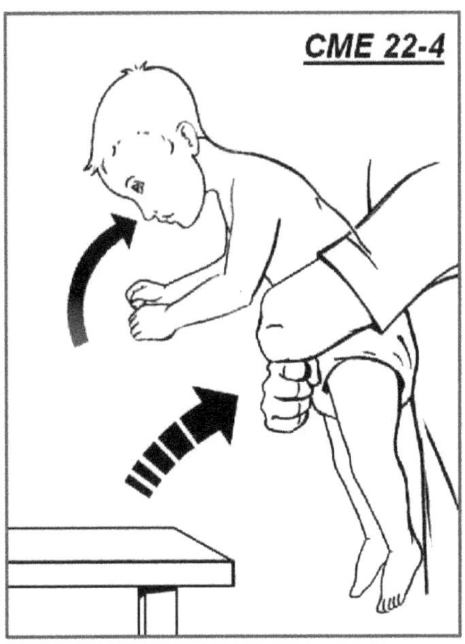

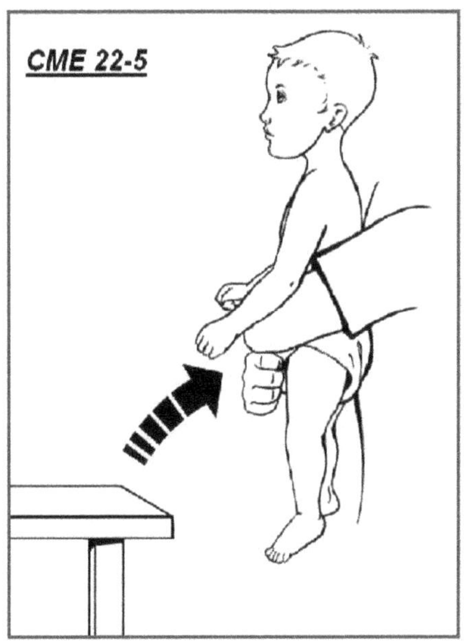

CME 23 SUPINE TO SITTING BY KNEES ON CHEST

MAIN GOAL: To provoke the sitting reaction against gravity with no hands support.

INITIAL POSITION: Child in supine position, (face up) perpendicular to you, with knees bent at the edge of the table as illustrated in the figure 1.

HOLD: With right hand on his/her ankles, tuck your left forearm behind the bent knees and compress his/her legs firmly against your mid trunk area, (fig. 2). Place your right hand on his/her left waist area as indicated by the curved arrow on figure 3. (NO DIGGING IN WITH THUMB ON ABDOMEN).

MOVEMENT: Turn the child completely on his/her side as illustrated in figure 4. (The turning movement involves a simultaneous lateral inclination of your trunk to the right side). Keep your legs apart and knees slightly bent

Continue on next page

CME 23 SUPINE TO SITTING BY KNEES ON CHEST

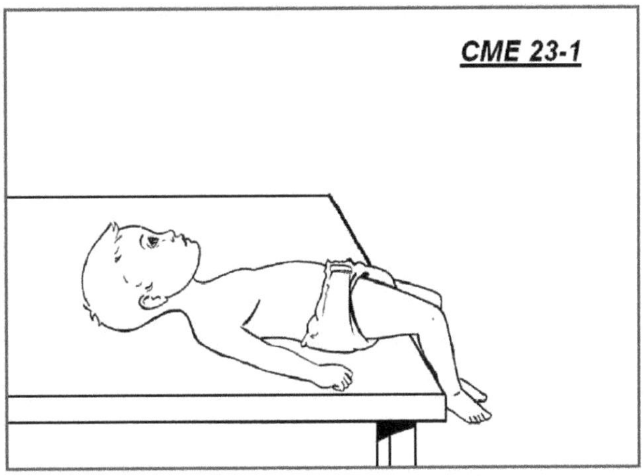

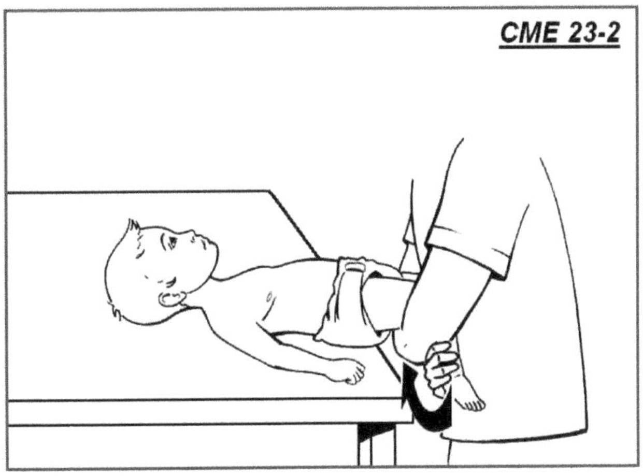

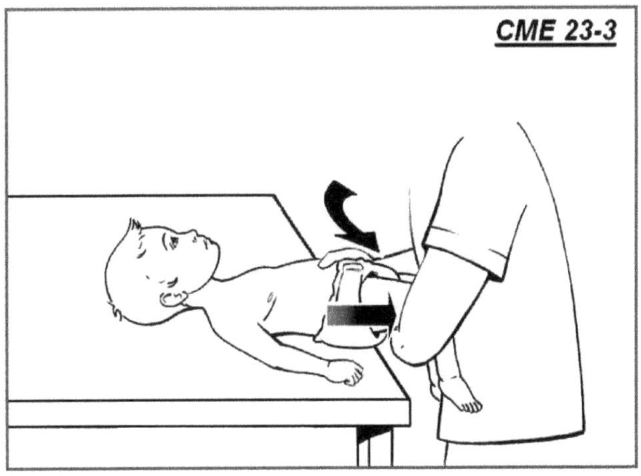

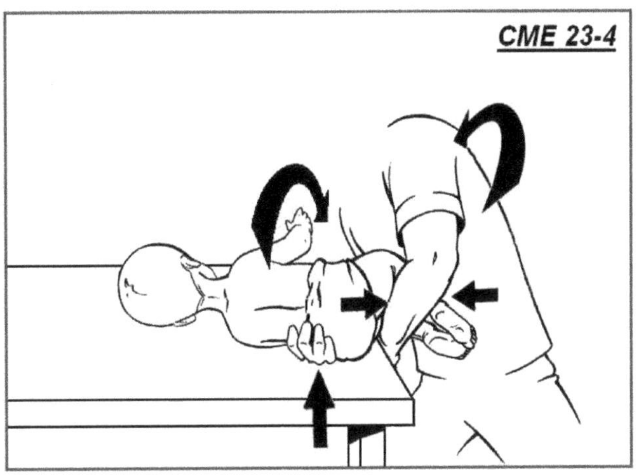

CME 23 SUPINE TO SITTING BY KNEES ON CHEST

Take steps slowly backwards, sliding the child off the table (keeping the sideways position at all times), until the shoulders reach the edge as illustrated in figure 5. At this time move your body to a vertical position keeping the compression of his/her legs on your lower chest and also firmly support your right hand on the child's lower trunk. This horizontal position suspended by the lower trunk with no support of the head and shoulder will compel the child to drive his/her trunk to a vertical position (sitting up reaction) as shown in figures 6 and 7. You should always push up softly by right hand and left forearm after the child triggers the upwards motion avoiding at all times any massive extension reaction. To repeat follow your CME practitioner instructions.

EXPECTED IMPROVEMENT: The child should react due to the force of gravity, and move up quickly to the sitting position on your forearm.

CME 23 SUPINE TO SITTING BY KNEES ON CHEST

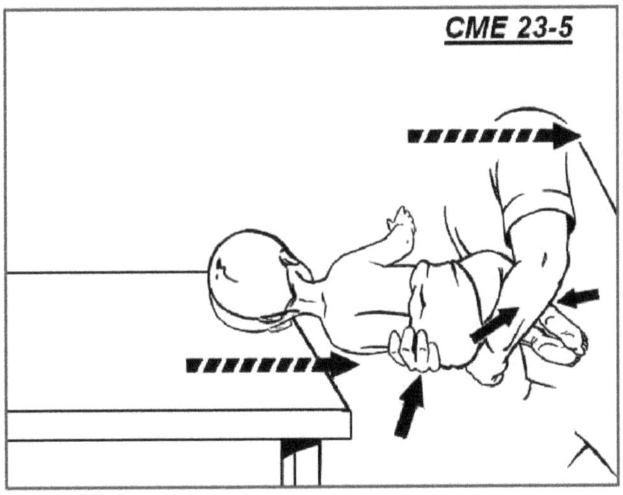

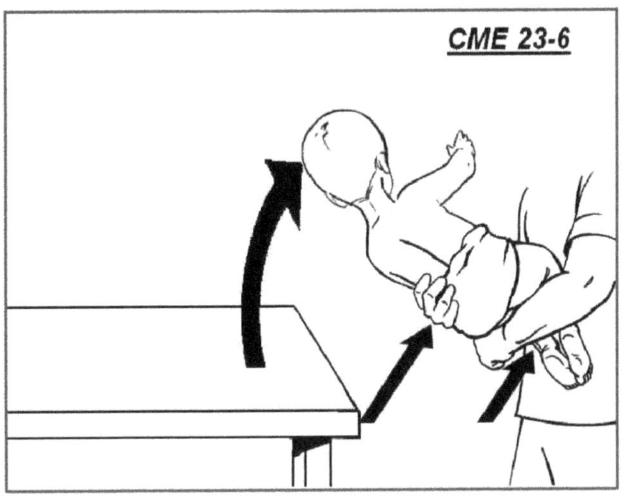

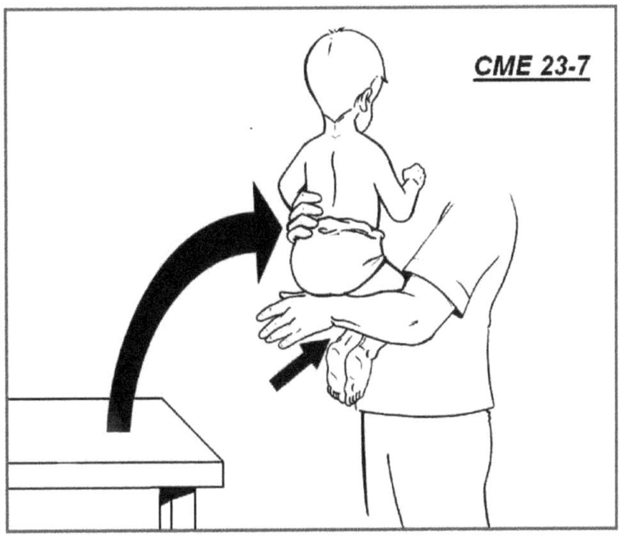

CME 24 SUPINE TO SITTING BY HORIZONTAL STRADDLE

MAIN GOAL: To provoke the sitting reaction and enhance strength of the trunk flexor muscles.

INITIAL POSITION: Child is placed face up on the table (in a supine position), perpendicular towards you, with knees bent at the edge of the table.

HOLD: Place legs around your waist and tuck tightly each leg with your forearms. Firmly hold the child by the mid trunk and strongly fix the pelvis (crotch area) against your abdomen as illustrated in figure 1. Keep your knees slightly bent.

MOVEMENT: Take steps backwards sliding the child off the table until the shoulders reach the edge as shown in figure 2. At this time turn the child totally sideways over his/her right side (fig. 3). Pull the entire body off the table to expose the trunk to the force of gravity as illustrated in figure 4. At this point the child begins to move his/her head up and the trunk towards a sitting up position assisted by you as illustrated in figure 5.

Continue on next page

CME 24 SUPINE TO SITTING BY HORIZONTAL STRADDLE

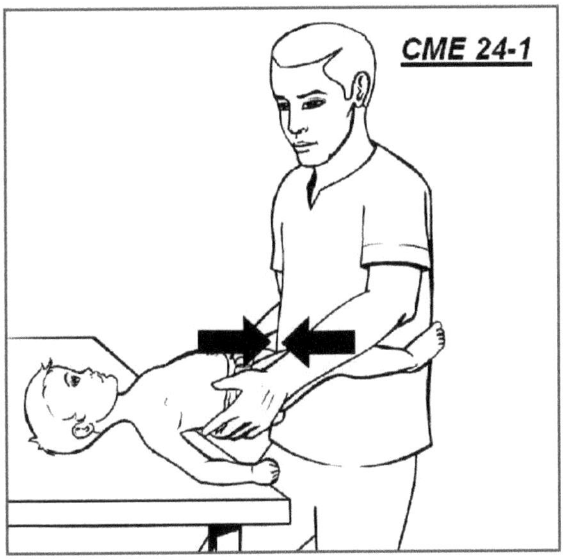

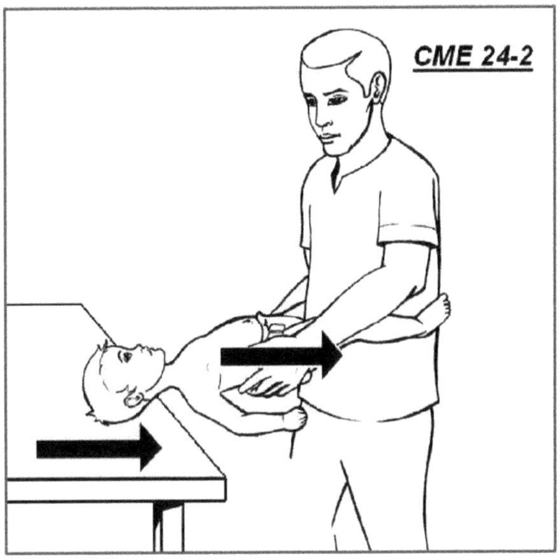

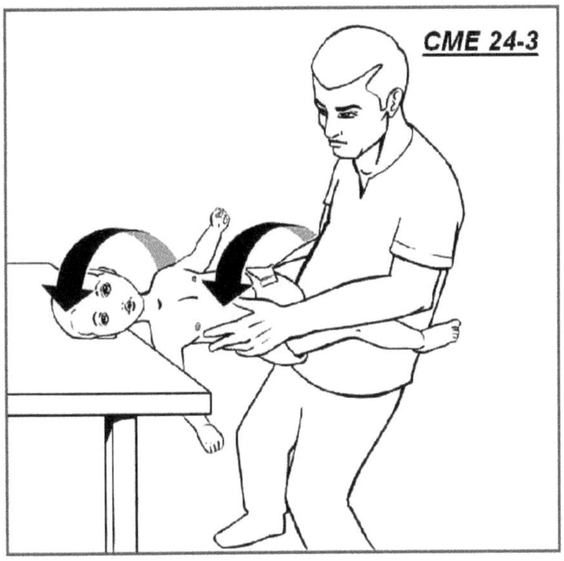

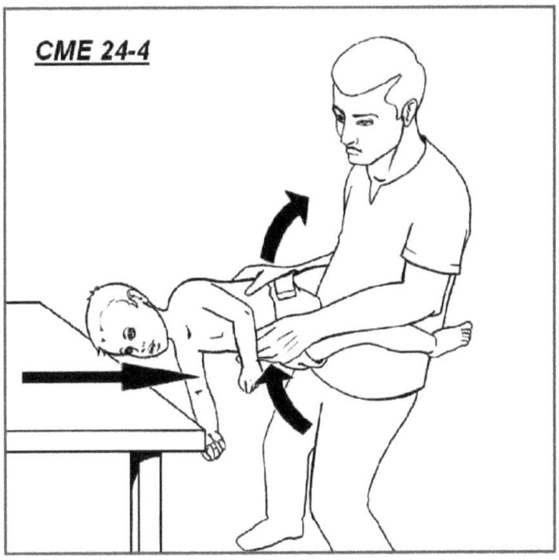

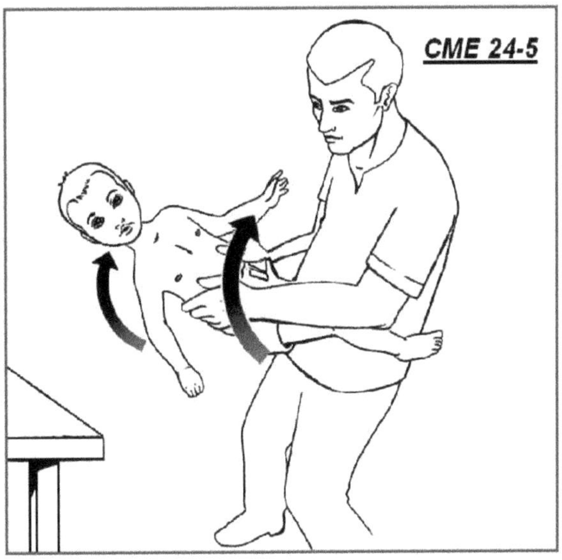

CME 24 SUPINE TO SITTING BY HORIZONTAL STRADDLE

Bend your knees and slant your trunk backwards while the child is moving to a sitting position in order to avoid premature grasping of the child over your chest as shown in figure 6. After reaching a sitting posture, place the right hand under buttocks as indicated in figure 7, and the other hand on nape and neck as illustrated in figure 8. Come close to the table (fig. 8), and carefully place the child in sitting position on the table as illustrated in figure 9. Repeat the exercise as recommended by your CME therapist.

EXPECTED IMPROVEMENT: The child should react due to the force of gravity, and move easily to the straddle sitting position.

CME 24 SUPINE TO SITTING BY HORIZONTAL STRADDLE

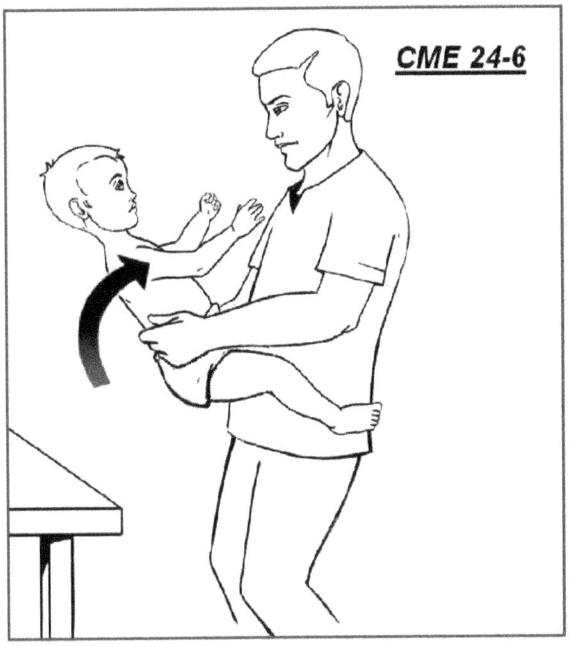

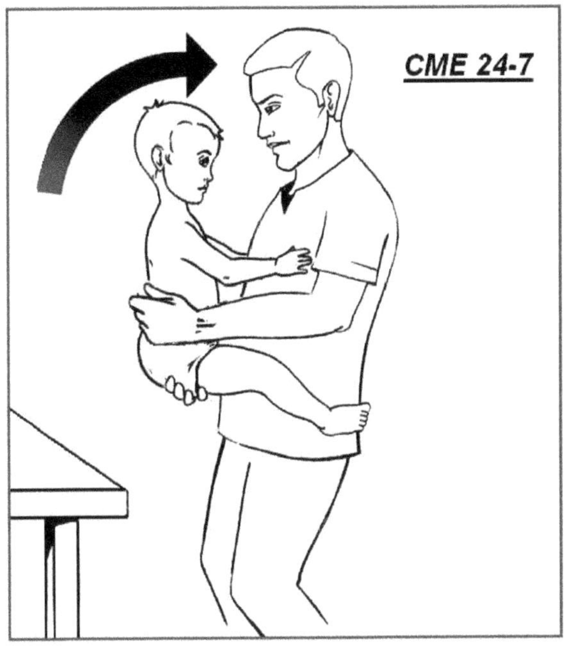

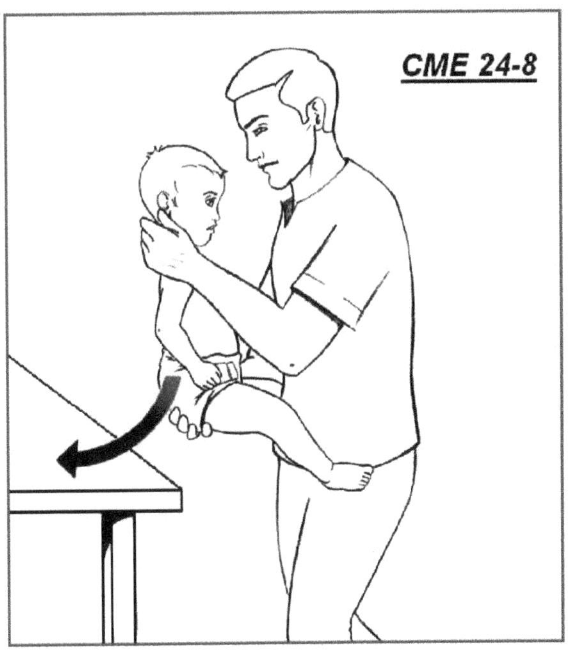

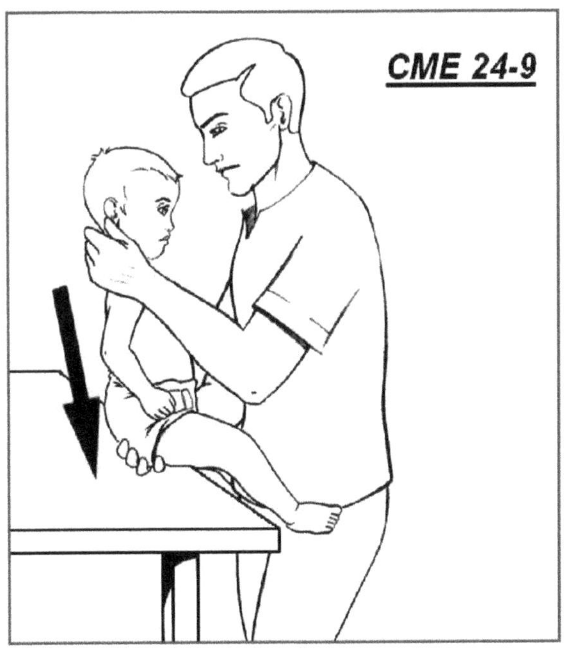

CME 25 SUPINE TO SITTING BY VERTICAL STRADDLE

MAIN GOAL: To provoke the sitting reaction and strengthen the trunk flexor muscles.

INITIAL POSITION: The child is placed face up on the table (in a supine position), perpendicular towards you, with legs spread out at edge of the table.

HOLD: Place the legs around your waist and tuck each leg tightly with your forearms, then firmly hold the child by the mid trunk and strongly fix the pelvis against your abdomen with no air between his/her crotch and your lower abdomen as illustrated in figure 1. Keep your knees slightly bent.

MOVEMENT: Take a step slowly backwards, sliding the child off the table until the shoulders reach the edge, (fig. 2), then turn the trunk completely sideways over the right side as shown in figure 3. Carefully pull the child off the table (fig. 4) moving his/her trunk softly to an upside down position as illustrated in figure 5.

Continue on next page

CME 25 SUPINE TO SITTING BY VERTICAL STRADDLE

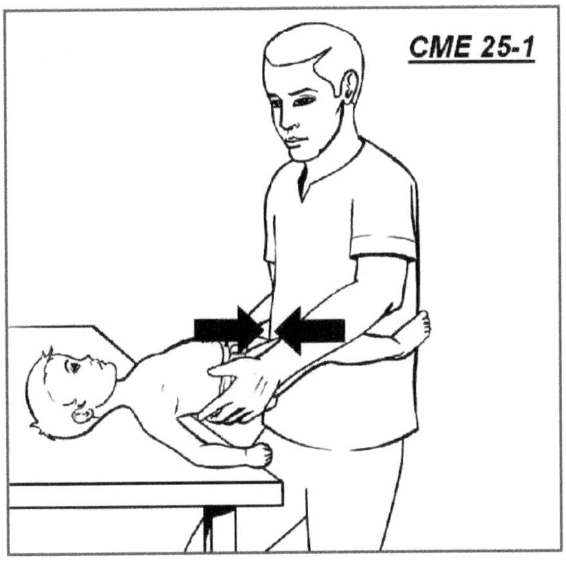

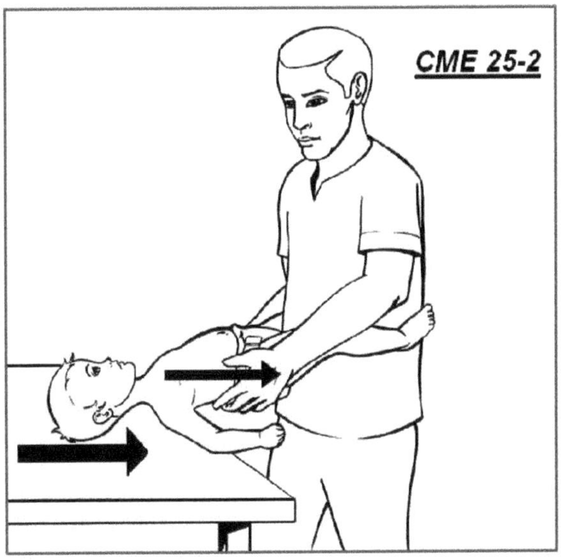

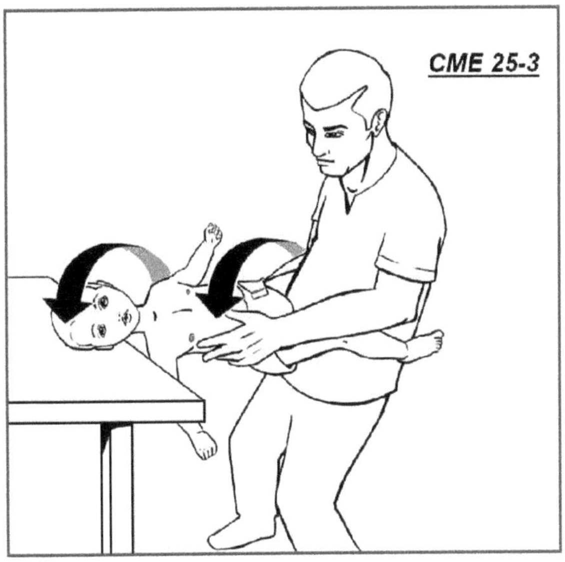

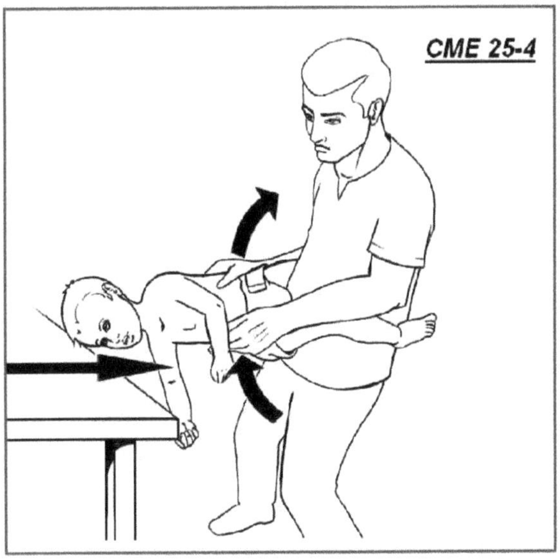

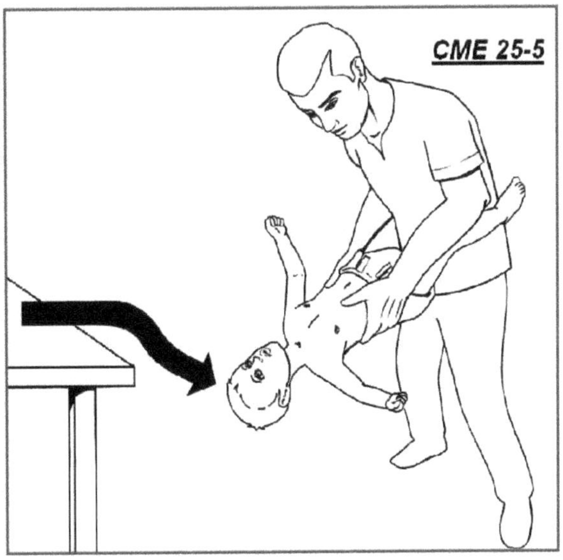

CME 25 SUPINE TO SITTING BY VERTICAL STRADDLE

Figure 6 illustrates exactly half of the exercise in a complete vertical upside down position. Your knees should be slightly bent and your trunk inclined forward. Hold the child strongly placed against to your lower abdomen at all times.

Rotate the child's trunk 90 degrees to a full lateral position as indicated in figure 7, and pull slowly up and steady toward the horizontal position as shown in figure 8. At this point, stop the movement and wait for the count of 10 for the automatic inner effort of the child to move upwards as illustrated in figure 9. Keep the sideways position steady while the child is moving up "***with your assistance***", and move your trunk backwards to avoid the child grasping onto your chest.

Continue on next page

CME 25 SUPINE TO SITTING BY VERTICAL STRADDLE

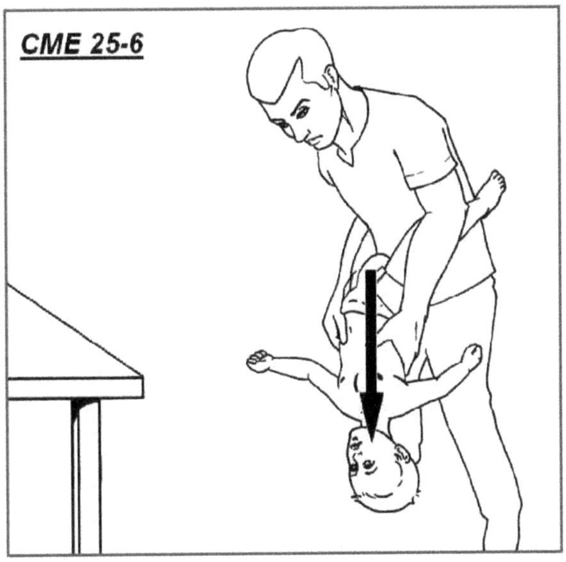

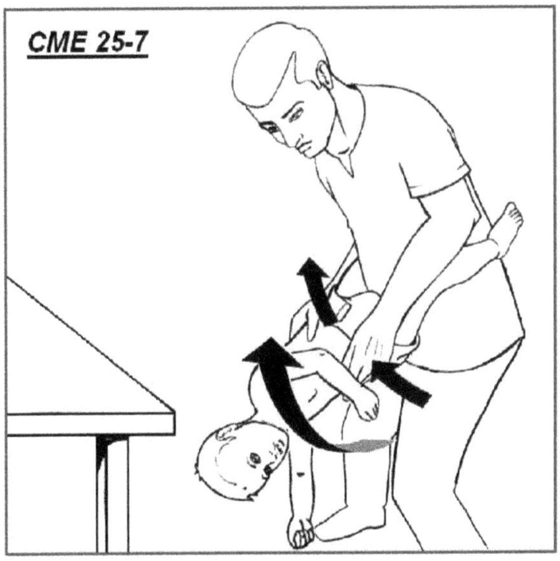

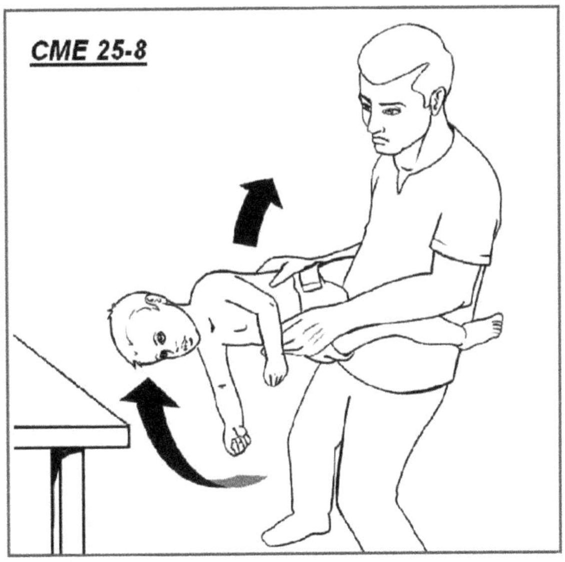

CME 25 SUPINE TO SITTING BY VERTICAL STRADDLE

The child moves 45 degrees up to a sitting posture as illustrated in figure 10, and finally succeeds to reach the straddle sitting position as shown in figure 11.

Place the right hand under buttocks and the other hand on the nape and neck as illustrated in figure 12. Come closer to the table (fig. 12), and carefully place the child into a sitting position on the table as illustrated in figure 13. Repeat the exercise as recommended by your CME therapist.

EXPECTED IMPROVEMENT: The child moves well to a straddle sitting position reacting due to the force of gravity, and try maintaining the free sitting position for a short span of time.

CME 25 SUPINE TO SITTING BY VERTICAL STRADDLE

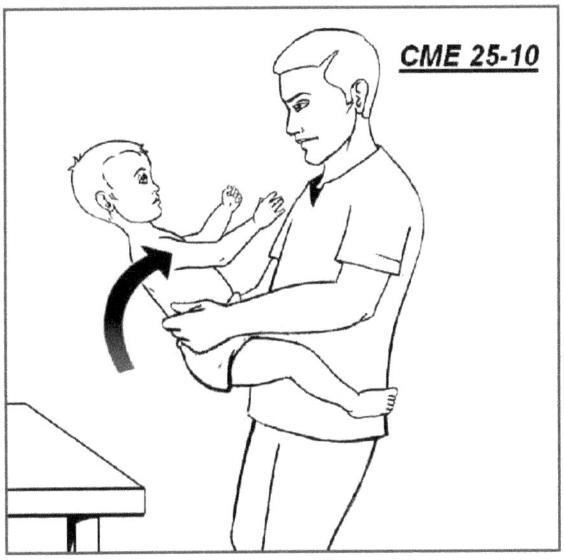

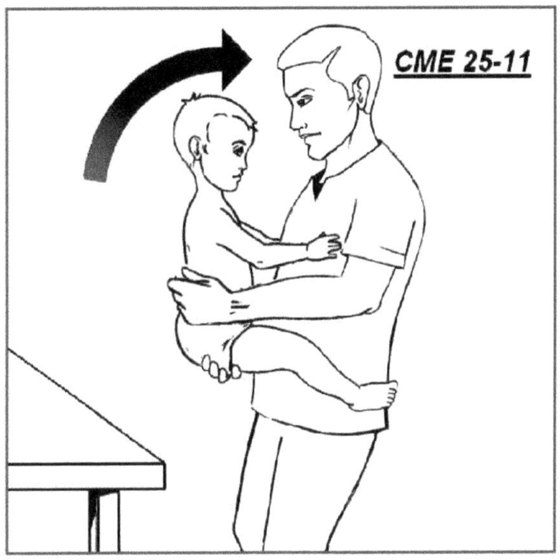

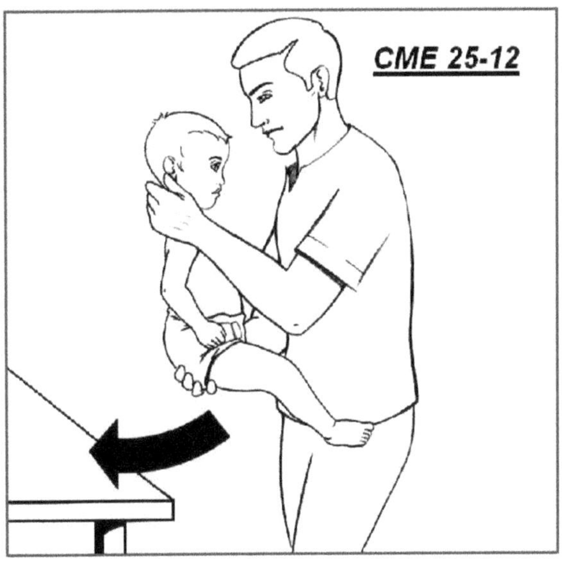

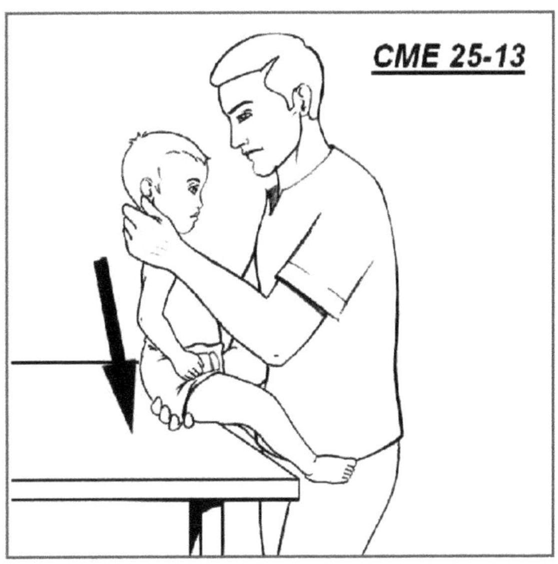

CME 26 AERIAL SITTING

MAIN GOAL: To provoke an upright sitting reaction and trunk stability in a vertical position. In addition, this exercise is indicated to decrease trunk hyperextension, and to desensitize primitive reflexes like the Moro and the TNR as well.

INITIAL POSITION: The child is lying in a supine position, crosswise (parallel to you), with his/her head to the right side of the table.

HOLD: With the left hand with both legs together above the ankles. Bend the knees, and with the right hand grab his/her right thigh (fig. 1), and compress thighs and legs together with both hands to place the heels to the buttock all along the exercise as illustrated in figure 2.

MOVEMENT: Slowly lift the legs diagonally up, with shoulders and head still on the table and turn the child to a full lateral position towards you supporting his/her body weight over your right forearm as shown in figure 3. Now push the trunk slowly up with your forearm while simultaneously moving down his/her legs and hips as shown in figure 4.

Continue on next page

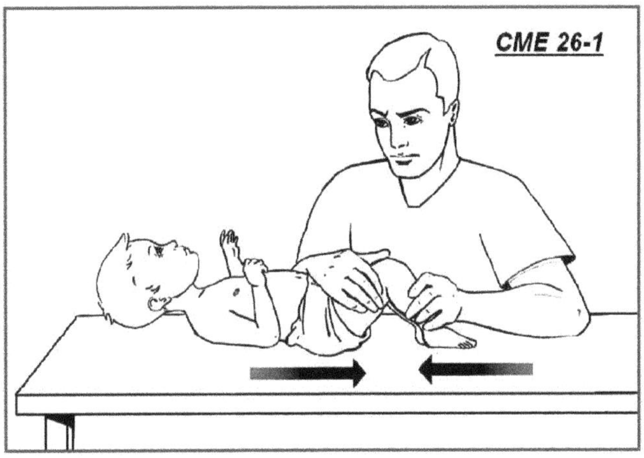
CME 26-1

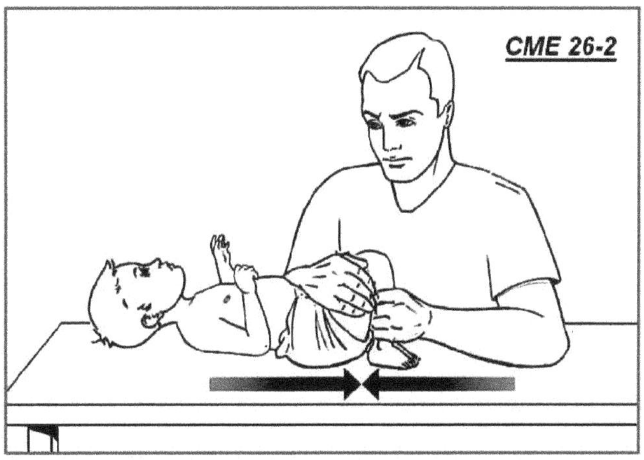
CME 26-2

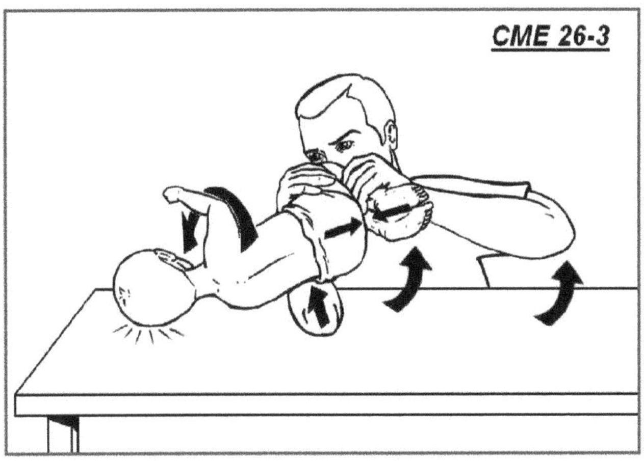
CME 26-3

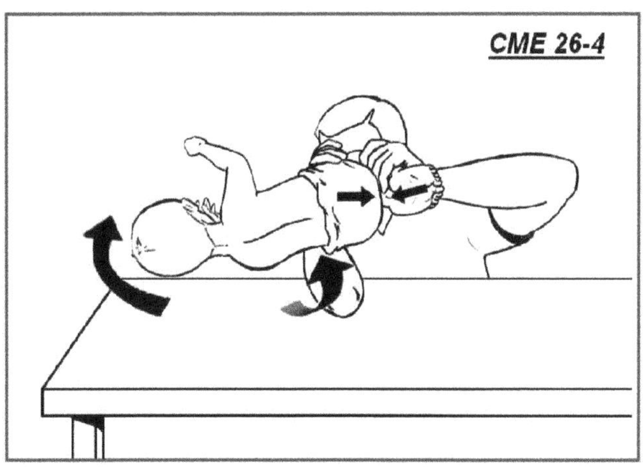
CME 26-4

CME 26 AERIAL SITTING

With a steady pace keep pushing the child's trunk up with your right forearm and move your left elbow to the table as indicated in figure 5. At the time the child is arriving to a vertical position your left arm should be completely flat on the table keeping the elbow bent at a 90 degrees angle (perpendicular to the table) as explained in figure 7. Hold the vertical aerial sitting position for the count of 10 as illustrated in figure 8. *YOUR CME PRACTITIONER MUST BE THE ONLY PERSON WHO CAN INDICATE HOW TO FINISH THE EXERCISE AND COME BACK TO THE TABLE*. Repeat the exercise as indicated by your CME practitioner.

EXPECTED IMPROVEMENT: The child moves quickly upward to the squat sitting position and stays steady with full vertical control.

CME 26 AERIAL SITTING

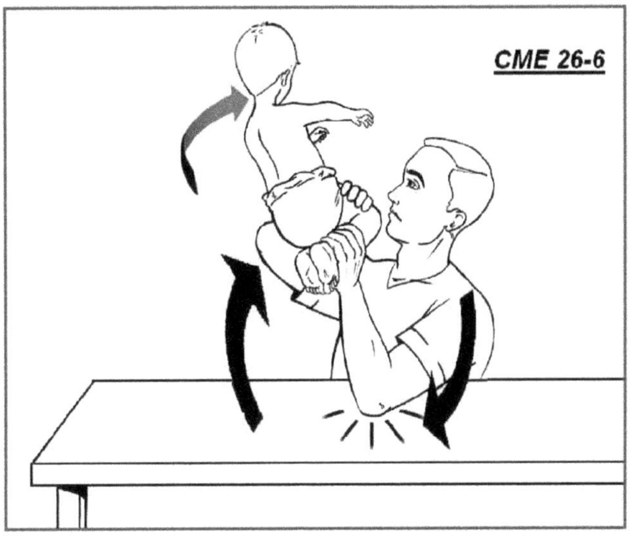

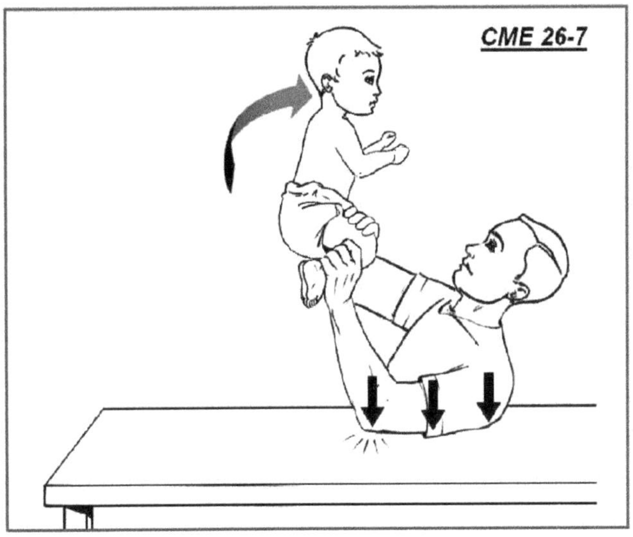

CME 27 CRAWLING POSITION BY ELBOWS

MAIN GOAL: To provoke control of the crawling position and weight bearing on the arms.

INITIAL POSITION: The child is lying in a prone position, on the table, placed perpendicular with the head towards you..

HOLD: With both hands locking elbows in extension as illustrated in figure 1. Your thumbs are extended along the arm which is pointing to the shoulder. The prominent bone of the elbow or point of the elbow called "olecranon" stays placed between the base of the ring finger and the middle finger.

MOVEMENT: Move arms downwards to the edge of the table in a diagonal position as shown in graph 2. Push his/her trunk up by the extended arms placing hands close to the edge and apply a soft rocking movement with a regular pace of 10 back and forth movements as illustrated in graph 3 in order to provoke the hip flexion to achieve a crawling position as shown in figure 4.

EXPECTED IMPROVEMENT: The child moves quickly to the crawling position as soon as his/her hands touch the table.

CME 27 CRAWLING POSITION BY ELBOWS

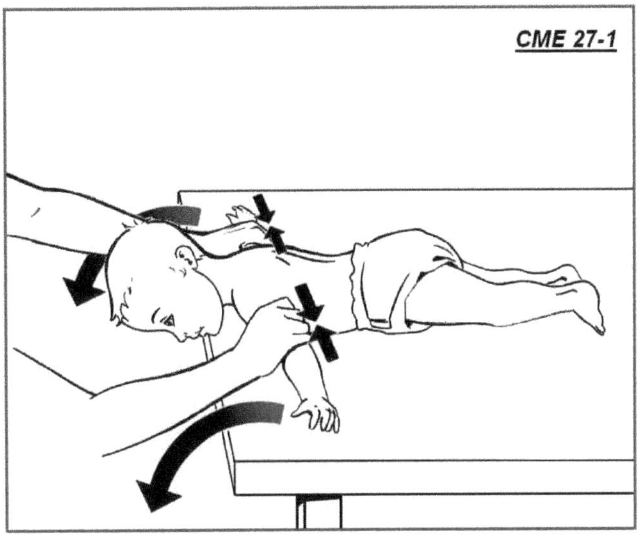

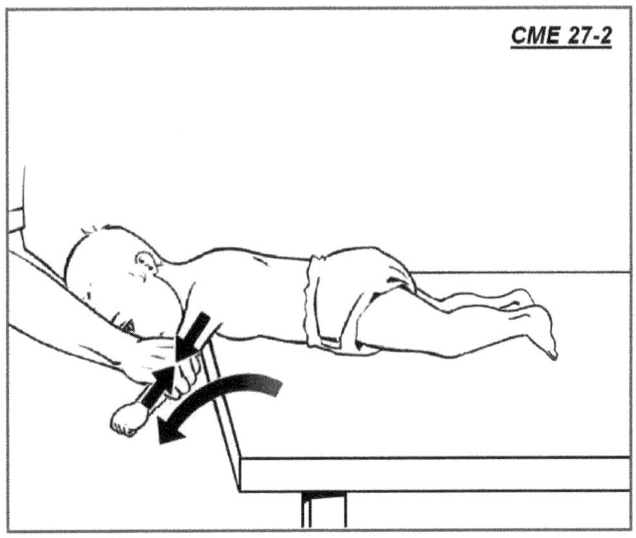

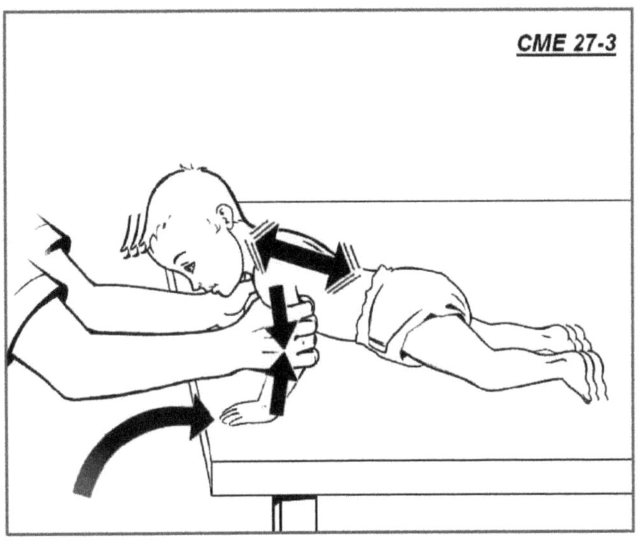

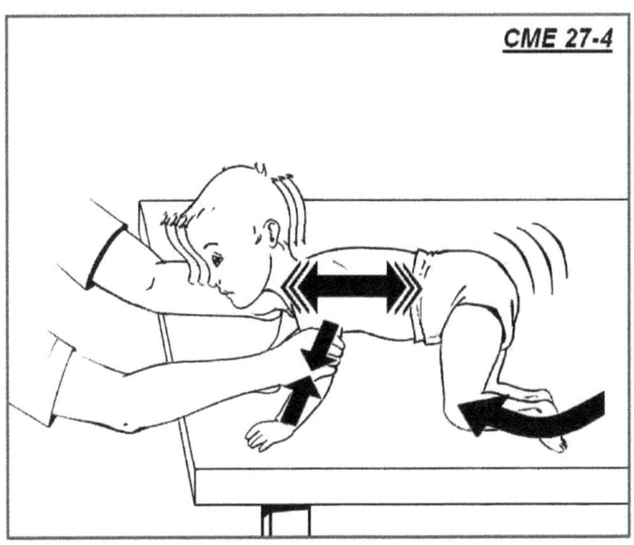

CME 28 CRAWLING TO SITTING BY ELBOWS

MAIN GOAL: To provoke control of the crawling position and weight bearing on the arms.

INITIAL POSITION: The child is lying in a prone position, on the table, placed perpendicular with his/her head towards you.

HOLD: With both hands locking elbows in extension as illustrated in figure 1. Your thumbs are extended along the arm which is pointing to the shoulder. The elbow or "olecranon" stays placed between the base of the ring finger and middle finger.

MOVEMENT: Move arms down to the edge of the table in a diagonal position as shown in graph 2. Push his/her trunk up by his/her extended arms, placing hands close to the edge and apply a soft rocking movement as illustrated in graph 3 in order to provoke the flexion of the hips (fig. 4), then, pivot the child sideways and back, over the right hand and right knee to achieve a sitting position as illustrated in figures 5 and 6. To repeat, follow the indications of your CME therapist.

EXPECTED IMPROVEMENT: The child moves quickly to the sitting position as soon as his/her hands touch the table.

CME 28 CRAWLING TO SITTING BY ELBOWS

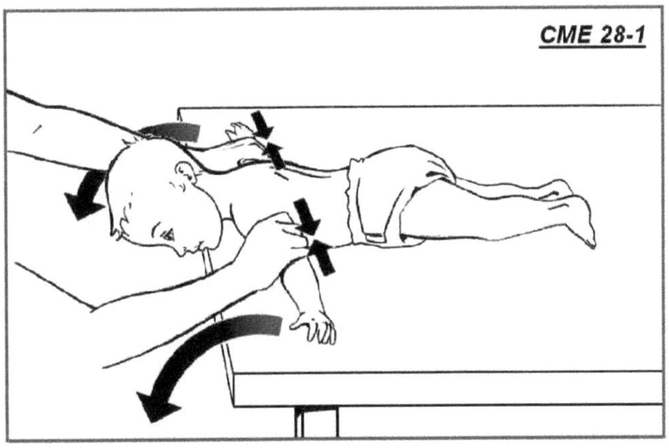

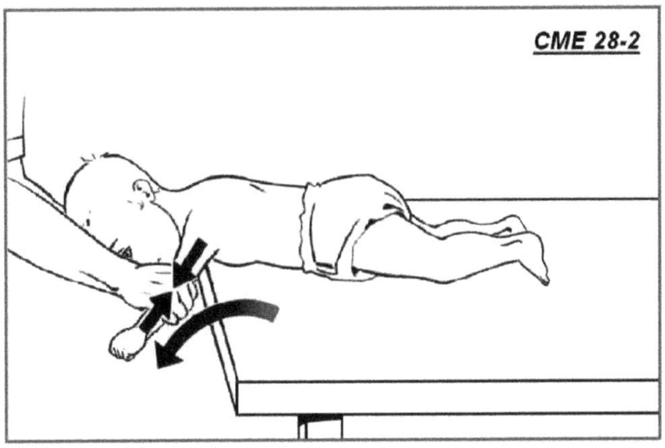

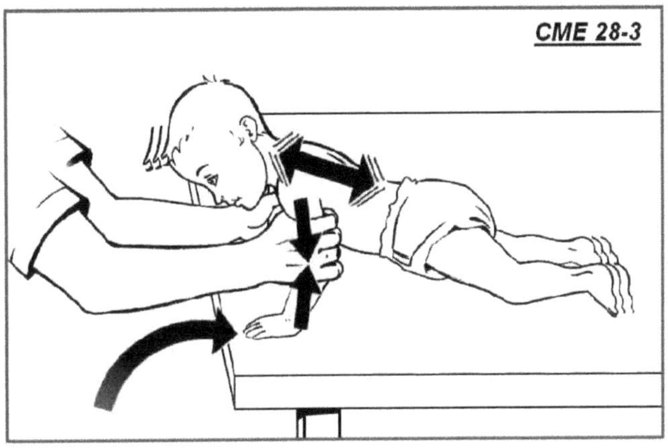

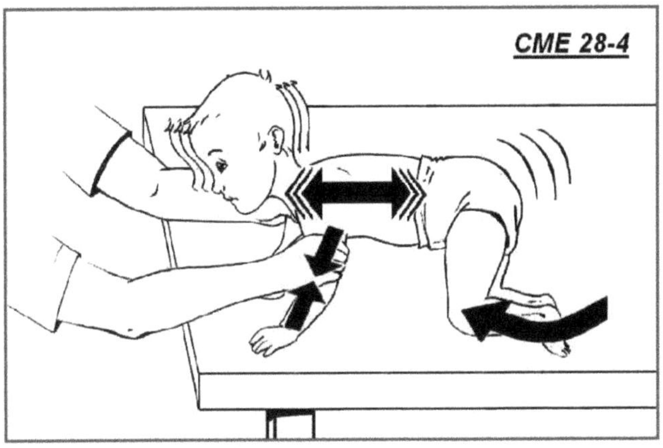

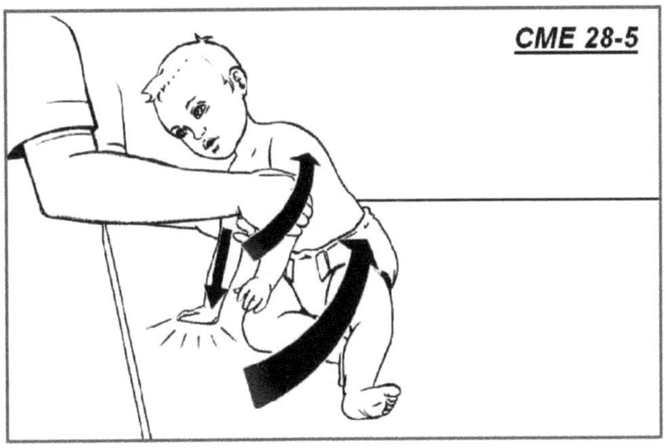

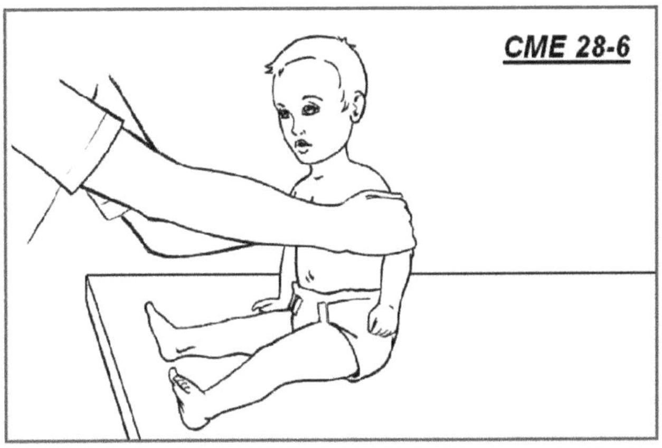

CME 29 UPSIDE-DOWN TO STANDING

MAIN GOAL: To provoke weight bearing reaction and trunk extension.

INITIAL POSITION: The child is face down on the table (prone position) crosswise with his/her head to your right side.

HOLD: With your left hand on the left knee with the thumb extended towards the buttock locking the joint, the kneecap stays located between the ring and middle finger. Firmly hold with your right hand on the right side of mid trunk (fig. 1).

MOVEMENT: Slowly lift the child pivoting his/her body towards an upside-down position as shown in figure 2, until reaching a full upside-down position with the child's head in contact with the table as illustrated in figure 3. Subsequently, apply a slow counter clockwise movement pivoting the child on your right hand as illustrated in figure 4.

Continue on next page

CME 29 UPSIDE-DOWN TO STANDING

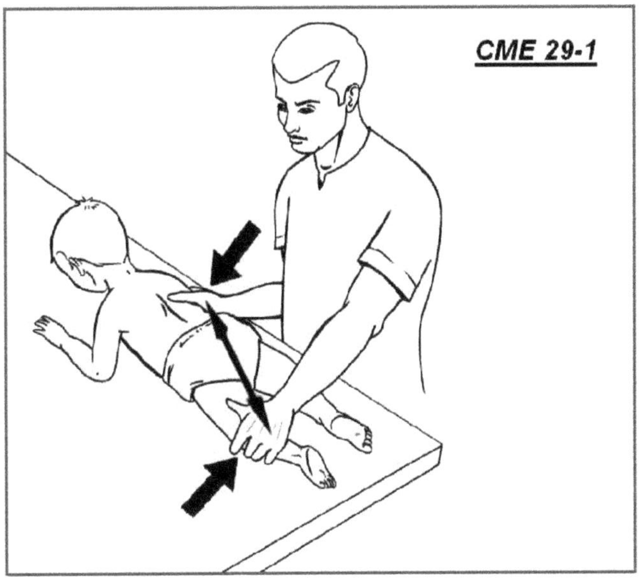
CME 29-1

CME 29-2

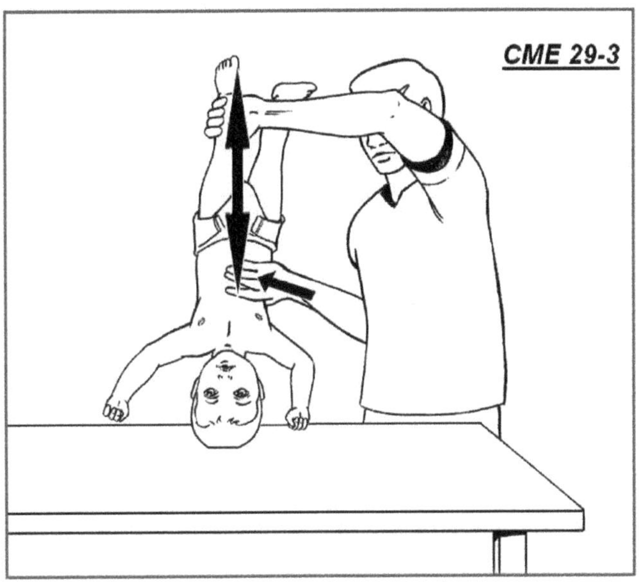
CME 29-3

CME 29-4

CME 29 UPSIDE-DOWN TO STANDING

With a non stop movement drive the child carefully (fig. 5), to a full aerial vertical posture as indicated in graph 6, in order to land firmly on the table in a full antigravity extension as illustrated in figure 7. Finally keep standing position for a count of 10. To repeat, follow the instructions of your CME therapist.

EXPECTED IMPROVEMENT: The child easily maintains the standing position.

CME 29 UPSIDE-DOWN TO STANDING

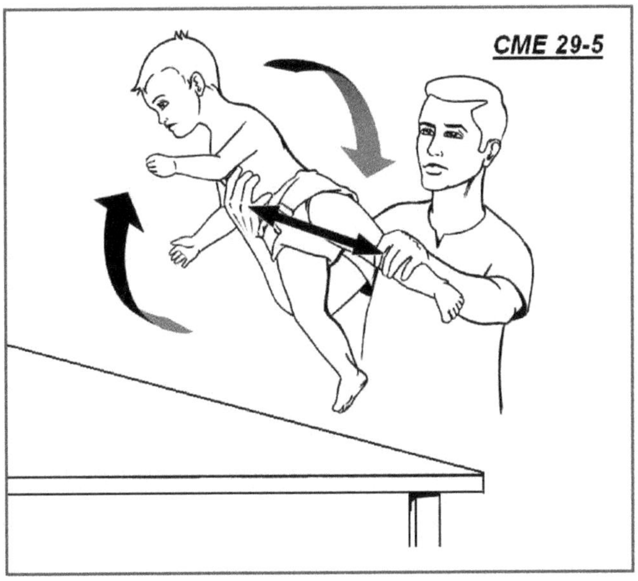

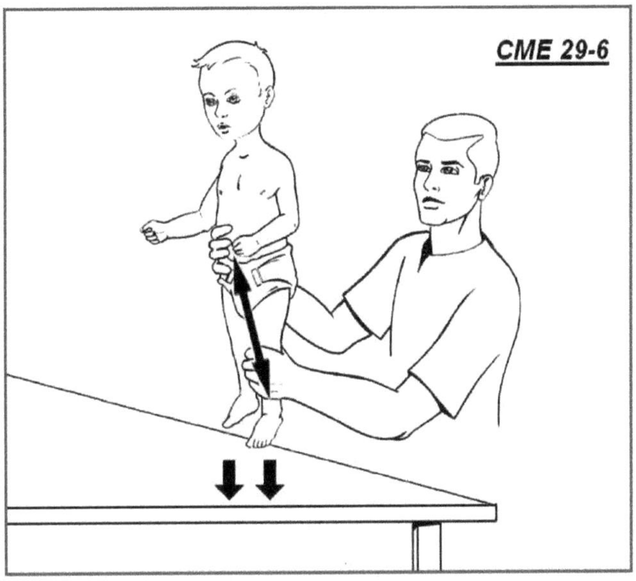

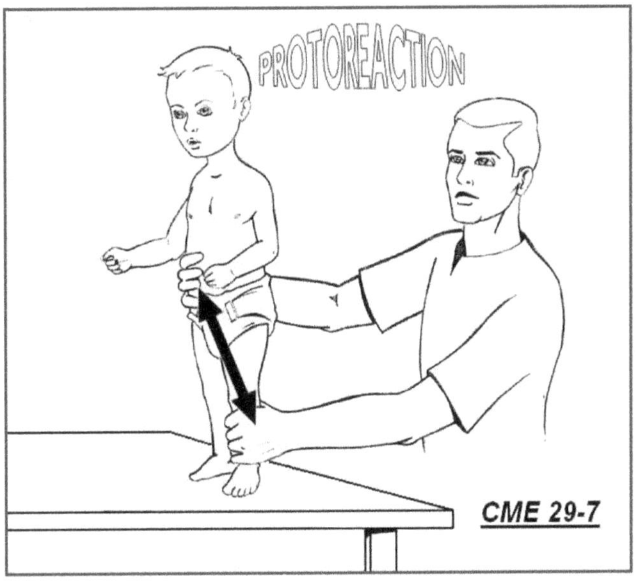

CME 30 PRONE TO STANDING BY ABDOMEN AND ANKLES

MAIN GOAL: To provoke trunk extension and weight bearing.

INITIAL POSITION: The child lies down on the table in a prone position.

HOLD: Firmly hold both ankles with your left hand and tuck the right hand under her/his low abdomen with fingers spread and thumb fully extended off of the abdomen, (fig. 1).

MOVEMENT: Slowly lift the hips and legs diagonally up until only the head and hands are touching the table (fig. 2), then hold the hips steady up bending down the hips with knees slightly bent, until both feet are in firm contact with the table (fig. 3). Maintain this posture steady until the child proceeds to respond with a full trunk extension as illustrated in figures 4, 5 and 6. Repeat the exercise as instructed by your CME practitioner.

EXPECTED IMPROVEMENT: The child does good trunk extension against gravity and maintains the standing position (assisted).

CME 30 PRONE TO STANDING BY ABDOMEN AND ANKLES

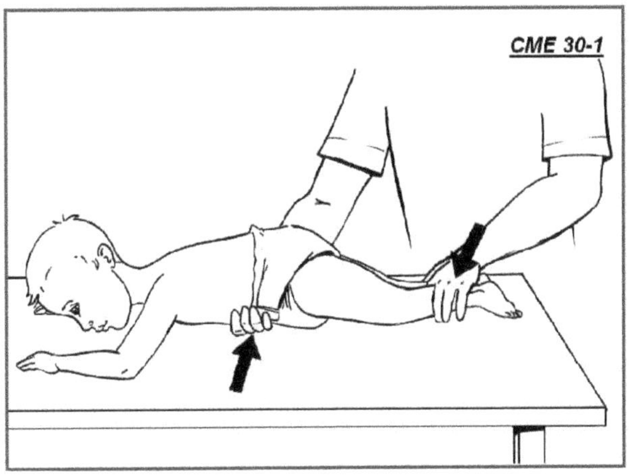

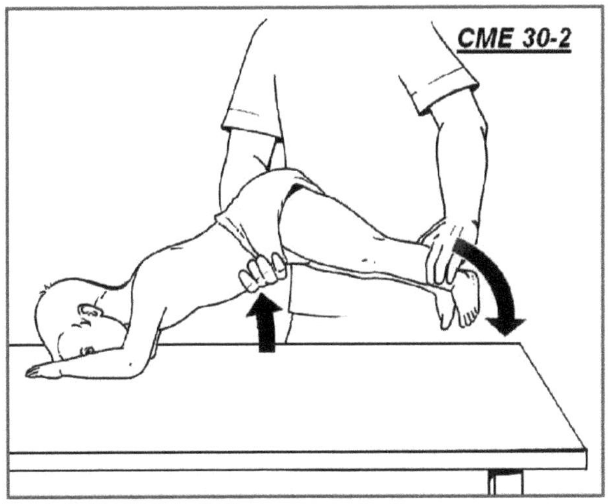

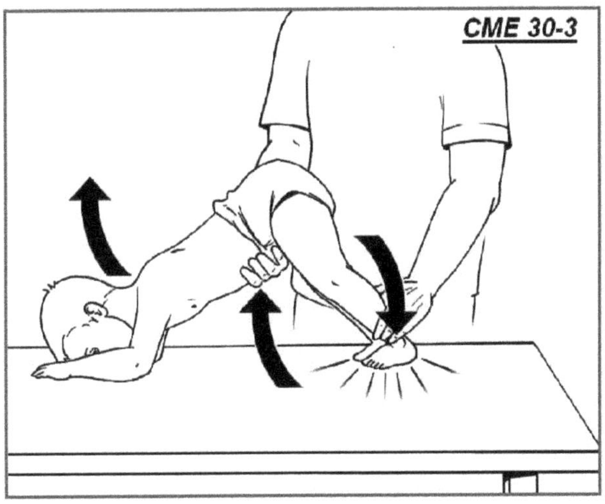

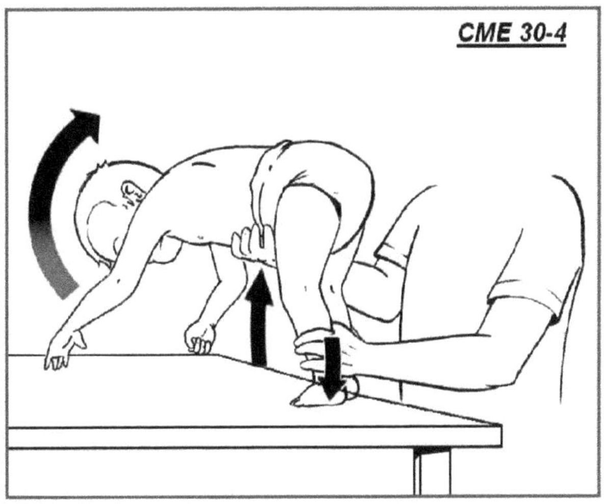

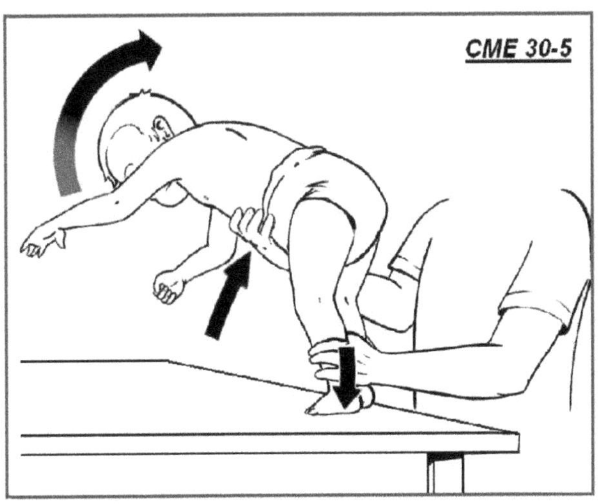

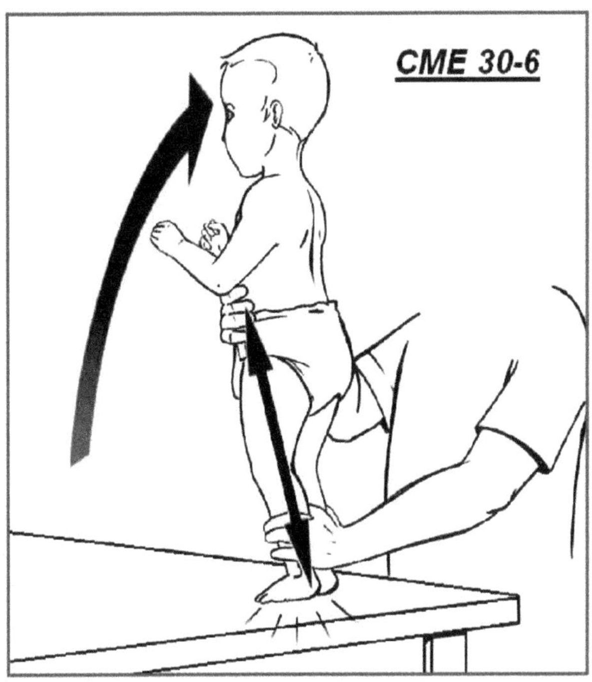

CME 31 SUPINE TO STANDING BY FOREARMS

MAIN GOAL: To provoke a standing reaction.

INITIAL POSITION: The child lies on the table diagonally to you, in a supine position.

HOLD: Firmly hold with your right hand by both forearms crossed together over chest, and with the left hand hold his/her ankles together (fig. 1), then bend the knees and place the feet firmly on the table (fig. 2).

MOVEMENT: Pull carefully by forearms diagonally up (fig. 2), until reaching the crouching position (fig. 3). At this point, hold the position for the count of 10, while waiting for the automatic standing reaction as shown in figure 4.

WARNING: DO NOT PULL FOREARMS UPWARDS.

EXPECTED IMPROVEMENT: The child responds by quickly pushing up to a standing position.

CME 31 SUPINE TO STANDING BY FOREARMS

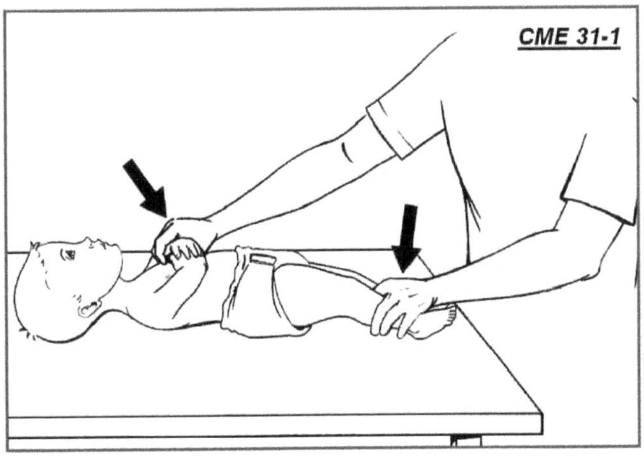

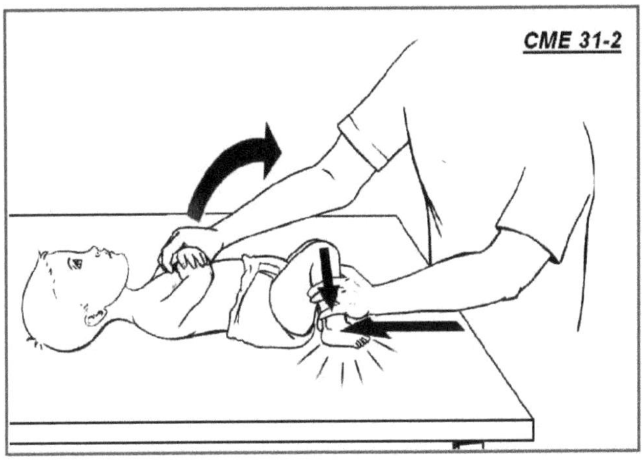

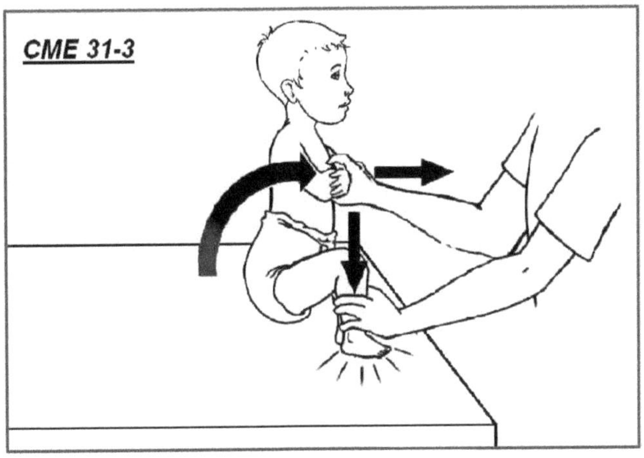

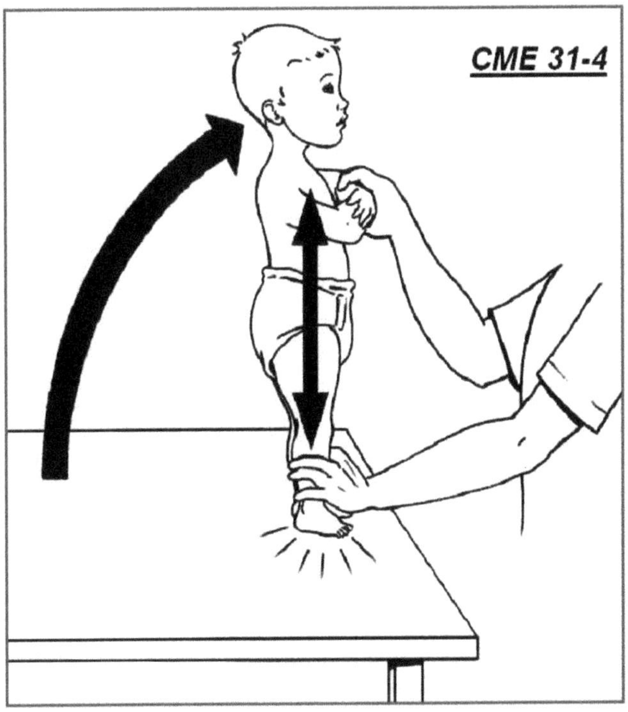

CME 32 KNEELING TO STANDING BY FOREARMS

MAIN GOAL: To provoke standing response

INITIAL POSITION: Child lies down in a prone position, perpendicular to you, with feet placed off the table.

HOLD: With both hands on the mid trunk (fig. 1), pull the child up to a kneeling position (fig. 2), then, hold with your left hand by both forearms close together, and with the right hand grasp his/her right ankle (fig. 3).

MOVEMENT: Place left foot firmly on the table and incline the child slightly forwards, then hold the posture steady for the count of 10 (fig. 4), waiting for the standing up automatic reaction (fig. 5).

EXPECTED IMPROVEMENT: The child will respond standing up easily.

CME 32 KNEELING TO STANDING BY FOREARMS

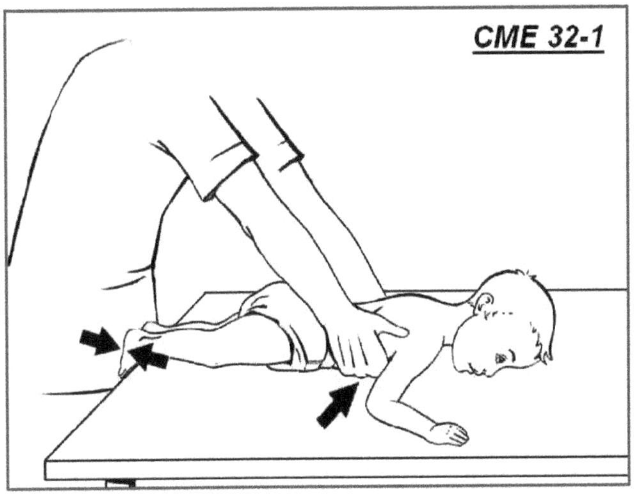

CME 32-1

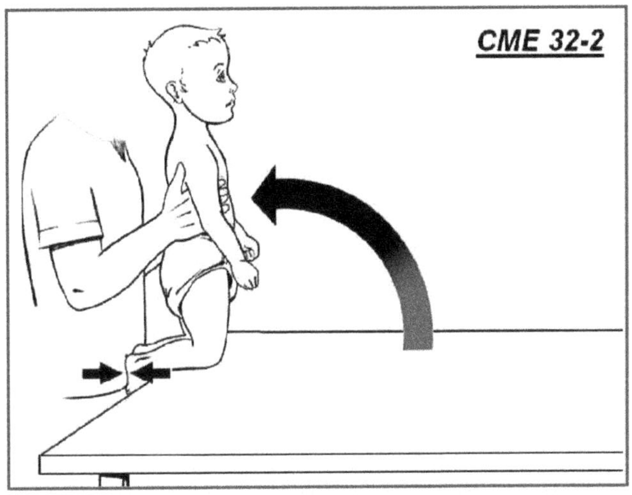

CME 32-2

CME 32-3

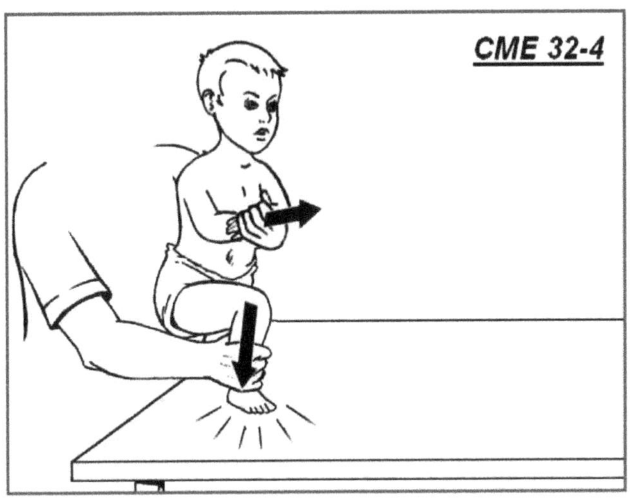

CME 32-4

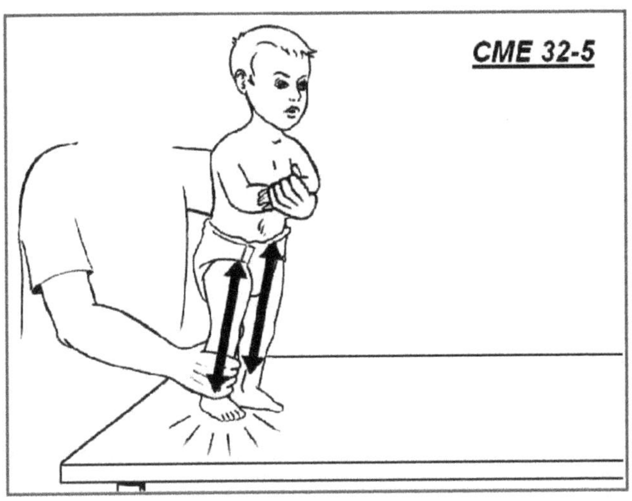

CME 32-5

CME 33 KNEELING TO STANDING BY ONE FOREARM

MAIN GOAL: To provoke a standing response.

INITIAL POSITION: Child lies down in prone position, perpendicular to you, with feet placed off the table.

HOLD: With both hands on the mid trunk (fig. 1), pull up the child to kneeling position (fig. 2), then, hold with your left hand by his/her left forearm and with right hand by right ankle as illustrated in figure 3.

MOVEMENT: Place his/her right foot firmly on the table and incline the child slightly forwards, then hold the posture for a count of 10 (fig. 4), waiting for the automatic standing up reaction as illustrated in figures 5 and 6.

EXPECTED IMPROVEMENT: The child responds by standing up easily.

CME 33 KNEELING TO STANDING BY ONE FOREARMS

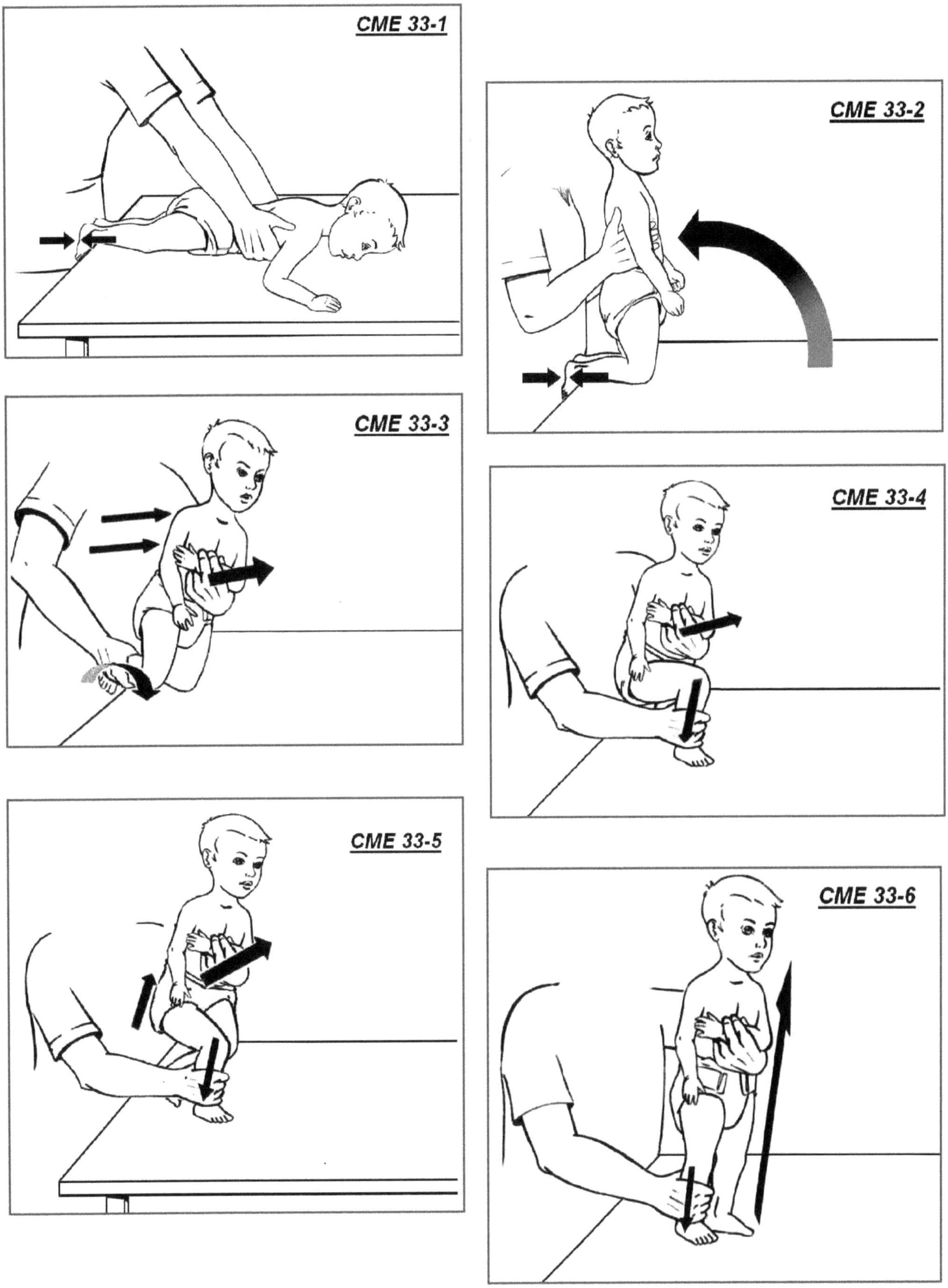

CME 34 SUPINE TO STANDING BY NAPE

MAIN GOAL: To provoke a standing up reaction.

INITIAL POSITION: Child lies on the table crosswise to you, in a supine position.

HOLD: Firmly hold with your left hand both ankles and bend the knees: Place the feet steady on the table and with your right hand cup the child's nape (fig. 1), fixing the point without digging into the soft tissue with the fingertips.

MOVEMENT: Push the child's body up carefully by the nape with a diagonal forwards direction, until reaching the crouching position (fig. 2), then hold the position for a count of 10, waiting for the standing up reaction (fig. 3). Repeat the exercise according to your CME practitioner instructions.

EXPECTED IMPROVEMENT: The child responds standing up easily.

CME 34 SUPINE TO STANDING BY NAPE

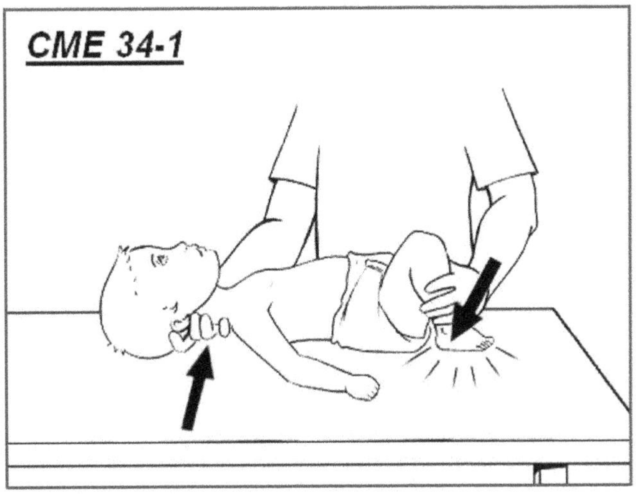

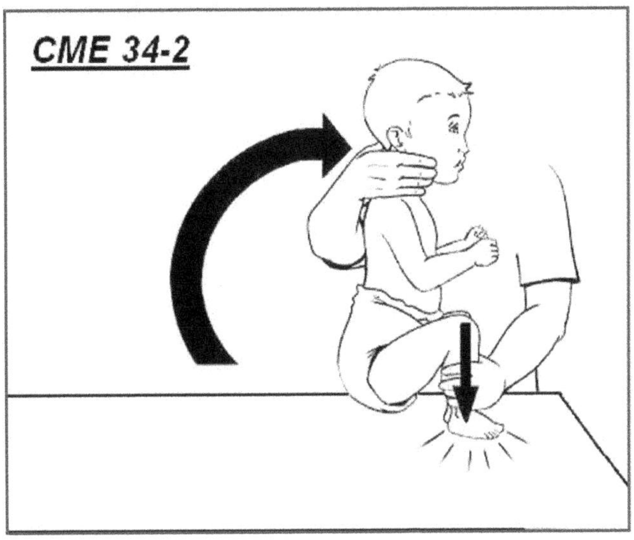

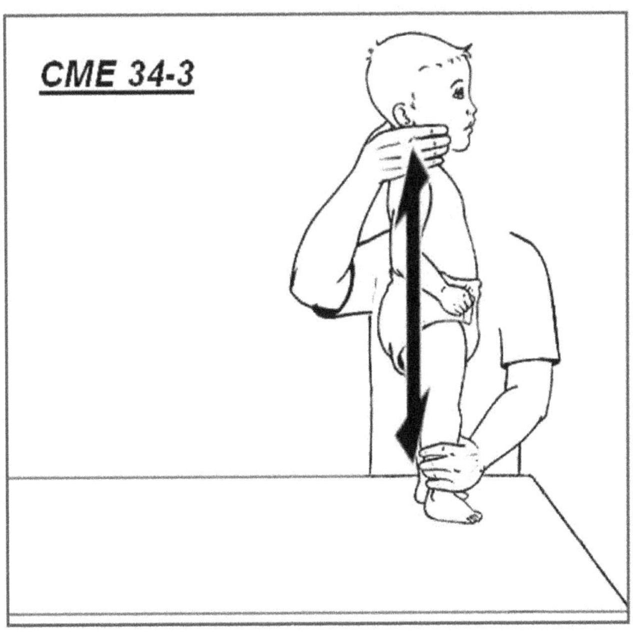

CME 35 KNEELING TO STANDING BY THIGHS

MAIN GOAL: To provoke a standing up reaction, trunk stability and weight bearing.

INITIAL POSITION: The child is face down on the table, with the feet off the edge of the table (fig. 1).

HOLD: With both hands on the mid trunk and tucking feet at the edge of the table as indicated in figure 1. Pull the child up to a kneeling position as shown in figure 2.

MOVEMENT: Move down both hands to grasp thighs and shift weight slightly over left knee as illustrated in figure 3. Place the right foot on the table by the tight as illustrated in figure 4, subsequently push up with right thumb under hip, to provoke the standing reaction (figures 5 and 6). Repeat the exercise following the instructions of your CME therapist.

EXPECTED IMPROVEMENT: The child will keep the trunk vertically and moves easily to a standing position.

CME 35 KNEELING TO STANDING BY THIGHS

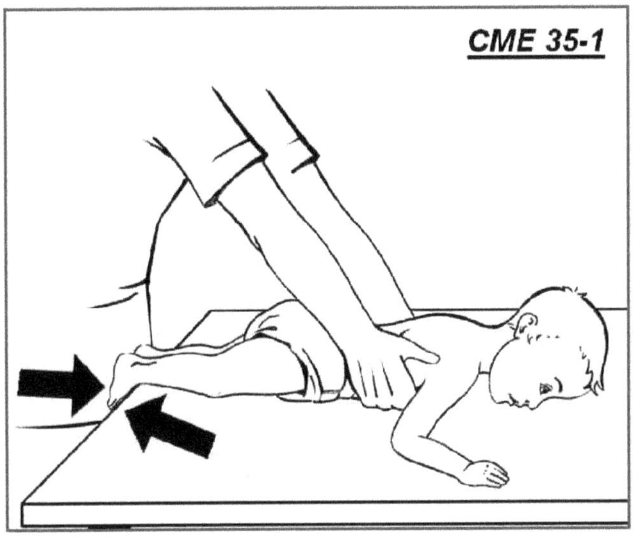

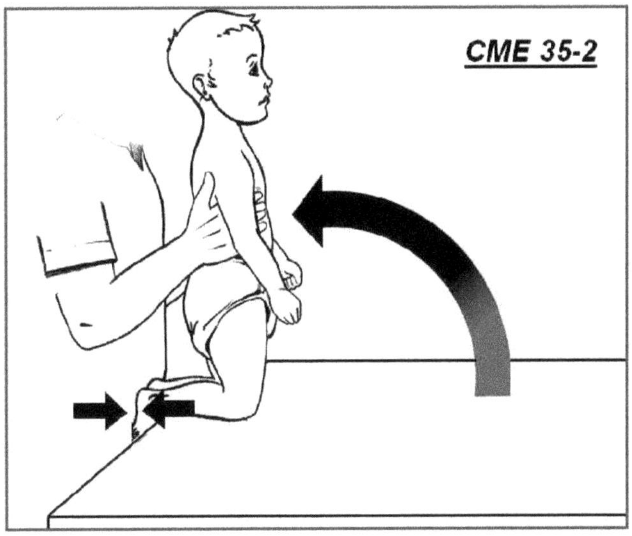

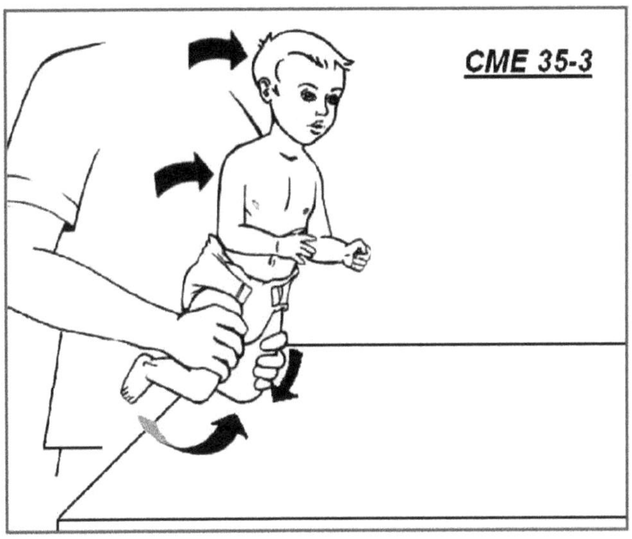

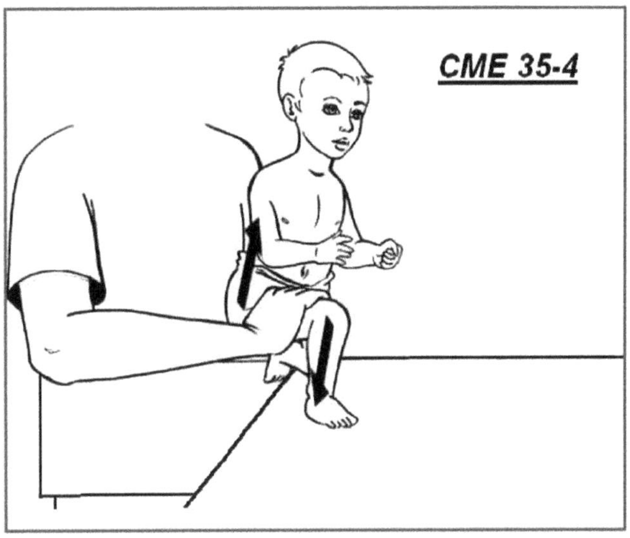

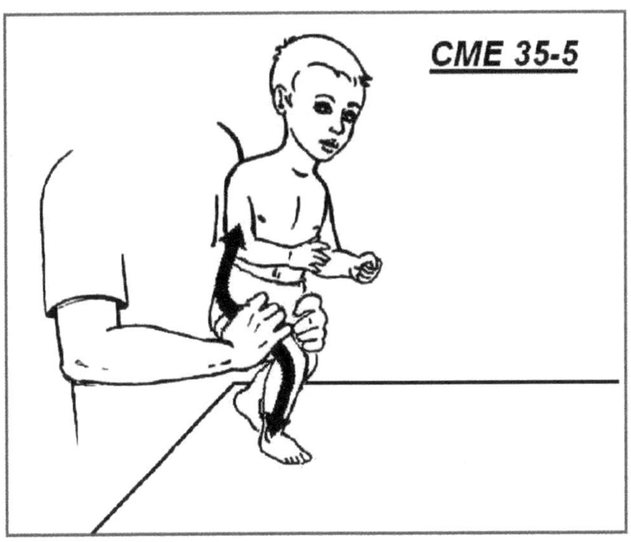

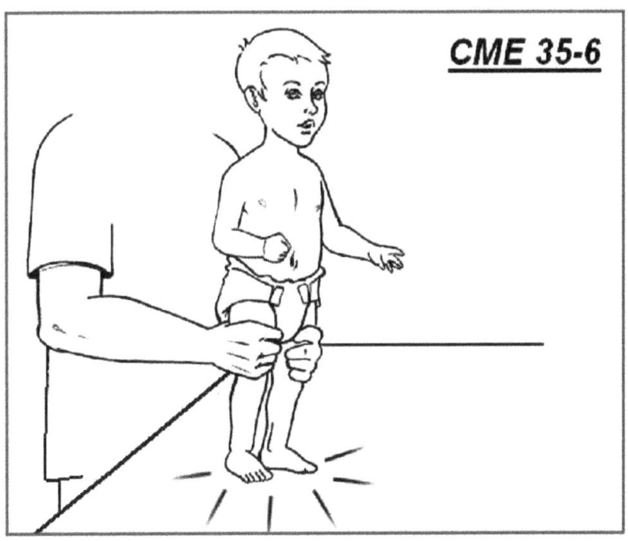

CME 36 4 POINTS TO STANDING BY THIGH AND OPPOSITE LEG

MAIN GOAL: To provoke dynamic standing up reaction.

INITIAL POSITION: Child is face down (prone position) on the table with legs towards you and feet off the edge of table (fig. 1).

HOLD: Hold firmly by thighs keeping feet tucked between your lower abdomen and the edge of the table as indicated in figure 1.

MOVEMENT: Bend hips until reaching a crawling position as illustrated in figure 2, then with right hand hold his/her right ankle and place the foot firmly on the table as illustrated in figure 3. Subsequently move your right hand up to grasp his/her right thigh as observed in the illustration number 4.

Continue on next page

CME 36 4 POINTS TO STANDING BY THIGH AND OPPOSITE LEG

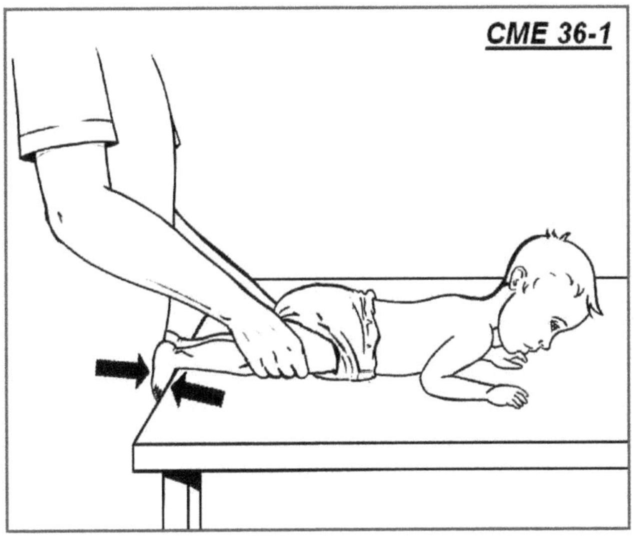

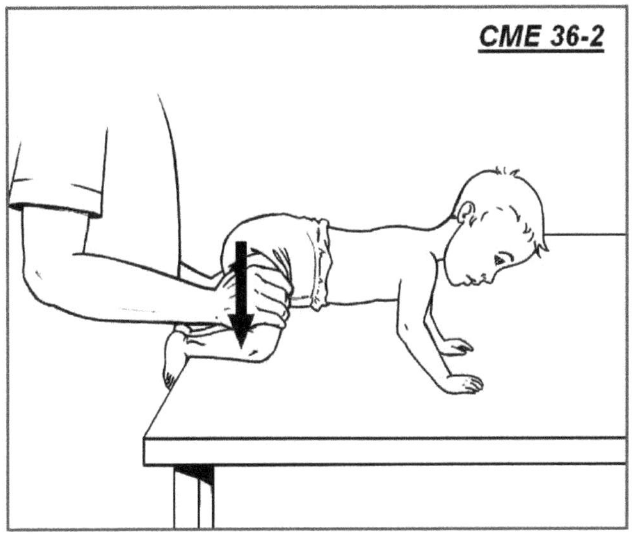

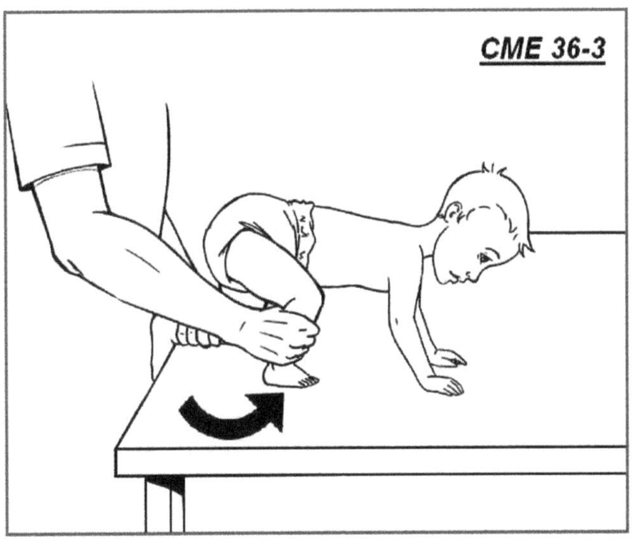

CME 36 — 4 POINTS TO STANDING BY THIGH AND OPPOSITE LEG

Move the left hand from his/her left thigh to his/her left ankle and place the left foot firmly on the table as shown in figure 5. Hold the quadruped position steady, and pull the pelvic area slightly backwards with your right hand as indicated by the curved arrow in figure 5, to provoke the automatic standing up reaction as illustrated in figures 6 and 7. Repeat the exercise as indicated by your CME practitioner.

EXPECTED IMPROVEMENT: The child moves up to a standing position rapidly and stays still.

CME 36 4 POINTS TO STANDING BY THIGH AND OPPOSITE LEG

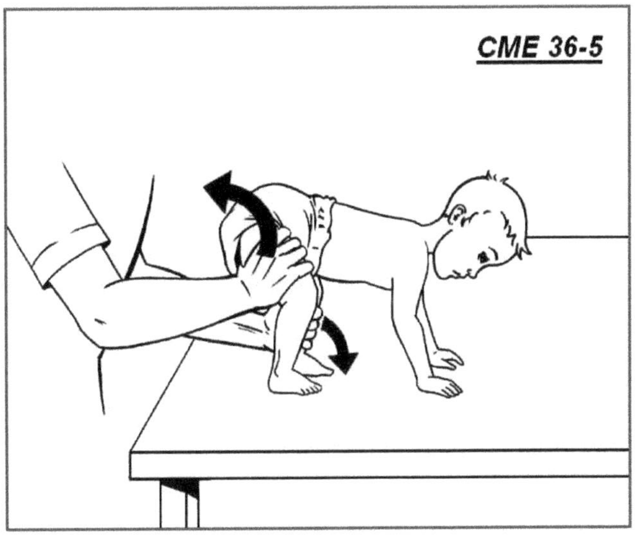

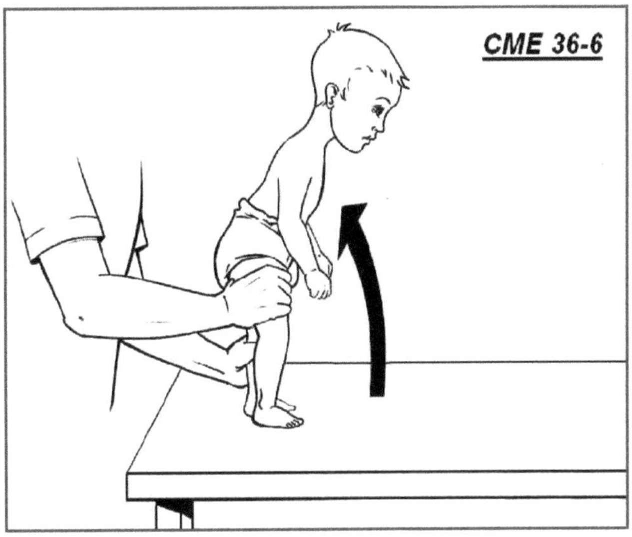

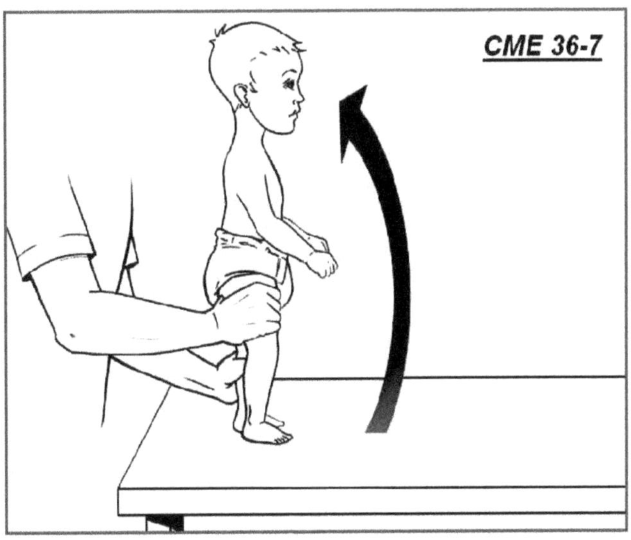

CME 37 SERIES OF 3

MAIN GOAL: To improve dynamic trunk stability and weight bearing.

INITIAL POSITION: The child is in a standing position supported against your trunk.

HOLD: The child with your right forearm, while holding the left ankle between the index and middle finger, locking the knee with the wrist, and fix the right ankle in the same way in order to achieve a steady standing posture as illustrated in figure 1.

MOVEMENT 1: Lift the child and fall to table with 10 consecutive bouncing movements, (figures 2 and 3).

MOVEMENT 2: Spread the child's legs, and incline your body side to side for the count of 10 sequential movements, (figures 4 and 5).

MOVEMENT 3: Put the child's legs together, and make slow regular steps on the spot, bending up each knee for a consecutive series of 10 steady steps (figs. 6 and 7). Follow your CME practitioner instructions.

EXPECTED IMPROVEMENT: The child maintains the trunk straight at all times.

CME 37 SERIES OF 3

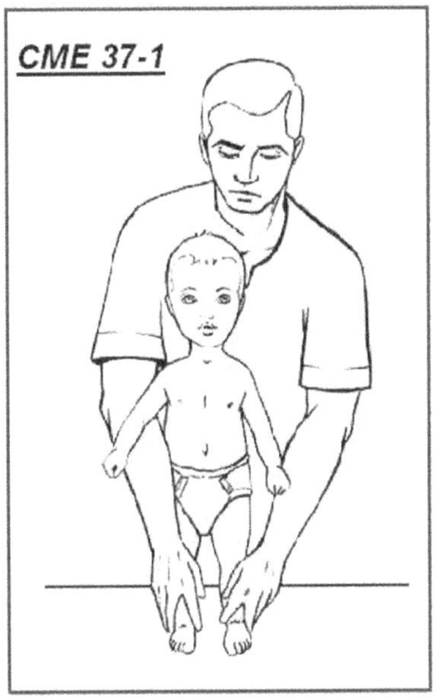

CME 37-1

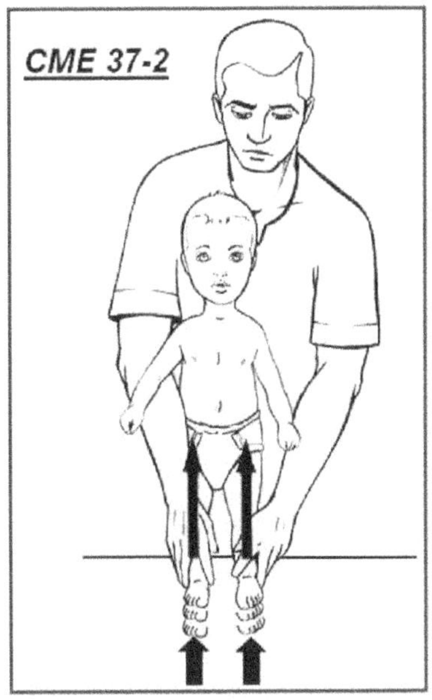

CME 37-2

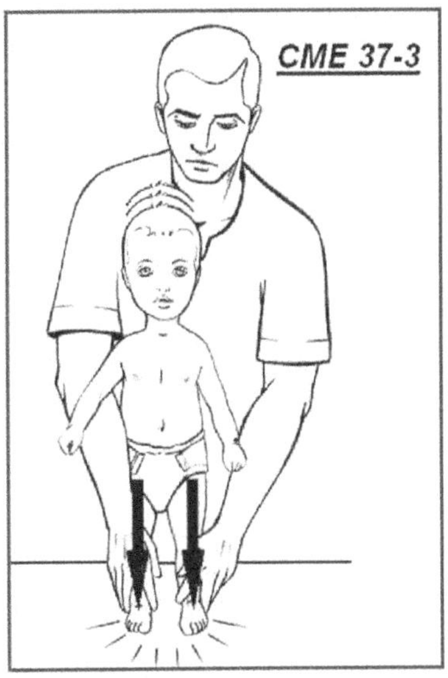

CME 37-3

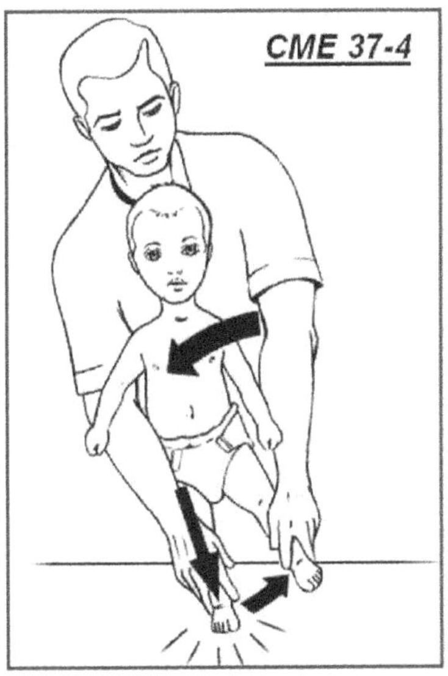

CME 37-4

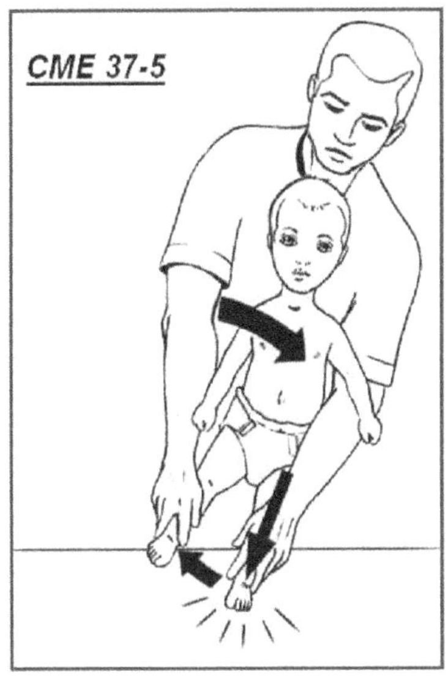

CME 37-5

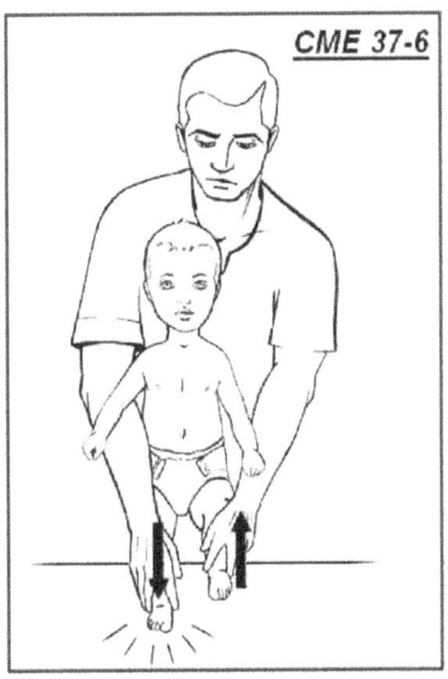

CME 37-6

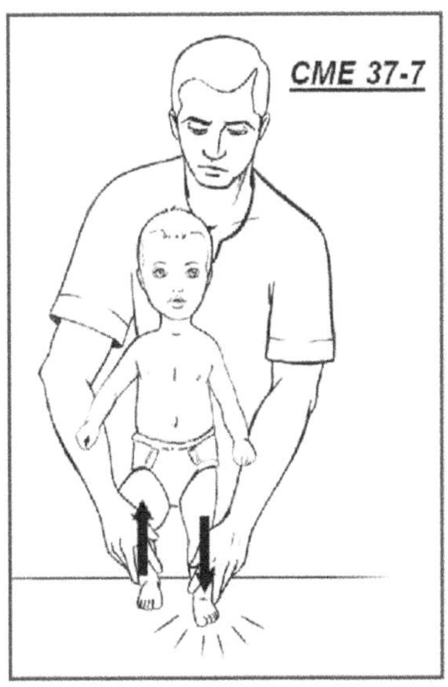

CME 37-7

CME 38 STANDING POSITION BY THIGHS

MAIN GOAL: Enhance vertical trunk stability and weight bearing.

INITIAL POSITION: Child is in standing position in front of you, held by mid trunk as shown in the figure 1.

HOLD: With right hand on his/her right thigh as indicated in the figure 2.

MOVEMENT: Move your left hand to his/her left thigh and hold the standing position for a count of 30 as illustrated in figure 3. Repeat the exercise as indicated by your CME therapist.

EXPECTED IMPROVEMENT: The child maintains a standing posture with trunk fully vertical.

CME 38 STANDING POSITION BY THIGHS

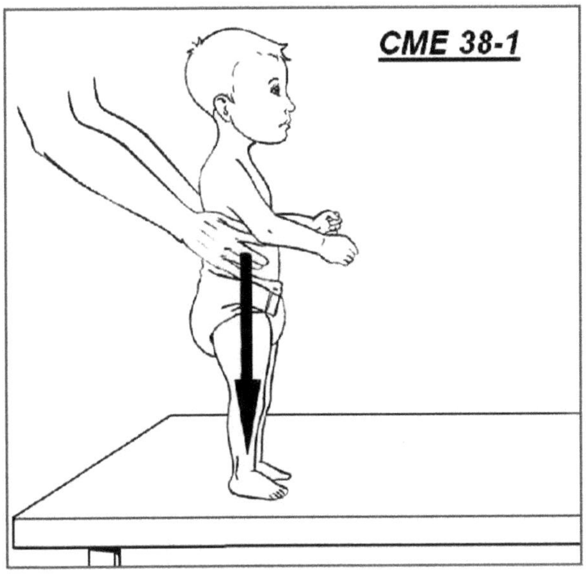

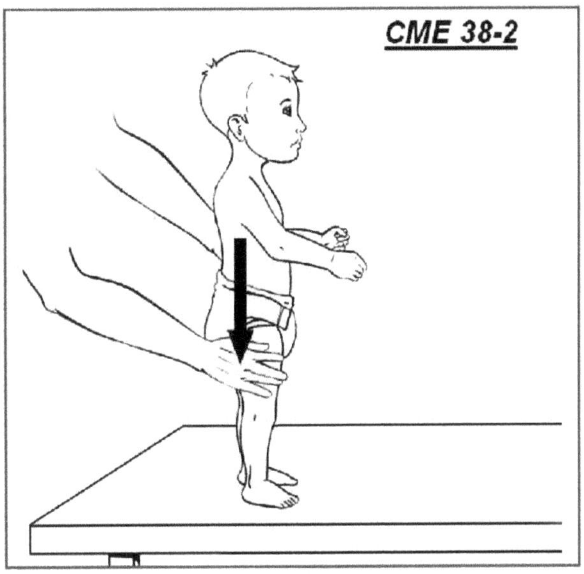

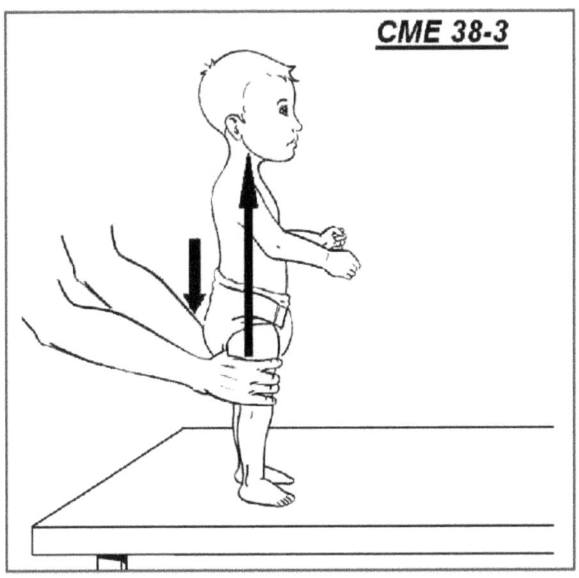

CME 39 ABD-ADD STANDING ON LAP BY THIGHS

MAIN GOAL: Enhance standing posture, and it is also a good exercise to increase hips range of motion.

INITIAL POSITION: Sit on the floor, your back to the wall, and the child stands on your lap supported by hips as illustrated in the figure 1.

HOLD: Firmly by his /her thighs as shown in figure 2.

MOVEMENT: Spread your legs apart slowly as indicated in figure 3, provoking the abduction of his/her legs in the maximum range that the child can resist. Subsequently close your legs together as illustrated in figure 4. (Immediately stop the abduction movement if the child begins to sit down on your hands or begin to incline the trunk forward). Repeat the ABD-ADD movement ten times. Apply the exercise as instructed by your CME practitioner.

EXPECTED IMPROVEMENT: The child opens and closes his/her legs without losing control standing.

CME 39 ABD-ADD STANDING ON LAP BY THIGHS

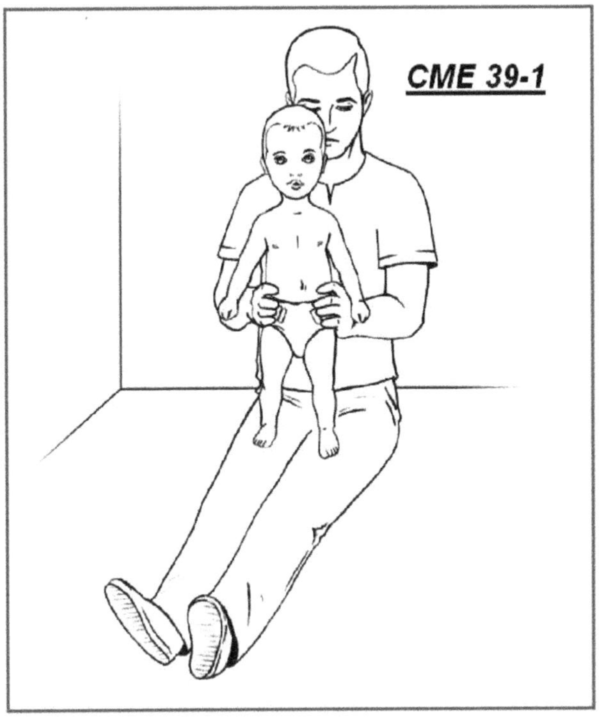

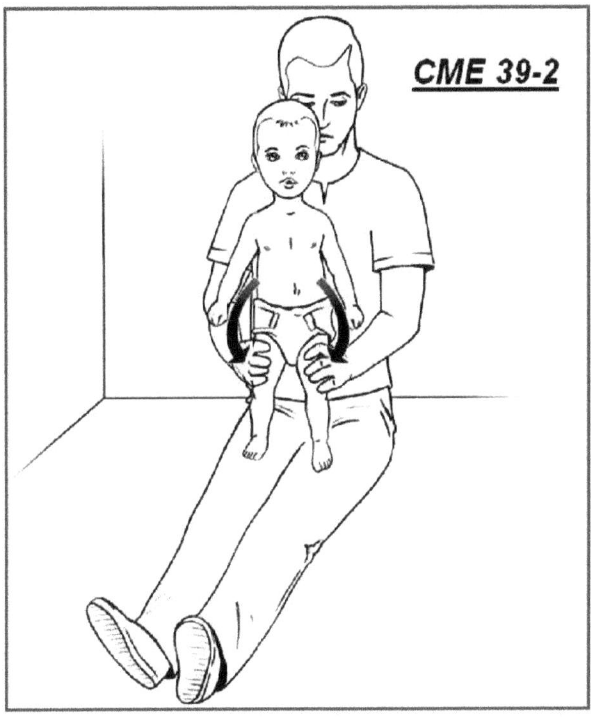

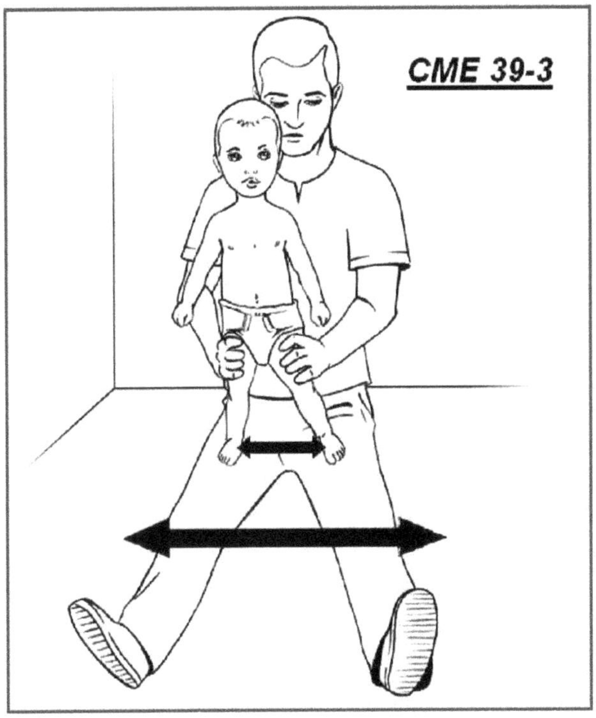

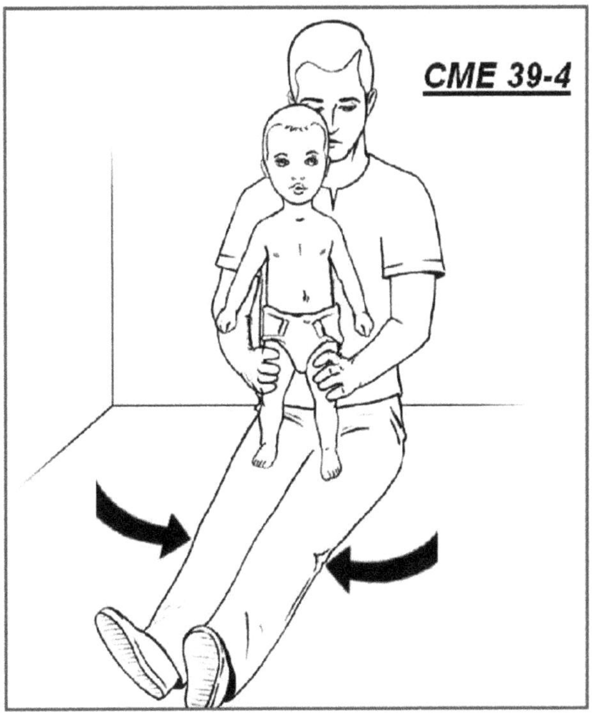

CME 40 COLLAPSE ON TOP OF 30 CM. BOX

Exercise Configuration: A 30 cm. box position: with slots up and toward you on wrapped balance bar, as illustrated in the figure 1.

MAIN GOAL: Strengthening of dynamic knees extension and trunk control in standing position.

INITIAL POSITION: The child is in a standing position in front of 30 cm. box; you are sitting on the floor.

HOLD: Hold his/her ankles firmly as shown in figure 1, and lift the child up until reaching the top of the box as illustrated in figure 2, with part of the feet off the edge of the box as shown in insert of figure 2.

MOVEMENT: Carefully and slowly pivot the box downwards, controlling the movement by both ankles with soles firmly in contact with the edge as shown in figures 3 and 4. Subsequently tuck the right heel under edge of box (fig. 5), and pull the child backwards and up on the box as illustrated in figures 6 and 7. Repeat as indicated by your CME therapist.

EXPECTED IMPROVEMENT: The child stays standing at all times.

CME 40 COLLAPSE ON TOP OF 30 CM. BOX

CME 40-1

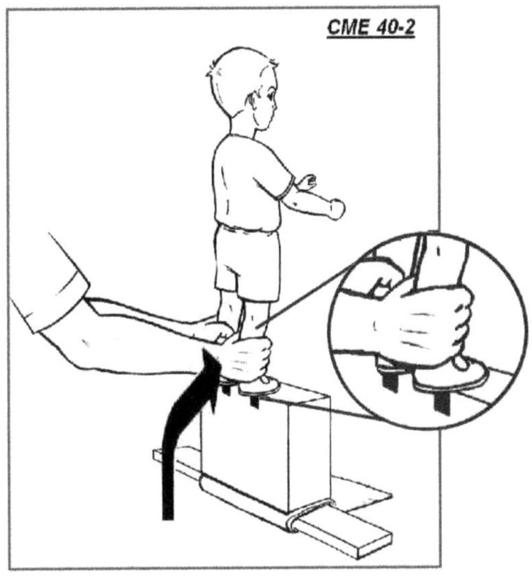
CME 40-2

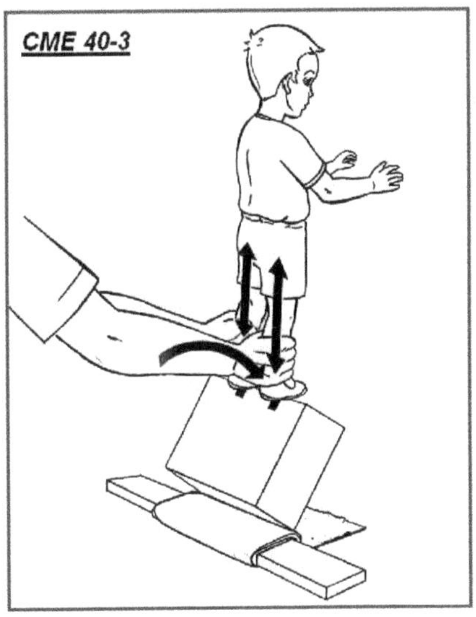
CME 40-3

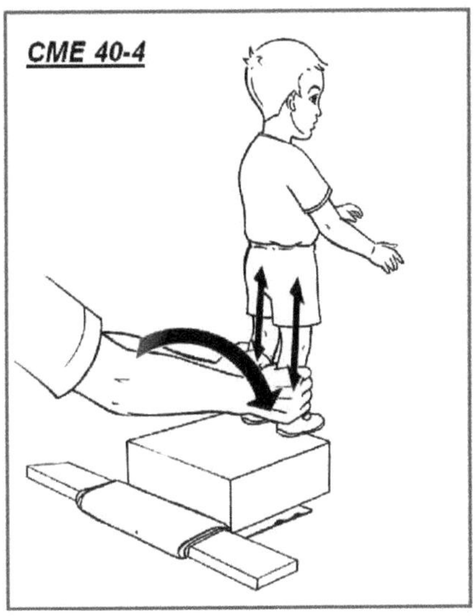
CME 40-4

CME 40-5

CME 40-6

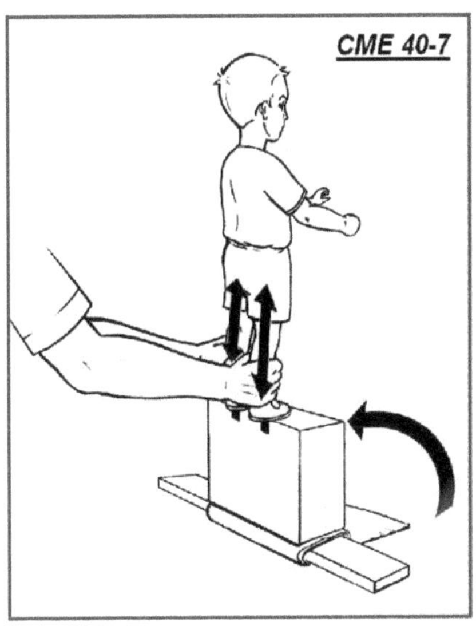
CME 40-7

CME 41 STANDING POSITION BY BELOW KNEES

MAIN GOAL: Enhance standing posture.

INITIAL POSITION: The child is in a standing position in front of you, held by mid trunk as shown in figure 1.

HOLD: With right hand on his/her right leg below the knee as indicated in figure 2.

MOVEMENT: Move your left hand to his/her lower left leg and hold the standing position for a count of 30 as illustrated in figure 3. To repeat, follow the instructions of your CME therapist.

EXPECTED IMPROVEMENT: The child maintains standing posture with trunk fully vertical.

CME 41 STANDING POSITION BY BELOW KNEES

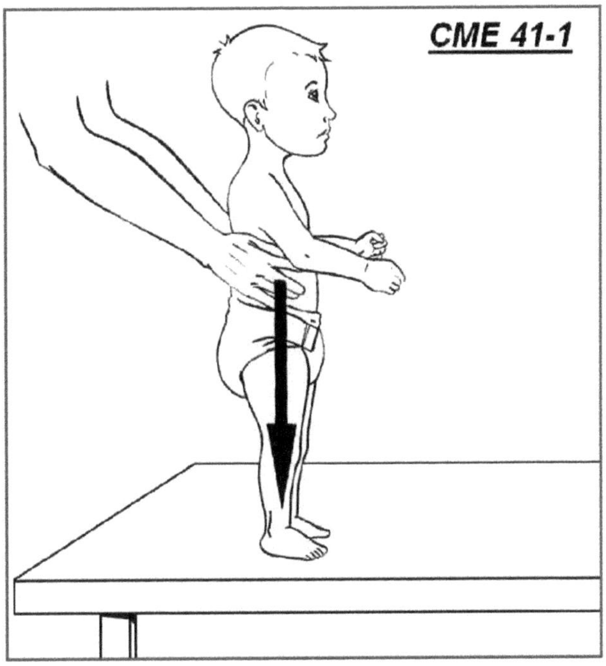

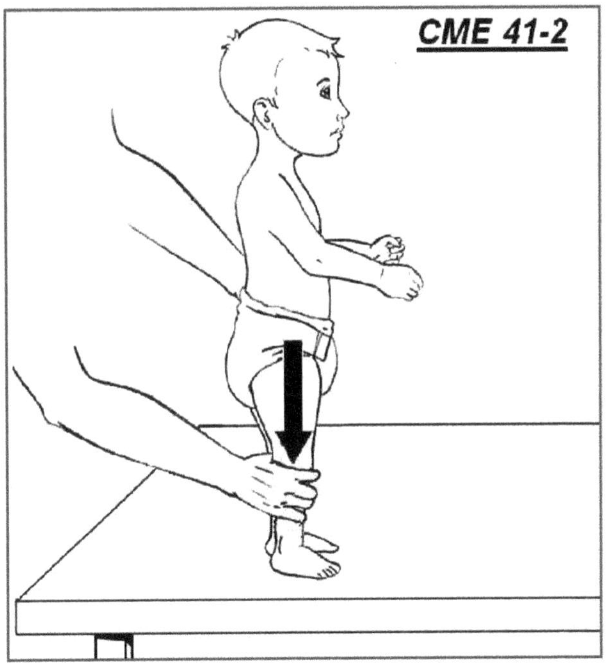

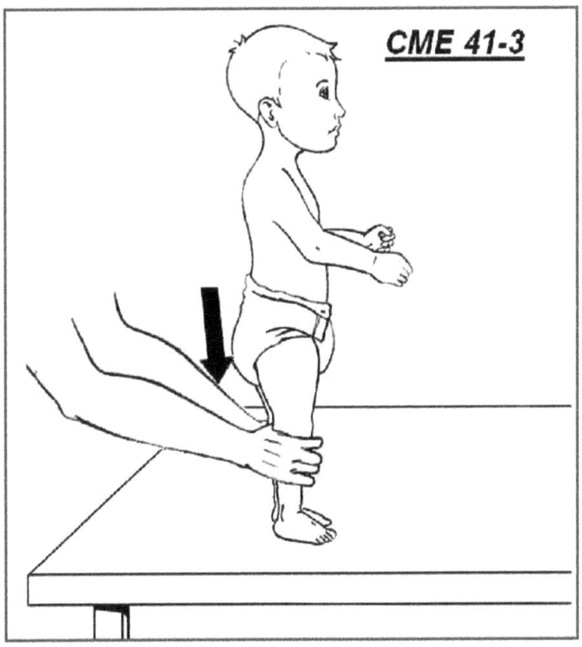

CME 42 AERIAL STANDING BY ONE LEG

MAIN GOAL: Enhance hip and trunk stability in standing position without soles contact.

INITIAL POSITION: Child is in a standing position in front of you, held by mid trunk as shown in the figure 1.

HOLD: With left hand on the inside of his/her right thigh as indicated in figure 2.

MOVEMENT: Move your right hand to his/her right ankle as shown in figure 3, and lift the child airborne, placing your right elbow on the table. Hold the suspended standing position as illustrated in figure 4 for a count of 20. Repeat the exercise as recommended by your CME therapist.

EXPECTED IMPROVEMENT: The child maintains a standing posture with trunk fully vertical.

CME 42 AERIAL STANDING BY ONE LEG

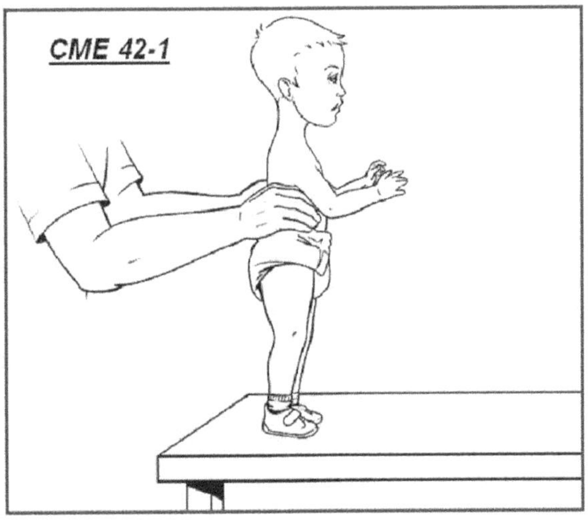
CME 42-1

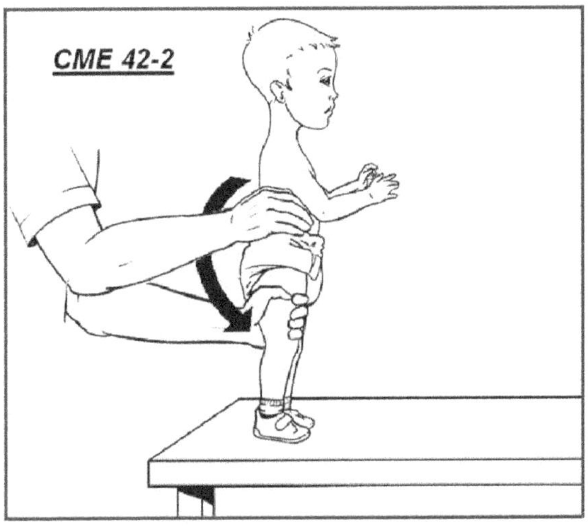
CME 42-2

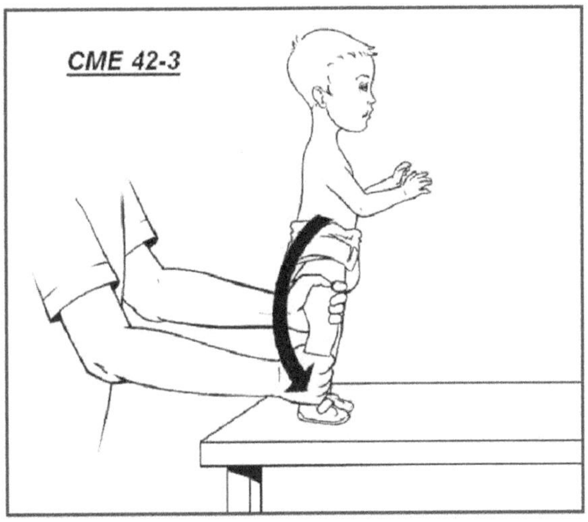
CME 42-3

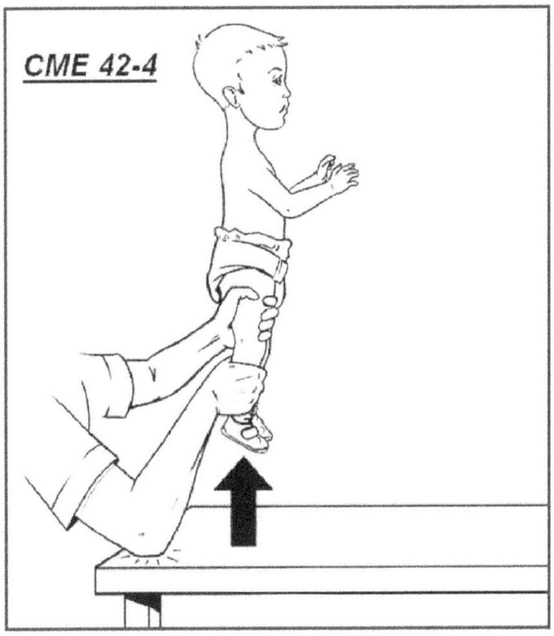
CME 42-4

CME 43 STANDING POSITION BY ANKLES

MAIN GOAL: Enhance functional hips and knees antigravity extension.

INITIAL POSITION: Child is in a standing position in front of you, held by thighs as shown in figure 1.

HOLD: With right hand on his/her right ankle as indicated in figure 2.

MOVEMENT: Move your left hand to his/her left ankle and hold the standing position for a count of 30 as illustrated in figure 3. To repeat, follow the instructions of your CME therapist.

EXPECTED IMPROVEMENT: The child maintains a standing posture with trunk fully vertical.

CME 43 STANDING POSITION BY ANKLES

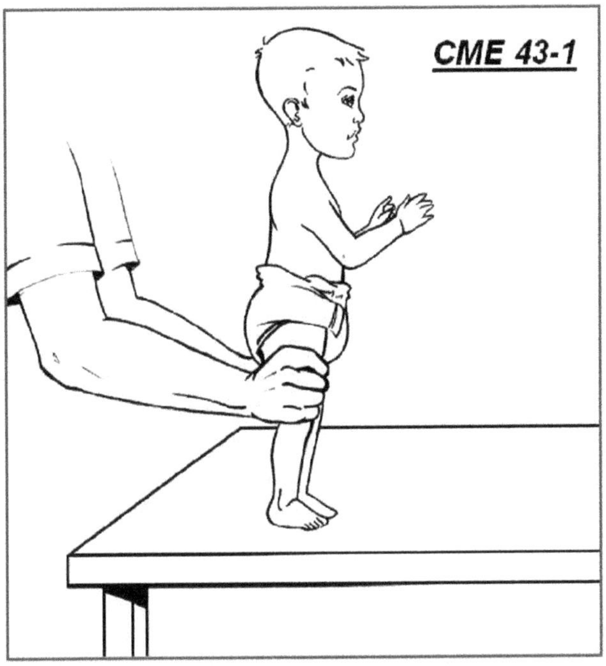

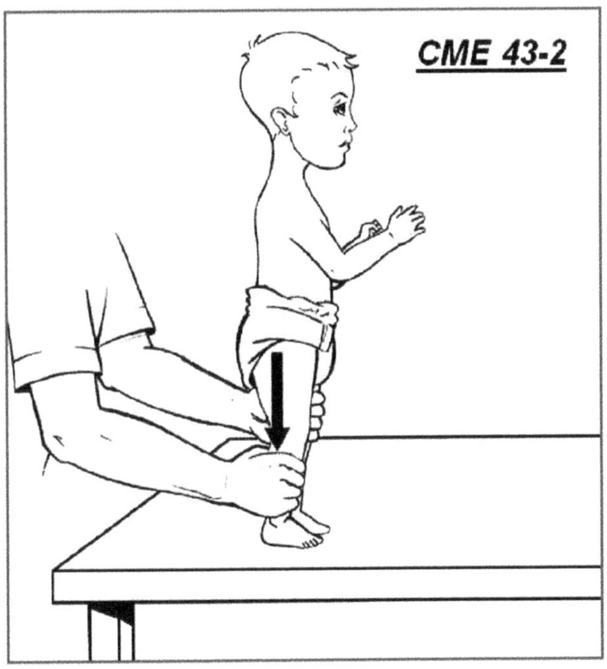

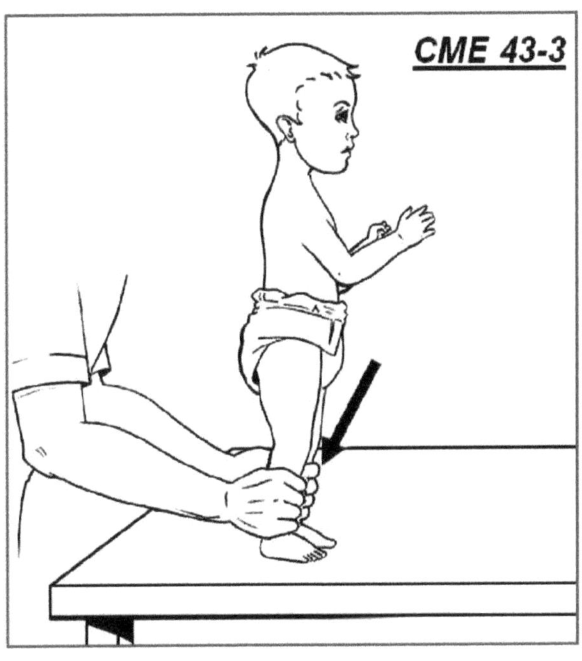

CME 44 STANDING UP BY PELVIS FROM BEHIND

Note: *this exercise was conceived especially to be used in cases of lack of hip extension (spina biphida), in order to try to reproduce the proprioception of legs and trunk in vertical position and some weight bearing as well.*

MAIN GOAL: Provoke a standing position.

INITIAL POSITION: Child is in sitting position on the table.

HOLD: With both hands at the stem of thighs with thumbs in full extension under buttocks as shown in figure 1.

MOVEMENT: Tilt his/her trunk slightly backwards as illustrate in figure 2, then make a leverage on thighs and hips, driving trunk straight upwards and legs down to the table as described in figure 3, ending the maneuver in full standing posture as indicated in figure 4. To repeat, follow the instructions of your CME therapist.

EXPECTED IMPROVEMENT: The child achieves a still standing posture.

CME 44 STANDING UP BY PELVIS FROM BEHIND

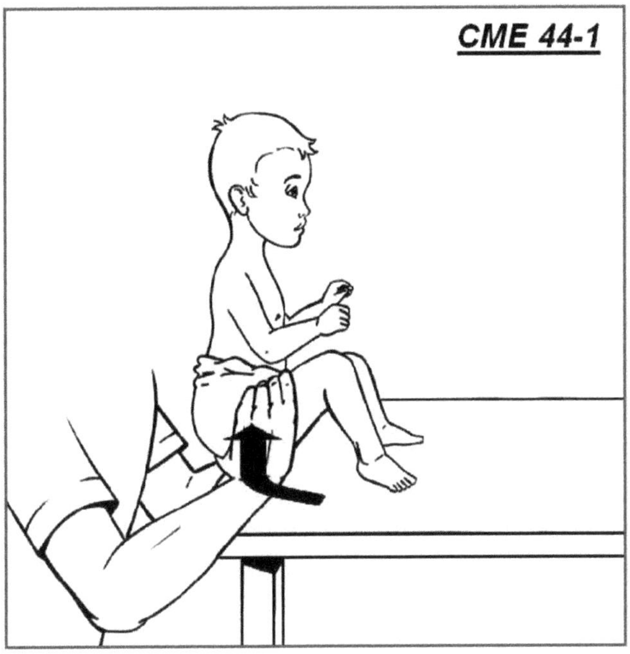

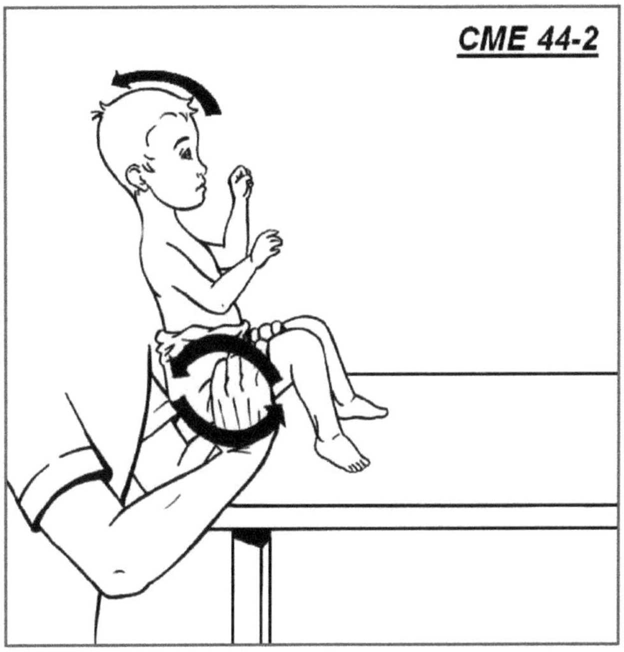

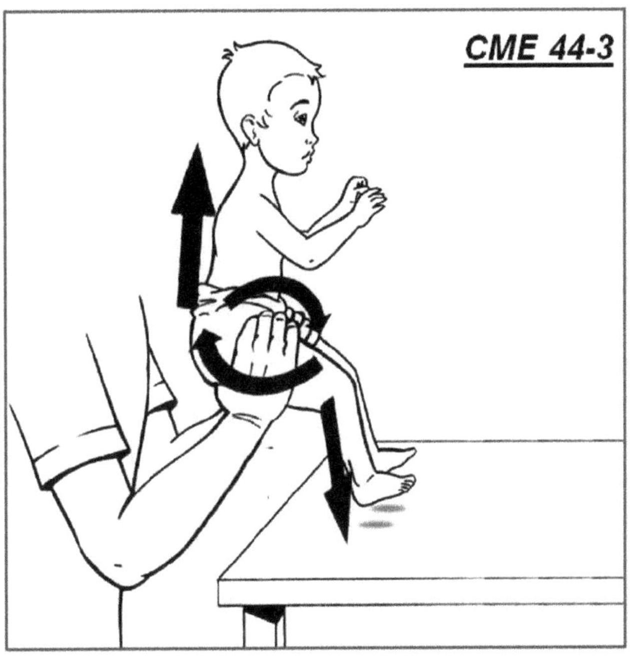

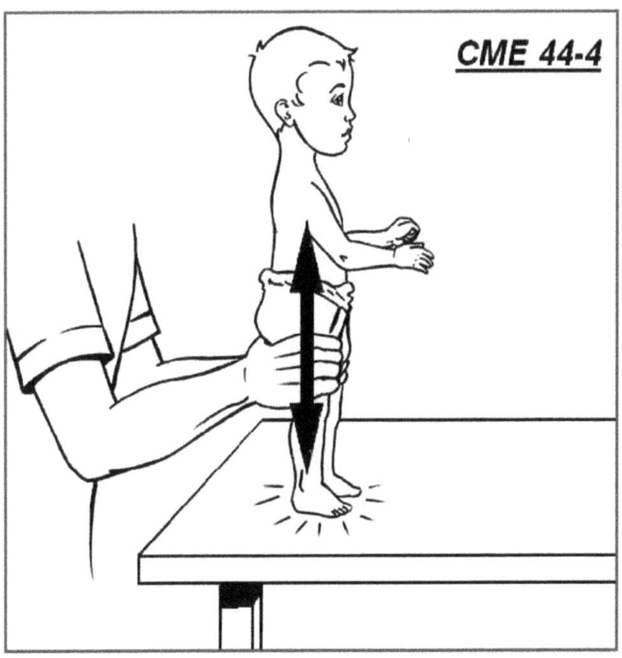

CME 45 HORSY STANDING

MAIN GOAL: Provoke weight bearing and standing position.

INITIAL POSITION: Child is in standing position on the table.

HOLD: With both hands by mid trunk as indicated in fig. 1.

MOVEMENT: Place your right hand between his/her legs as shown in figure 2, then place the left hand in the same position with palm vertically and thumb pointing up as illustrated in figure 3. Hold the standing position for the count of 30. Repeat as indicated by your CME therapist.

EXPECTED IMPROVEMENT: The child stays still in a standing posture.

CME 45 HORSY STANDING

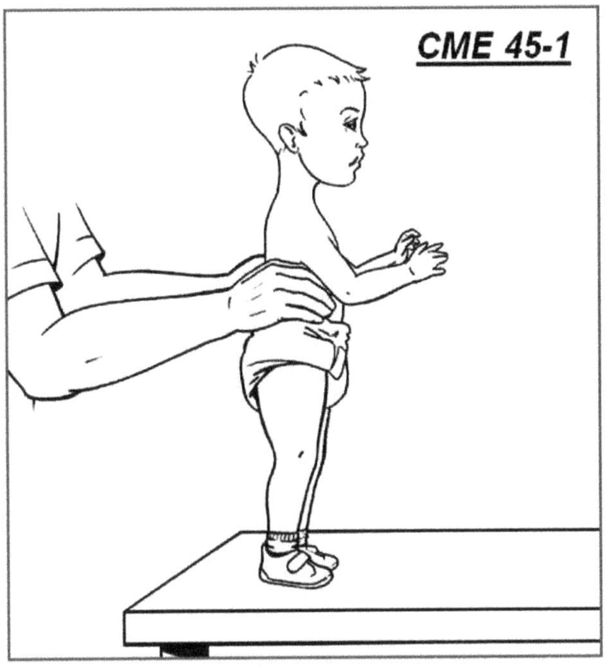

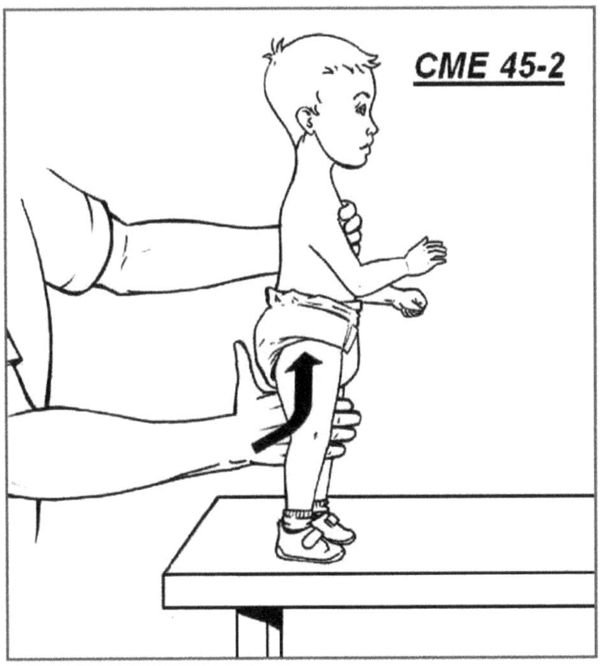

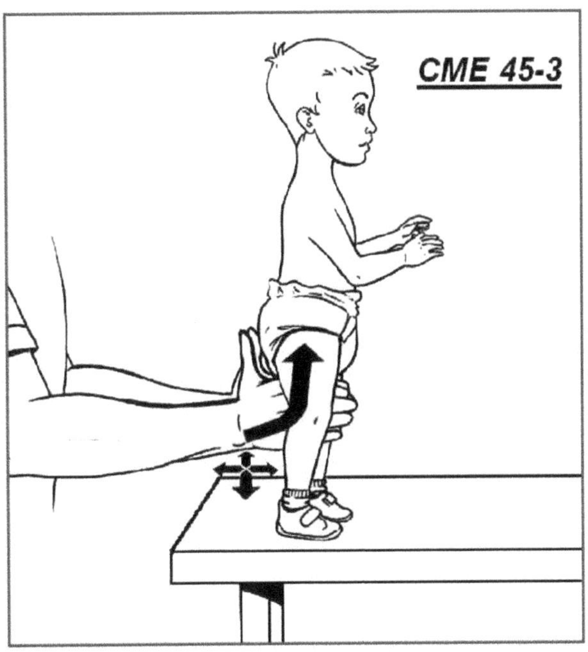

CME 46 HORSY STANDING BY BALANCE BAR

Exercise Configuration: *One end of the balance bar is wrapped with 2 full length grip linner material.*

MAIN GOAL: Provoke weight bearing and standing position.

INITIAL POSITION: Child is in a standing position on the table.

HOLD: With both hands by mid trunk as indicated in the fig. 1.

MOVEMENT: Place the wrapped end of the balance bar between his/her legs. between as illustrated in figure 2. Hold the bar with both hands (fig. 3) and keep the standing position for the count of 30. At the beginning it is necessary to move the bar properly to any direction to compensate the balance reactions. Repeat as indicated by your CME therapist.

EXPECTED IMPROVEMENT: The child stays still in a standing posture.

CME 46 HORSY STANDING BY BALANCE BAR

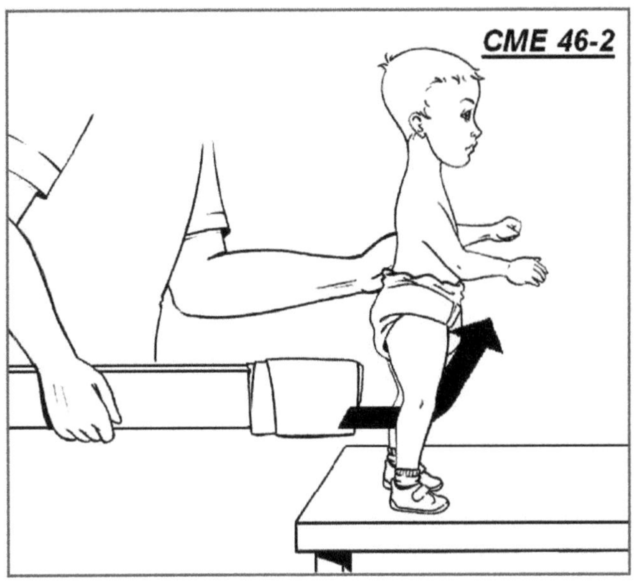

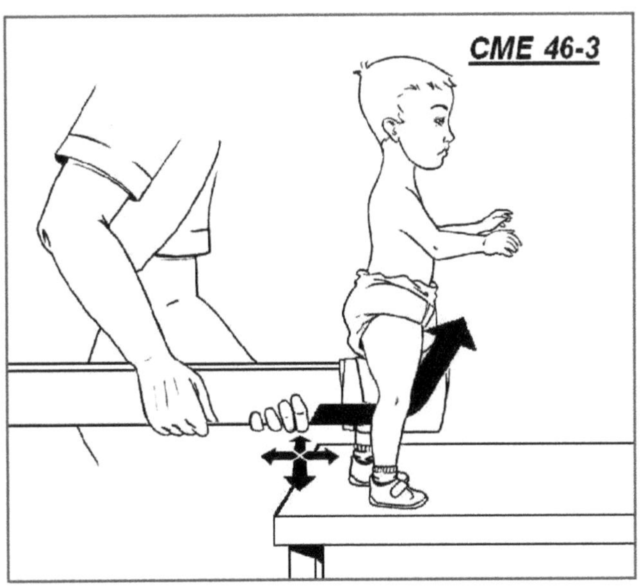

CME 47 STANDING POSITION BY CROSSED STRAPS

Exercise Configuration: *To cross the straps under crotch area as indicated in the circled insert in the figure one.*

MAIN GOAL: Provoke weight bearing and standing position.

INITIAL POSITION: Child is in standing position on the table leaning on your chest.

HOLD: Pull straps horizontally in the opposite direction as illustrated in figure 1.

MOVEMENT: Remove your chest support and let the child stand only by strap tension and hold the standing position for a count of 30 as shown in figure 2. Repeat as indicated by your CME therapist.

EXPECTED IMPROVEMENT: The child stays still in a standing posture.

CME 47 STANDING POSITION BY CROSSED STRAPS

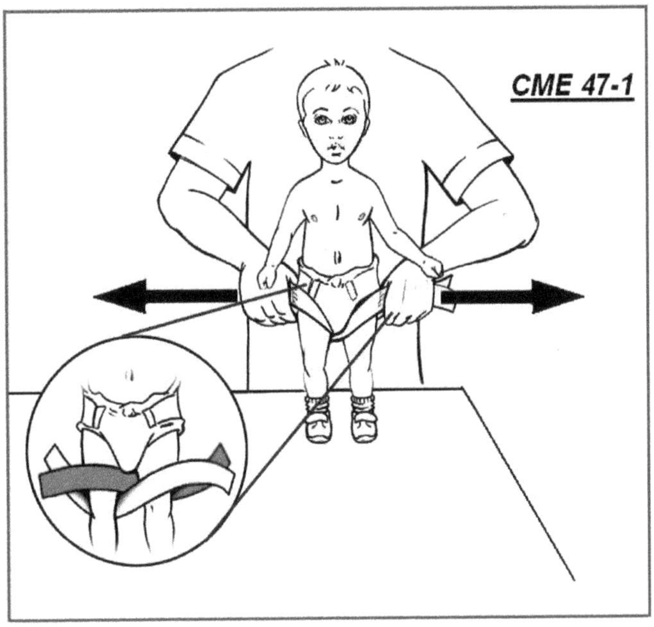

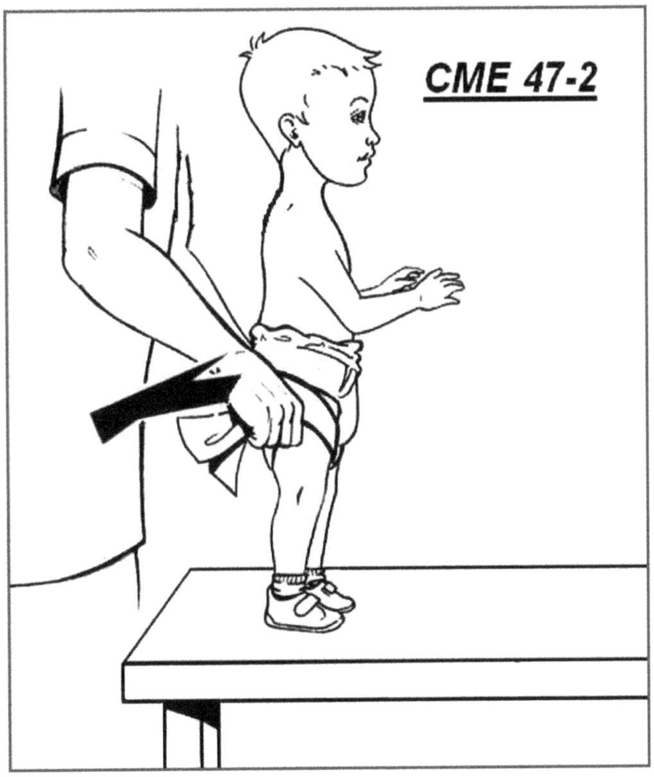

CME 48 STANDING POSITION BY STRAP ON WAIST

MAIN GOAL: Provoke weight bearing and enhance stability control in standing posture.

INITIAL POSITION: The child is in a standing position on the table supported by your chest.

HOLD: By strap wrapped around his/her lower abdomen and pull apart in the horizontally opposite direction as illustrated in figure 1.

MOVEMENT: Move your chest support away and leave the child standing only by the tension of the strap and hold the position for the count of 30. Repeat as indicated by your CME therapist.

EXPECTED IMPROVEMENT: The child stays still in a standing posture.

CME 48 STANDING POSITION BY STRAP ON WAIST

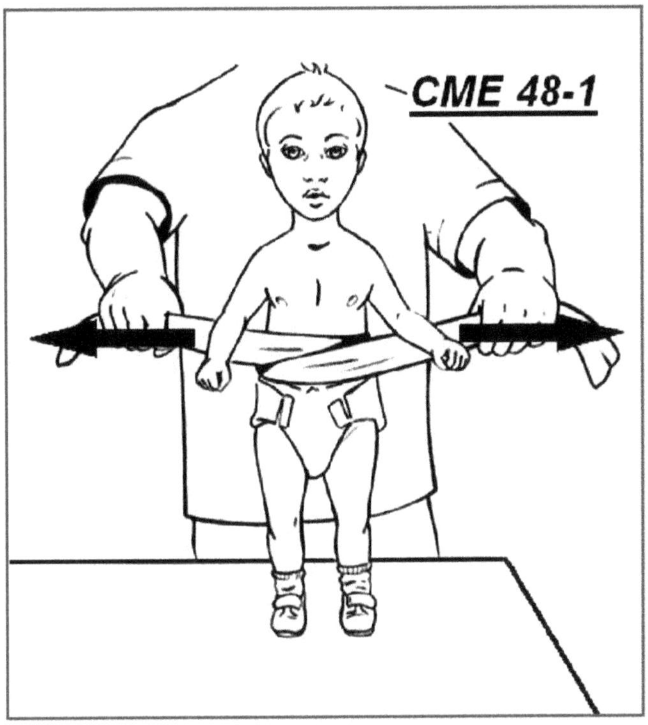

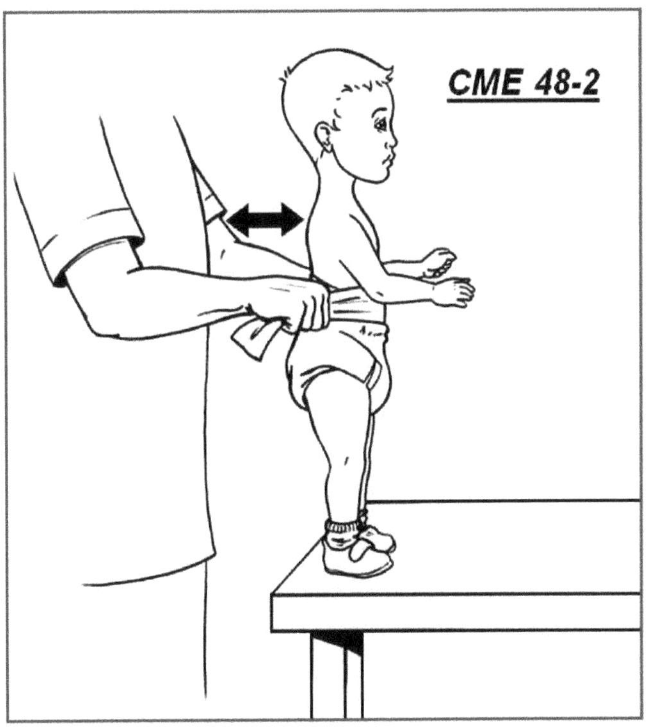

CME 49 STANDING POSITION BY STRAPS ON ANKLES

Exercise Configuration: *Put one strap on each ankle and twist them very tightly like a tourniquet.*

MAIN GOAL: Provoke weight bearing and balance reactions in a standing position.

INITIAL POSITION: The child is in standing position on the table supported by your chest.

HOLD: By a strap wrapped around his/her ankles.

MOVEMENT: Move your chest support away and leave the child standing only by the tension of the straps and hold the standing position as illustrated in figure 2, for the count of 30. At the beginning it is mandatory to apply compensatory pulses in any necessary direction, as suggested by multi directional arrows in figures 1 and 2, in order to compensate the lack of stability responses. Repeat as indicated by your CME therapist.

EXPECTED IMPROVEMENT: The child stays still in a standing posture.

CME 49 STANDING POSITION BY STRAPS ON ANKLES

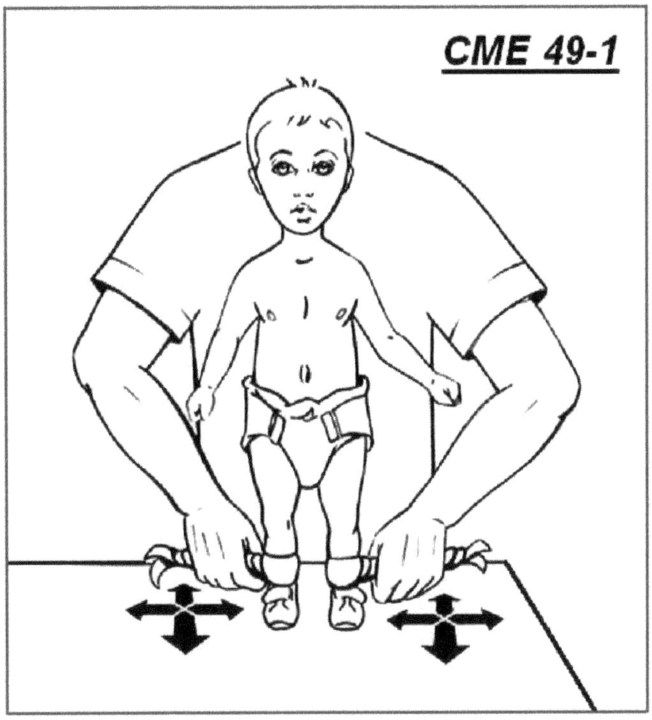

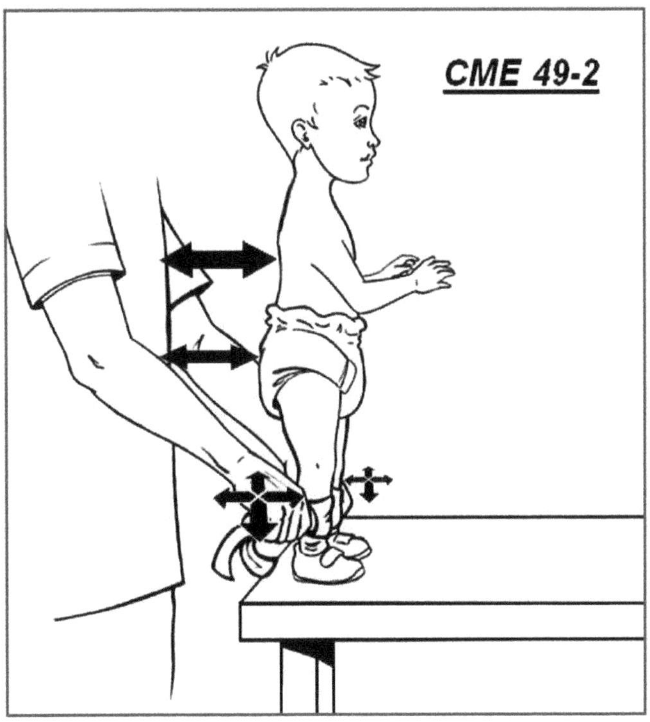

CME 50 STANDING ON TRAPEZE

Exercise Configuration: *Put small hard cover book of approx. 15x25 centimeters at the center of two full length grip linner straps. Fold the straps on the book and hold up by both ends (fig.1).*

MAIN GOAL: Provoke balance reactions in a standing position.

INITIAL POSITION: The child lies down in a prone position with hips bent at the edge of table as illustrated in figure 2.

HOLD: Move up the trapeze configuration until making solid contact with his/her feet as indicated in figures 2 and 3.

MOVEMENT: The child responds by making a complete trunk extension with full weight bearing on the trapeze as shown in figure 4. At this time, lift the trapeze up again to move the child completely off the table and hold the free standing for the count of 20 as observed in figure 5. To compensate the balance responses you need to move the trapeze to any convenient direction. Repeat as indicated by your CME therapist.

EXPECTED IMPROVEMENT: The child stays still in standing posture.

CME 50 STANDING ON TRAPEZE

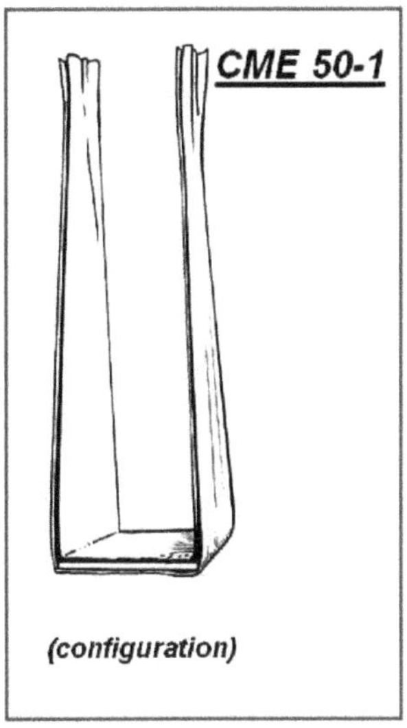

CME 50-1

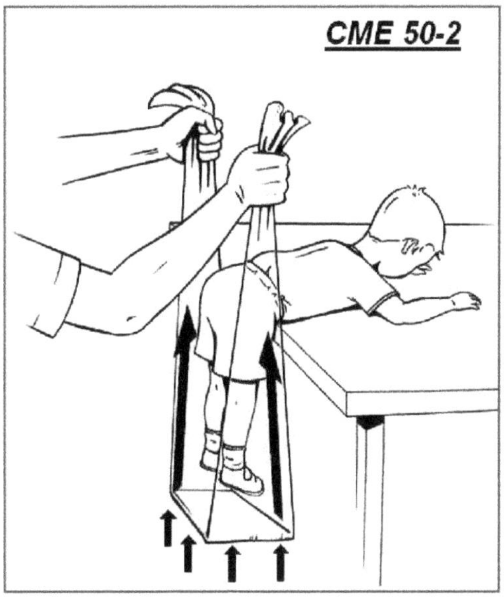

CME 50-2

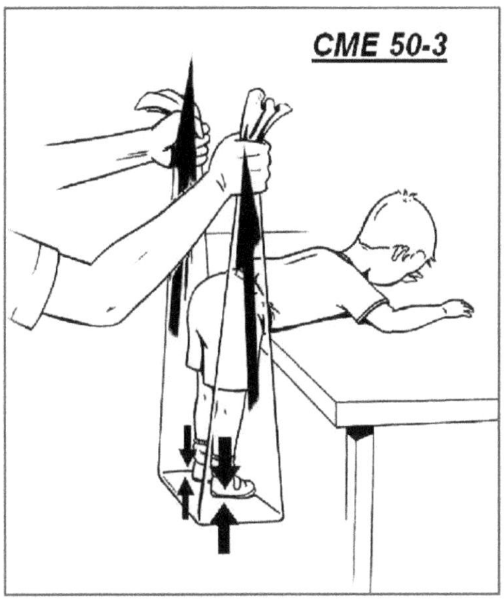

CME 50-3

CME 50-4

CME 50-5

CME 51 STANDING UP FROM PRONE BY LONG STRAP

MAIN GOAL: To provoke a standing up reaction from a quadruped posture.

INITIAL POSITION: The child lies down in a prone position on the floor.

HOLD: By a long strap (1.80 cm or 6 feet) under his/her hips as indicated in figure 1.

MOVEMENT: Pull slowly up and backward by the ends of the straps to provoke a quadruped posture (fig, 2), keep pulling steady until the child places feet on the floor (fig. 3 and 4) and begin to move up to a standing posture as illustrated in figures 4, 5, 6 and 7. Repeat the exercise as indicated by your CME therapist.

EXPECTED IMPROVEMENT: The Child moves quickly from a prone position to a standing position and stays still.

CME 51 STANDING UP FROM PRONE BY LONG STRAP

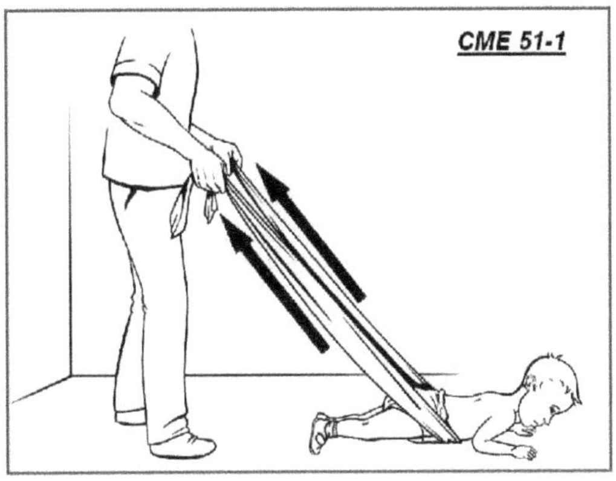
CME 51-1

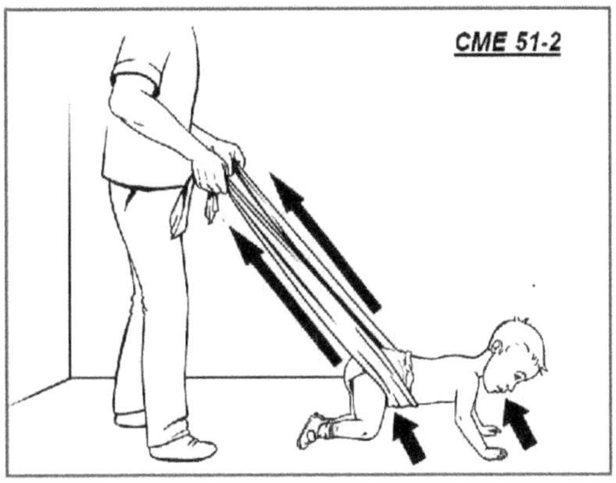
CME 51-2

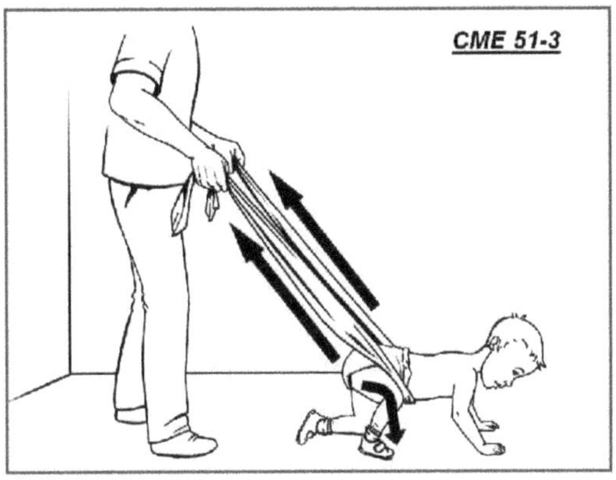
CME 51-3

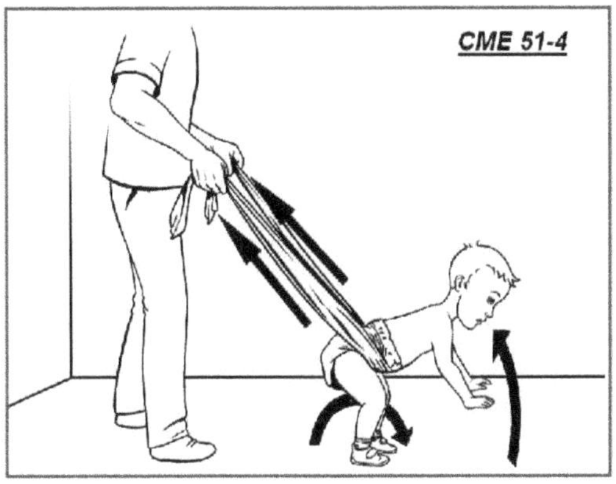
CME 51-4

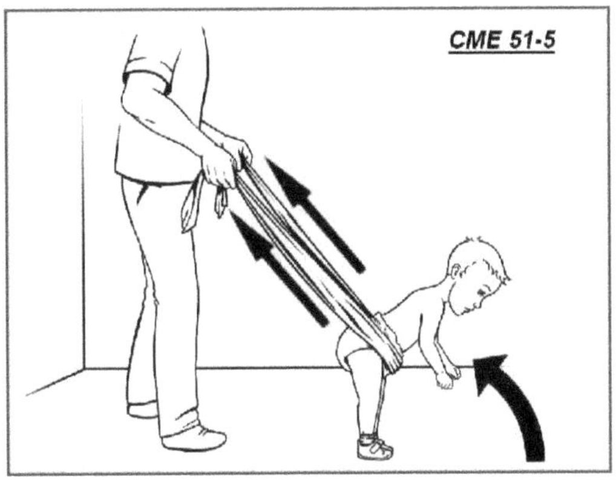
CME 51-5

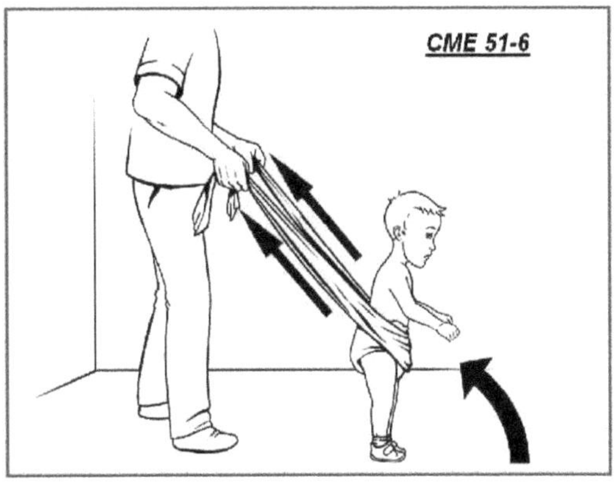
CME 51-6

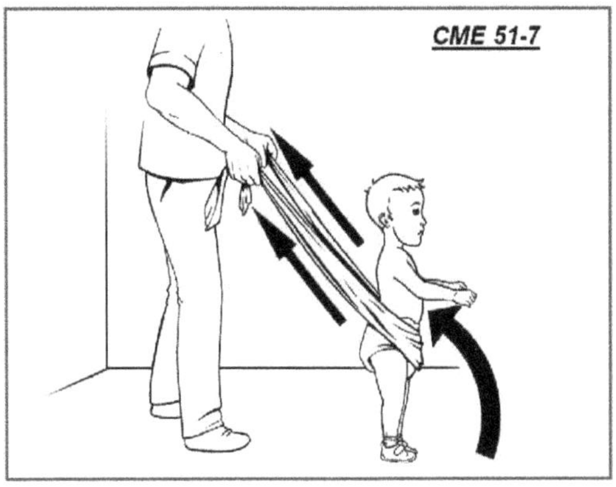
CME 51-7

CME 52 STEPS BY SERIES OF 3 POSITION

MAIN GOAL: To improve the dynamic trunk stability and step awareness.

INITIAL POSITION: Standing on the table.

HOLD: To fix the position place the index and middle finger around the joint of the ankle of the right foot, and lock the knee with the wrist and then do the same on the left ankle (fig. 1).
Subsequently scoot sideways to the right, placing the child between the left side of your trunk and your left arm as illustrated in figure 2.

MOVEMENT: Drive the child forward taking small steps along the edge of the table as shown in figures 3, 4, and 5, until you complete 50 steps. Repeat the exercise as indicated by your CME therapist.

EXPECTED IMPROVEMENT: The child should respond taking steps easily and holding the trunk upright at all times.

CME 52 STEPS BY SERIES OF 3 POSITION

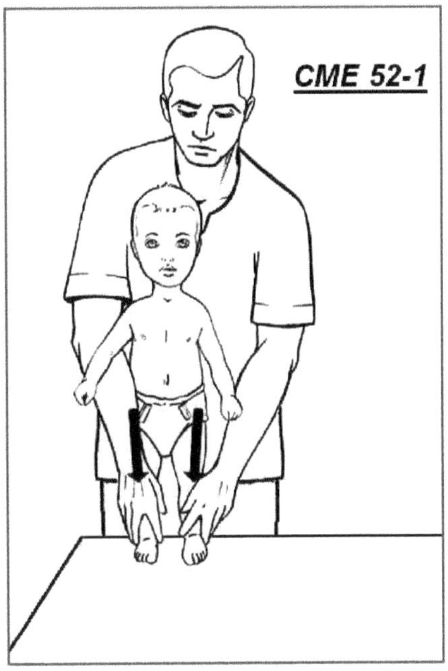
CME 52-1

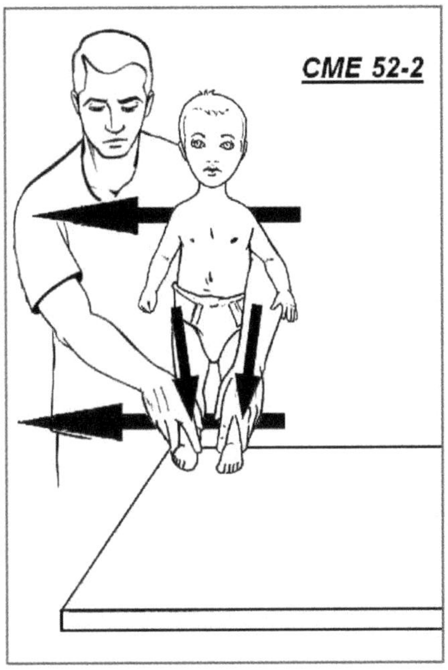
CME 52-2

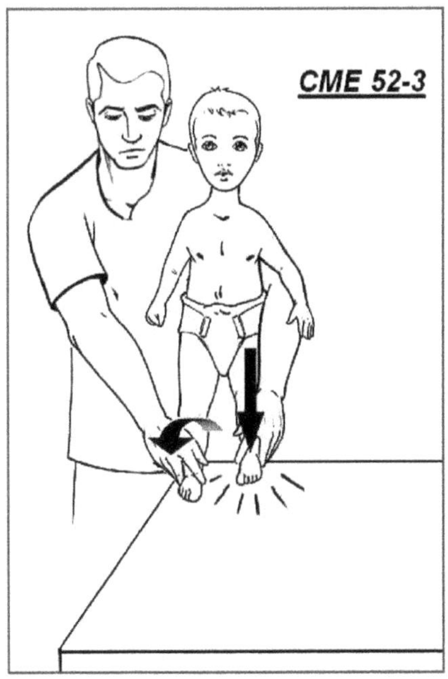
CME 52-3

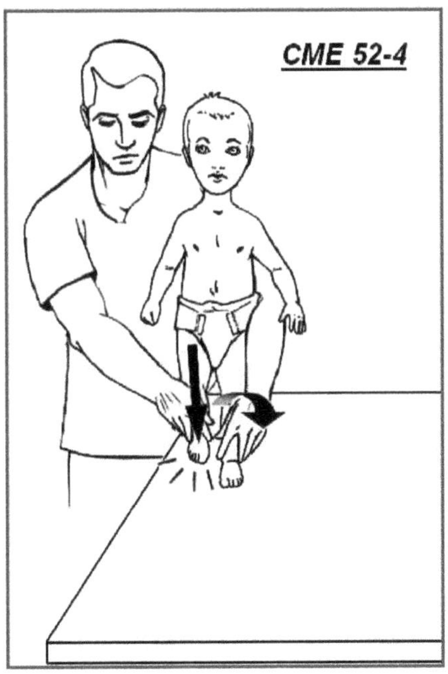
CME 52-4

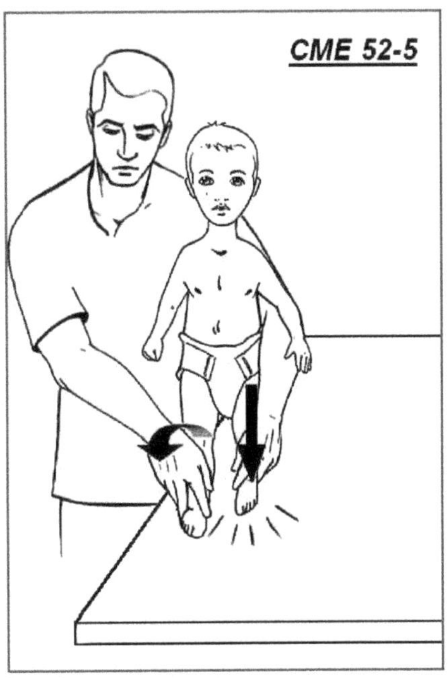
CME 52-5

CME 53 PLACING

MAIN GOAL: To provoke automatic steps and dynamic posture correction of standing posture.

INITIAL POSITION: Child is in a standing position with heels at edge of the table (fig. 1).

HOLD: With both hands by the thighs.

MOVEMENT: While holding the standing position firmly by his/her left thigh, place your right hand in front of the child's right ankle as illustrated in figure 1. Subsequently swift backwards sliding the left foot completely off the table (fig. 2), and the child moves the leg back to the table to recover stability, as illustred in the figure 3. Follow the instructions of your CME therapist.

EXPECTED IMPROVEMENT: The child keep his/her body upright at all times.

CME 53 PLACING

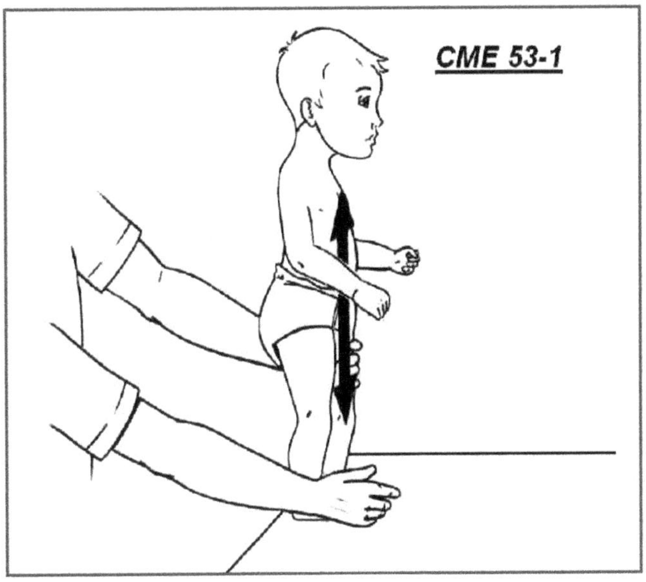

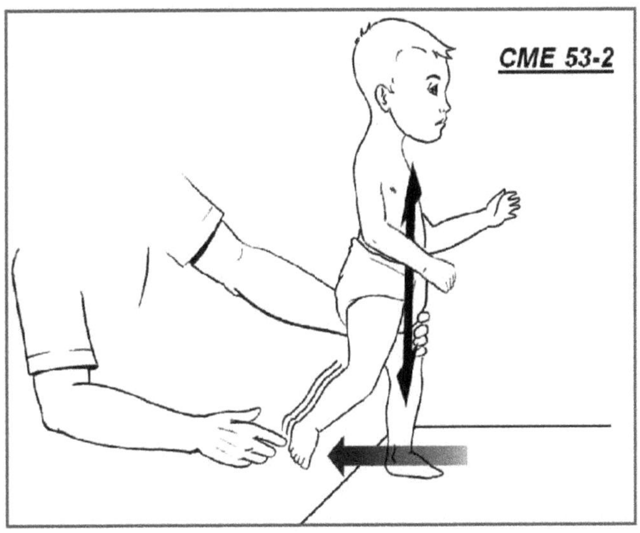

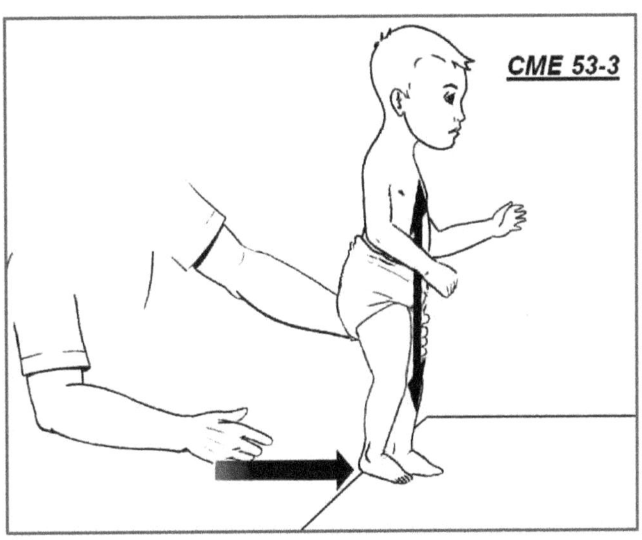

CME 54 STEPS BY THIGHS

MAIN GOAL: To provoke vertical trunk stability and walking pattern.

INITIAL POSITION: The child lies down in a prone position on the table.

HOLD: With both hands on the mid trunk and pulling up the child to a standing position away from you (fig. 1), then move your hands to the thighs and hold the position firmly as shown in figure 2.

MOVEMENT: Move ahead one leg at a time with a regular pace, making small steps until you reach the other edge of the table as suggested in figures 3 and 4. Follow the instructions of your CME practitioner.

EXPECTED IMPROVEMENT: The child follows the steps and also maintains the trunk vertical and stable at all times.

CME 54 STEPS BY THIGHS

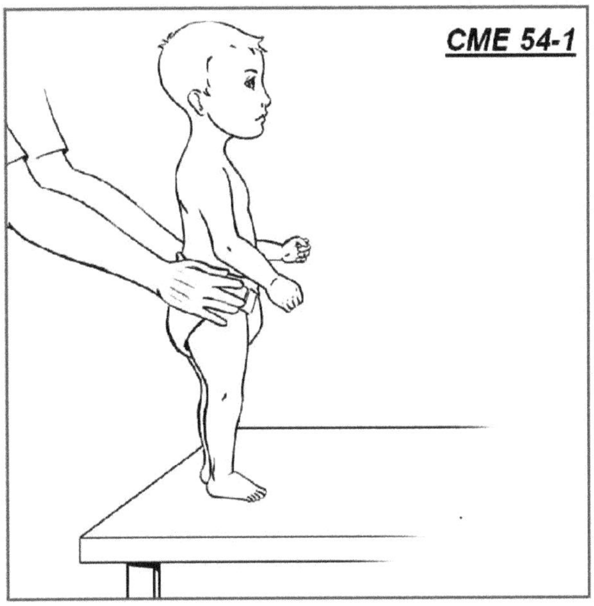

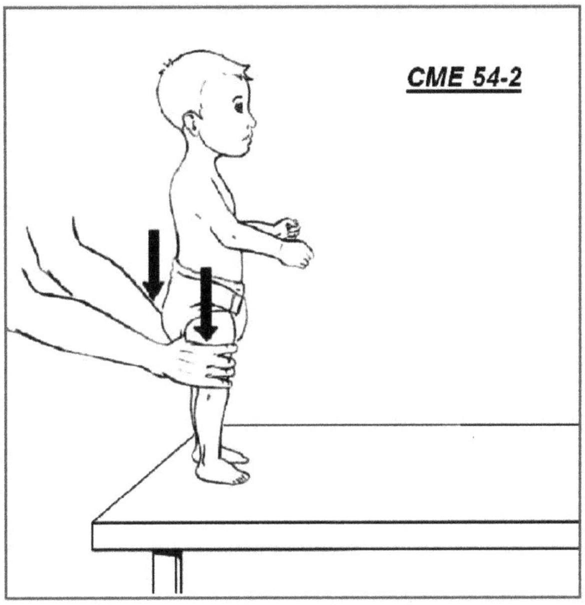

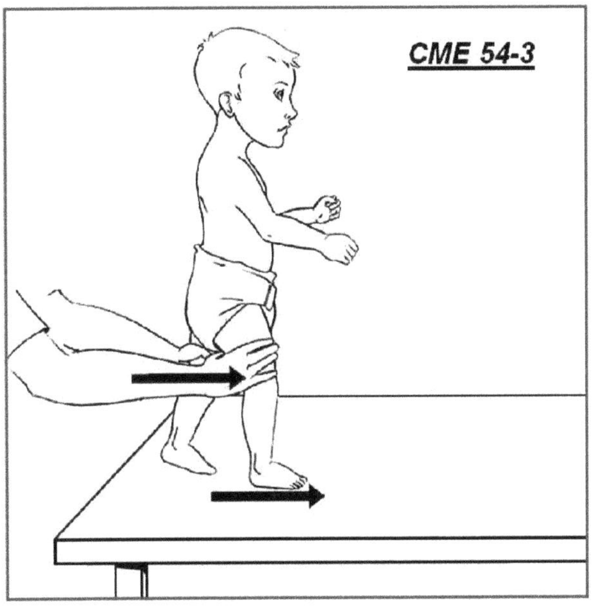

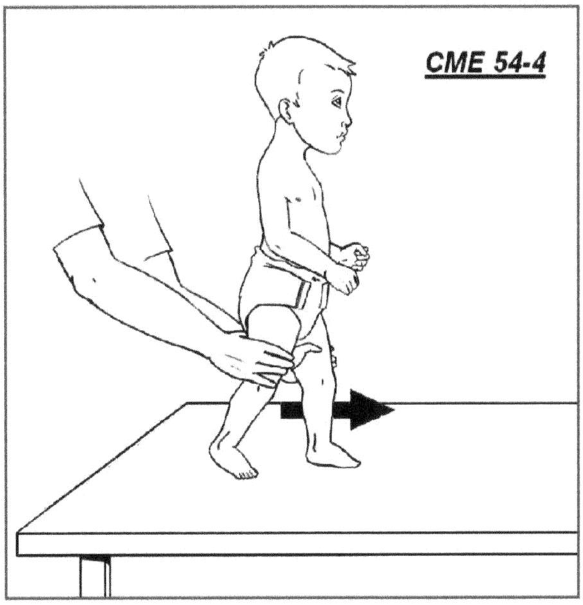

CME 55 STEPS ON EDGE OF TABLE

MAIN GOAL: To improve the dynamic trunk stability and steps awareness.

INITIAL POSITION: Standing on the table.

HOLD: With the index and middle finger around the joint of the ankle of the right foot, lock the knee with the wrist, and tuck his/her trunk between your arm and your trunk as illustrated in figure 1.

MOVEMENT: Push the child forward taking small steps along the edge of table as shown in figures 2, 3 and 4 until 30 steps are completed. Repeat the exercise as indicated by your CME therapist.

EXPECTED IMPROVEMENT: The child should respond taking steps easily and holding the trunk upright at all times.

CME 55 STEPS ON EDGE OF TABLE

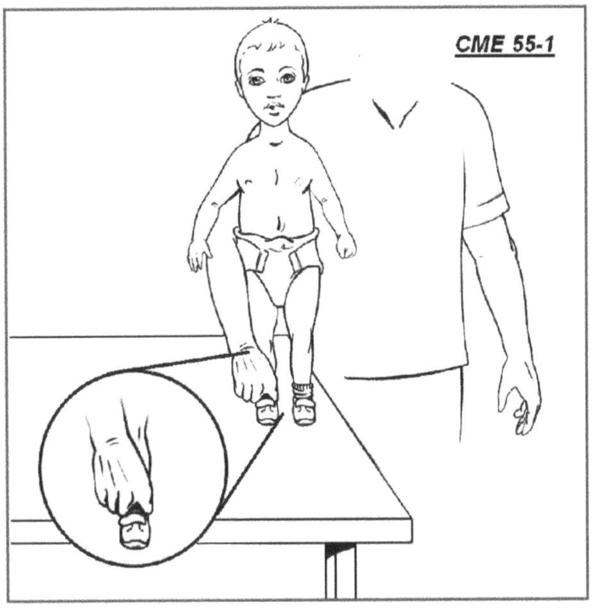

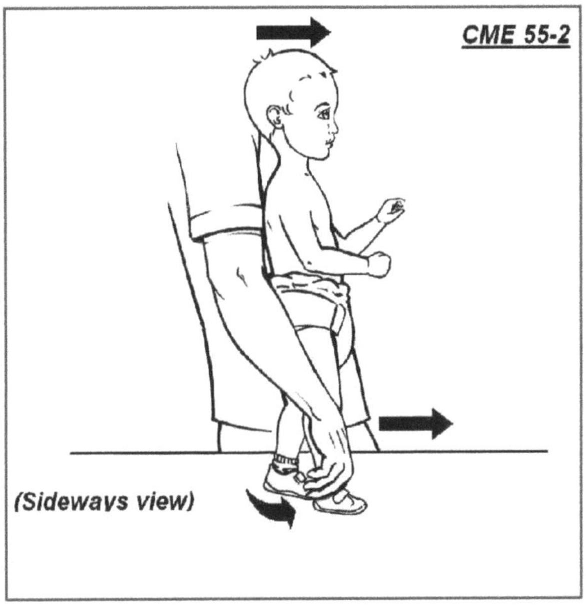

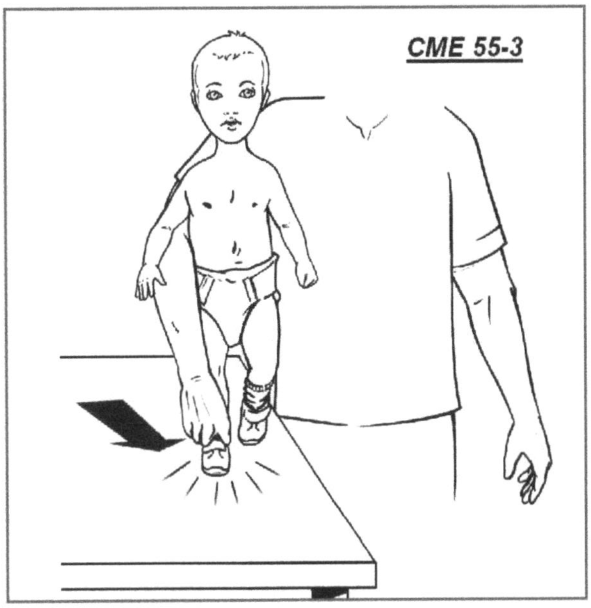

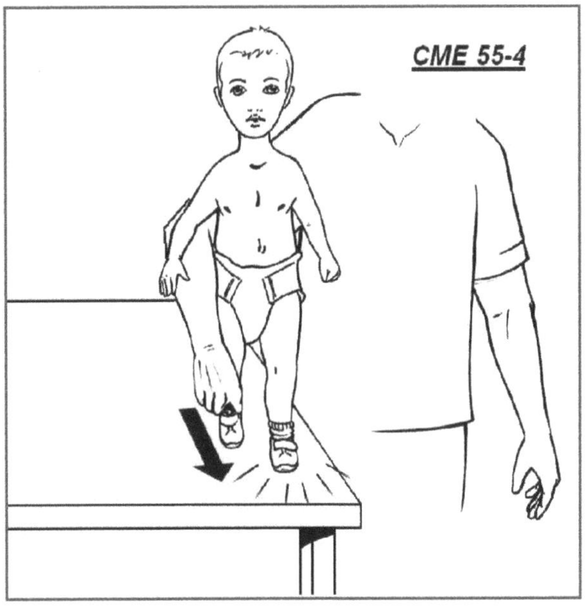

CME 56 STEPS BY ONE LEG

MAIN GOAL: To provoke vertical trunk stability and walking pattern.

INITIAL POSITION: Child is in the standing position.

HOLD: With left hand positioned on the inside of his/her right thigh and with your right hand by his/her right ankle as illustrated in figure 1.

MOVEMENT: Move the right leg ahead and take small steps as shown in figure 2. Wait until the child moves the left leg himself/herself taking the step as indicated in figure 3. Repeat the cycle one step at a time with a regular pace, making small steps until you reach the other edge of the table. Repeat this exercise as instructed by your CME therapist.

EXPECTED IMPROVEMENT: The child follows the steps and also maintains the trunk vertically and stable at all times.

CME 56 STEPS BY ONE LEG

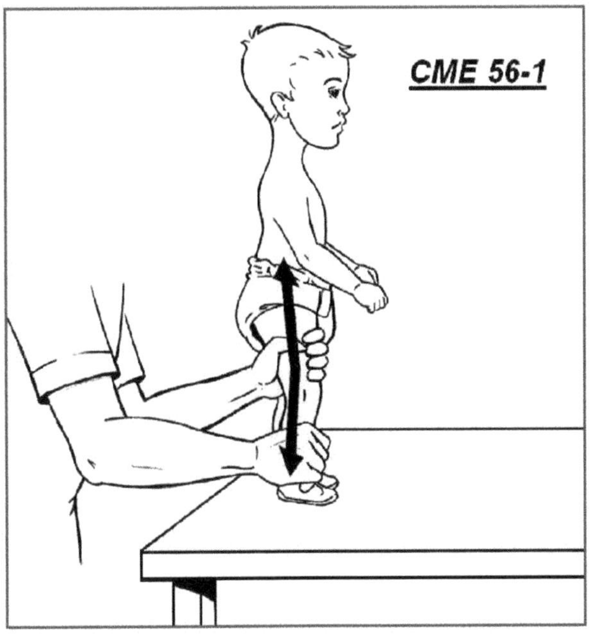

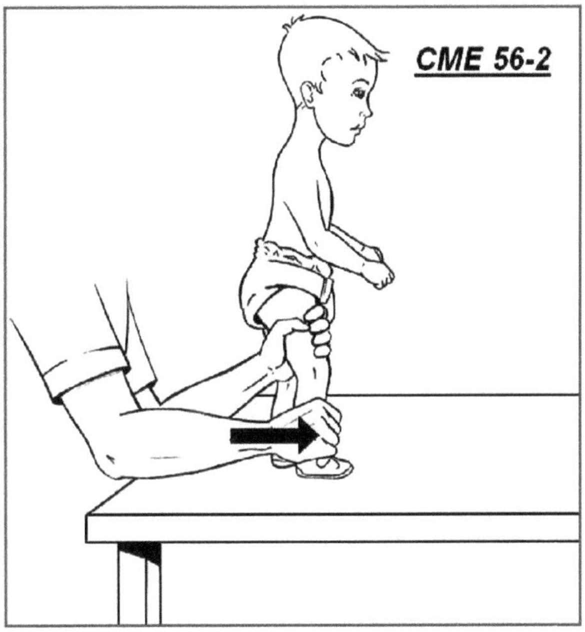

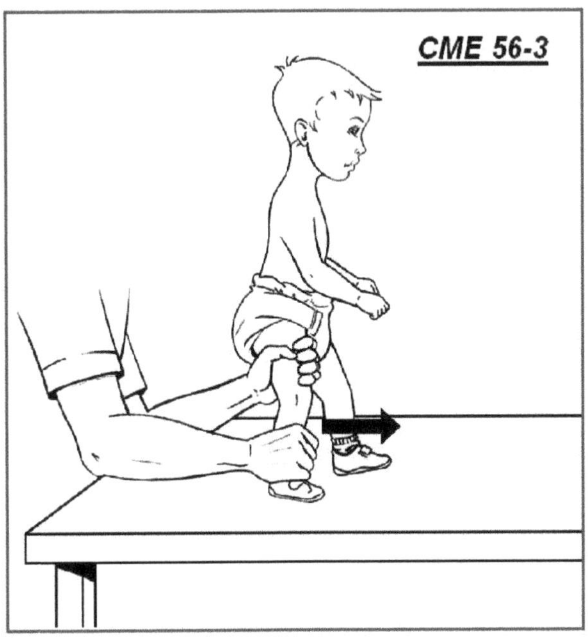

CME 57 HORSY STEPS BY ONE ANKLE

MAIN GOAL: Provoke steps awareness and dynamic stability in a standing position.

INITIAL POSITION: The child is in a standing position on the table.

HOLD: With your right hand between his/her legs and holding the left leg by the ankle as illustrated in figure 1.

MOVEMENT: Take steps with his/her left leg as illustrated in figure 2, then switch the left hand to his/her right ankle and take a step as indicated in figure 3, subsequently hold his/her left ankle and take a step (fig. 4). Repeat the cycle until reaching the end of the table. Repeat the exercise as indicated by your CME therapist.

EXPECTED IMPROVEMENT: The child takes steps easily all along the table.

CME 57 HORSY STEPS BY ONE ANKLE

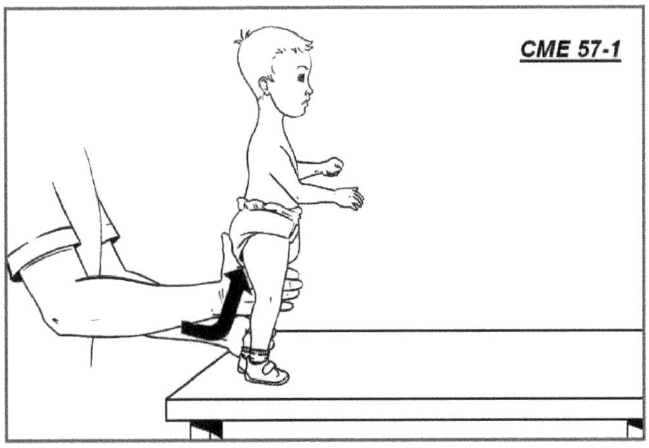

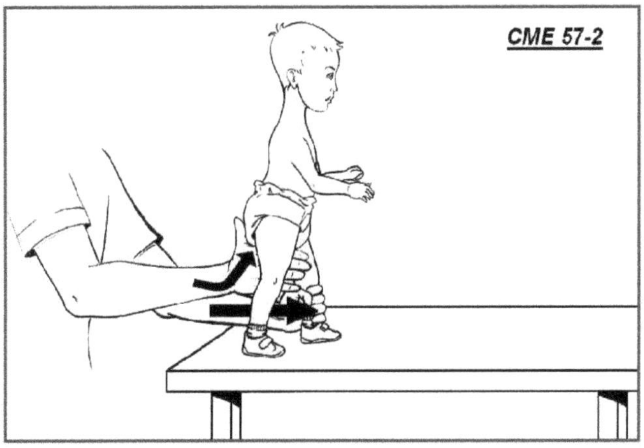

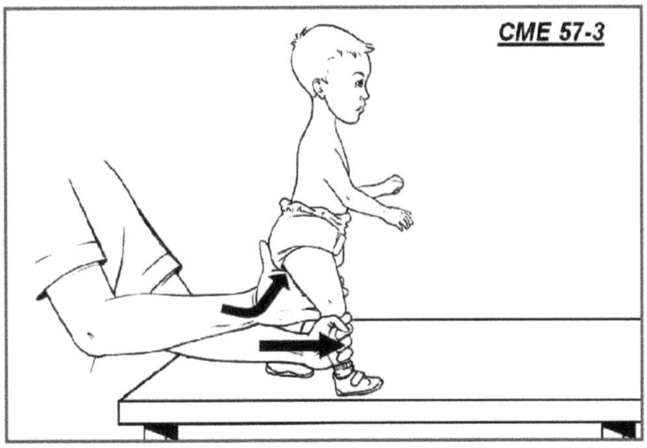

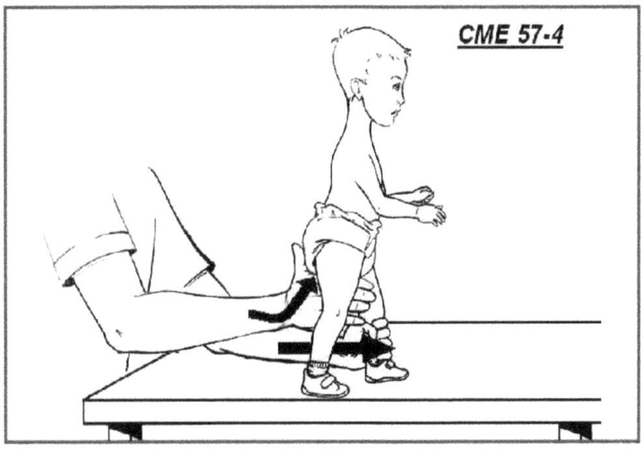

CME 58 STEPS BY OPPOSITE PULLING STRAPS

MAIN GOAL: Provoke steps awareness and dynamic stability.

INITIAL POSITION: Standing on the table.

HOLD: By one strap around the hips in front of the child, and by another strap around his/her chest behind the child as illustrated in figure 1.

MOVEMENT: Pulling the straps simultaneously in opposite directions, the previous push prevailing straps on hips, in order to provoke small steps forward as shown in figures 2 and 3. Take steps all along the table. Repeat as instructed by your CME therapist.

EXPECTED IMPROVEMENT: The child walks easily until the end of the table.

CME 58 STEPS BYOPPOSITE PULLING STRAPS

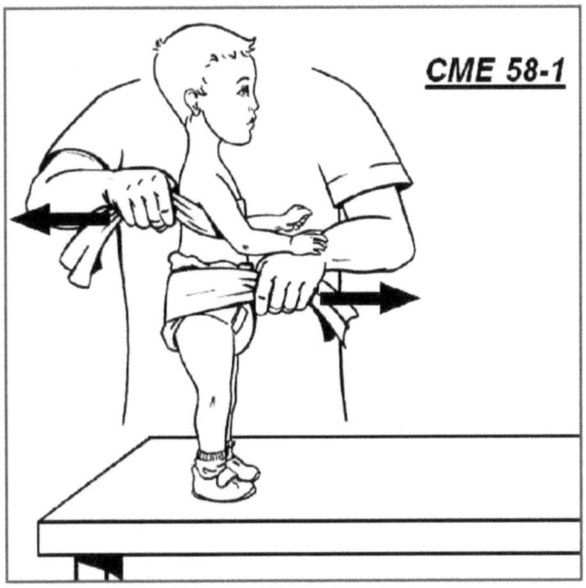

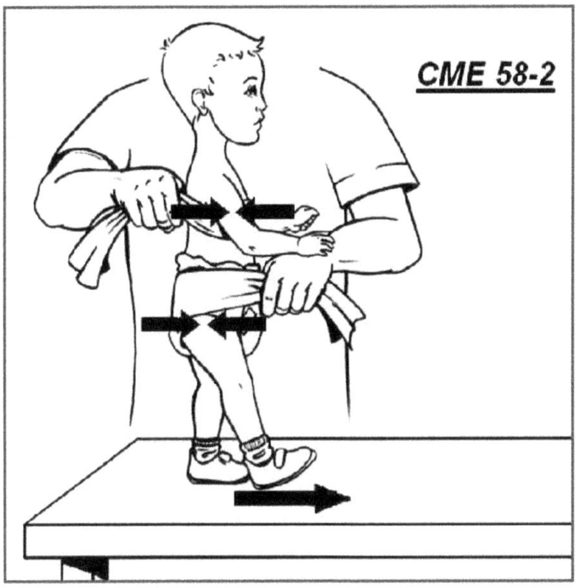

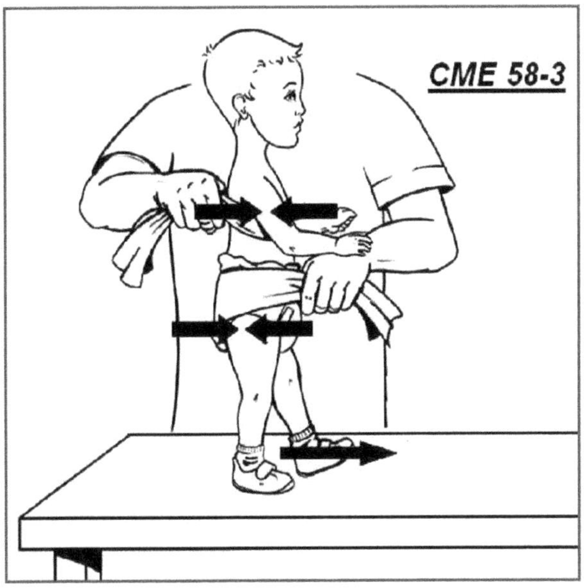

CME 59 STEPS BY PARACHUTE STRAPS

MAIN GOAL: To provoke weight bearing, balance reactions and spontaneous steps.

INITIAL POSITION: Standing on the floor supporting his/her back on your legs.

HOLD: By 2 long straps placed under crotch and supported upwards on a "V" shape, higher than his/her head, as illustrated in figure 1.

MOVEMENT: Move his/her back away from your legs as indicated in figure 2. Subsequently drive the child forward in order to provoke steps as shown in figures 3 and 4. Keep on going until 10 steps are completed. Repeat as instructed by your CME practitioner.

EXPECTED IMPROVEMENT: The Child walks easily with full weight bearing control.

CME 59 STEPS BY PARACHUTE STRAPS

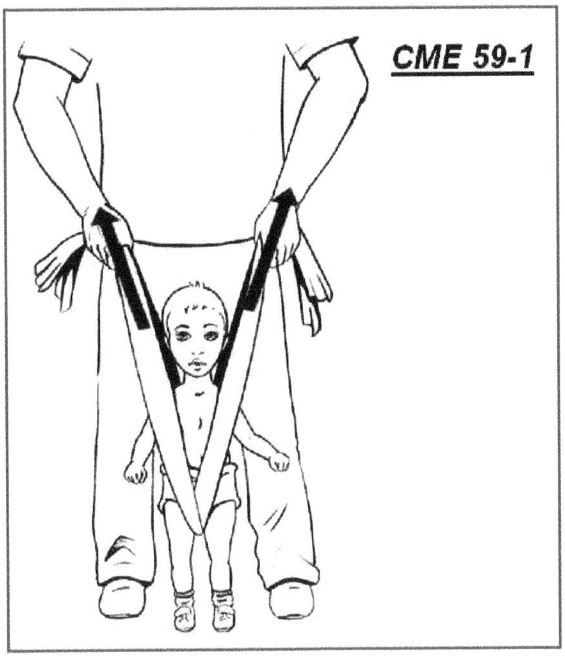

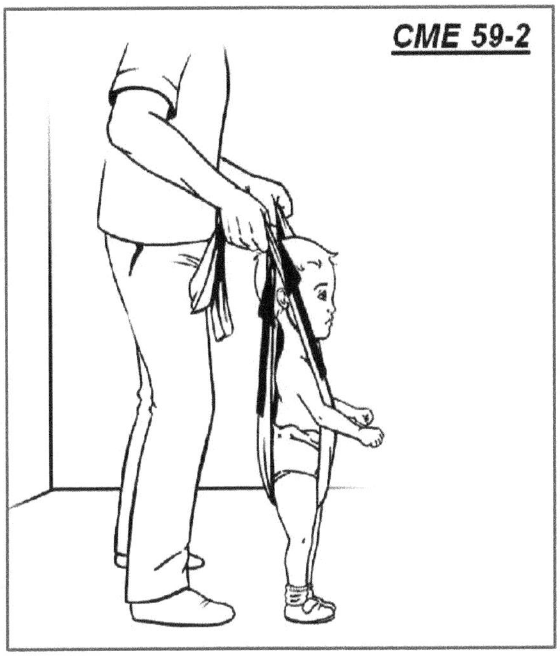

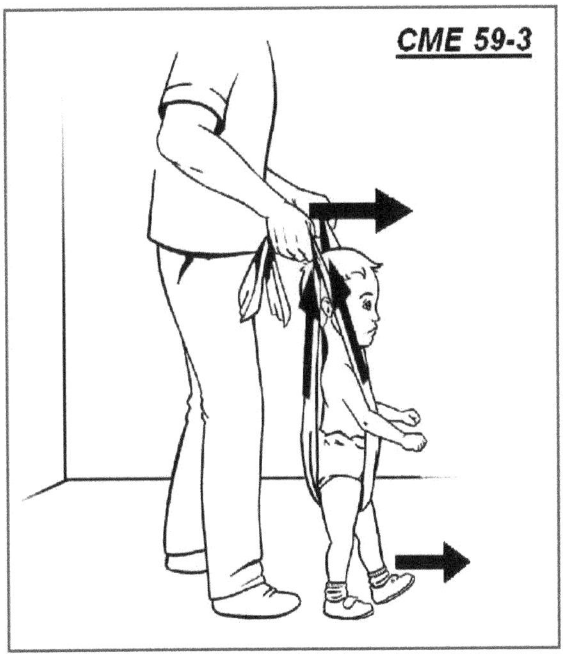

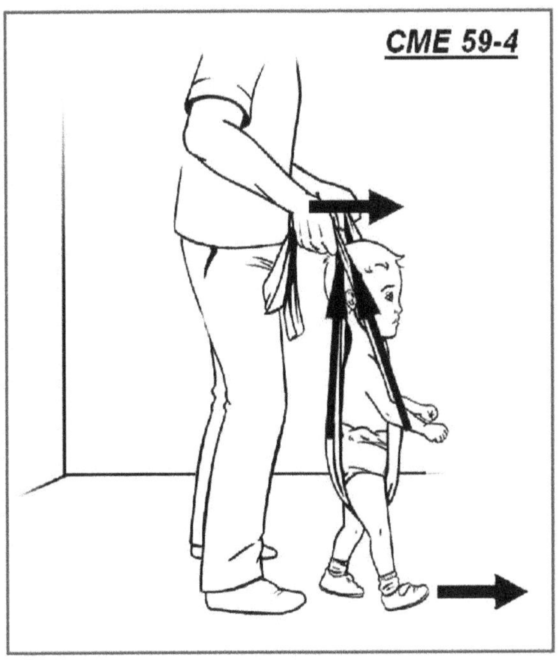

CME 60 STEPS ON BALANCE BOARD BY THIGHS

MAIN GOAL: Reinforce trunk vertical stability and provoke balance reactions.

INITIAL POSITION: Standing at edge of the balance board supported by hips (fig. 1).

HOLD: By thighs as illustrated in figure 2.

MOVEMENT: Take small steps across the balance board until reaching the opposite end as shown in figures 3 and 4. Repeat the exercise as recommended by your CME therapist.

EXPECTED IMPROVEMENT: The child moves across the balance board with good stability.

CME 60 STEPS ON BALANCE BOARD BY THIGHS

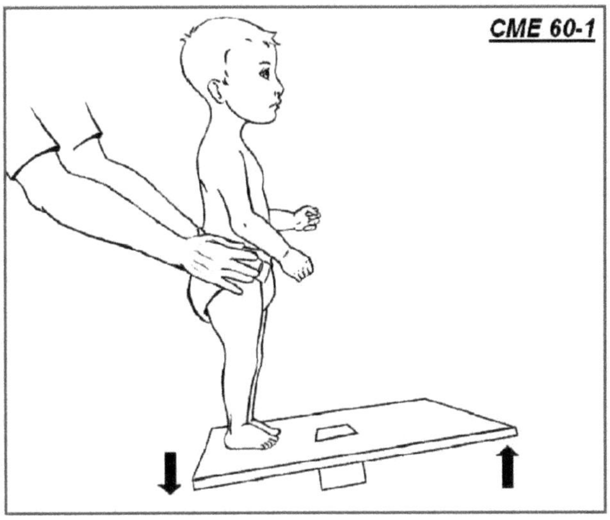

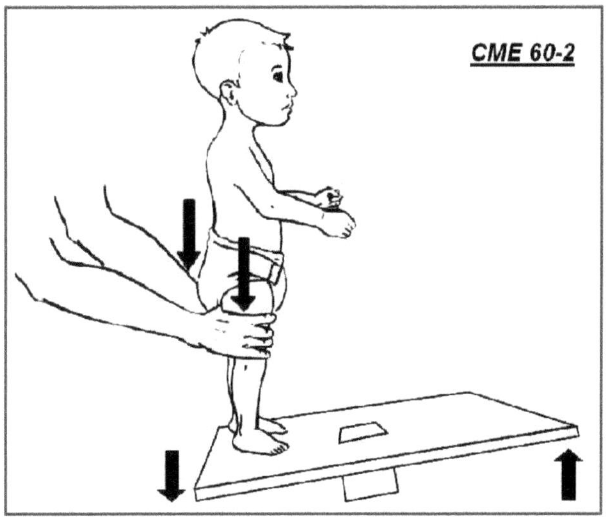

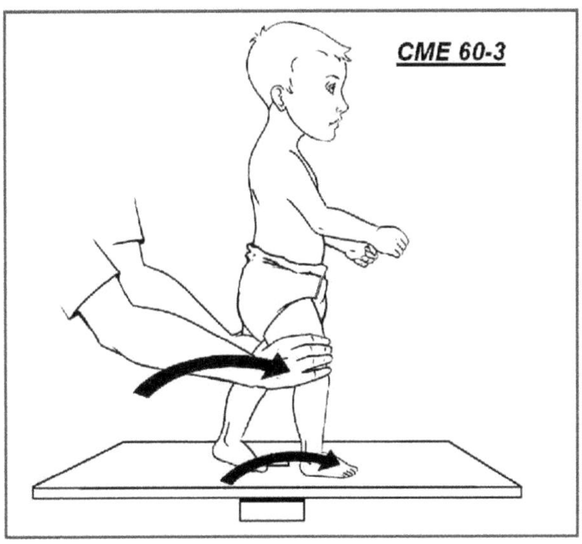

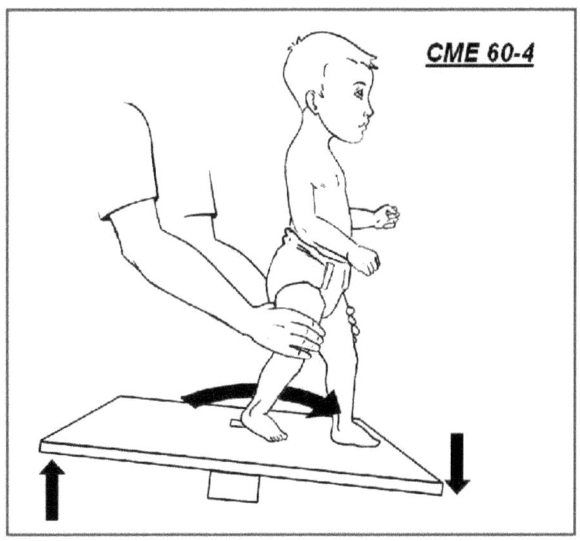

CME 61 STEPS ON BALANCE BAR BY ONE THIGH AND OPPOSITE LOW LEG

MAIN GOAL: To provoke balance reactions.

INITIAL POSITION: In a standing position on the box in front of the balance bar.

HOLD: By both hands below his/her knees as illustrated in figure 1.

MOVEMENT: Move up your right hand to his/her right thigh and take one step as illustrated in figure 2, then move your right hand quickly to his/her lower leg and your left hand to his/her left thigh and take the step as indicated in figure 3. (Stance leg is held by thigh and swing leg held by lower leg). Repeat the sequence until he/she reaches the end of the balance bar. Repeat the exercise as instructed by your CME therapist.

EXPECTED IMPROVEMENT: The child takes assisted steps along the balance bar with no difficulty.

CME 61 STEPS ON BALANCE BAR BY ONE THIGH AND OPPOSITE LOW LEG

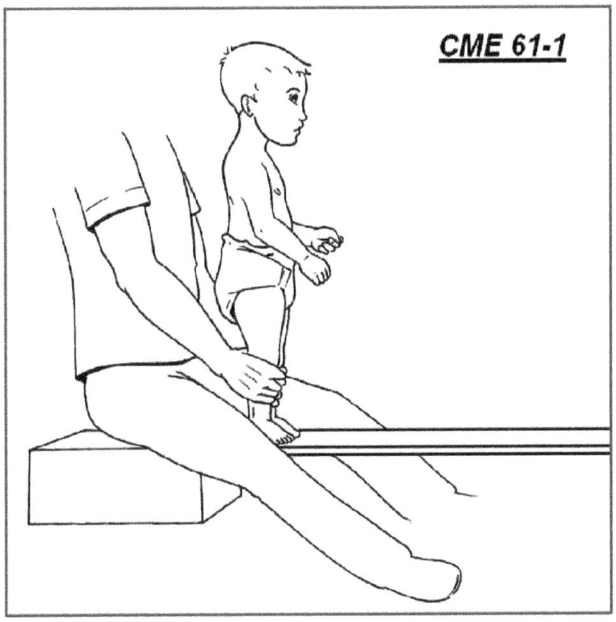

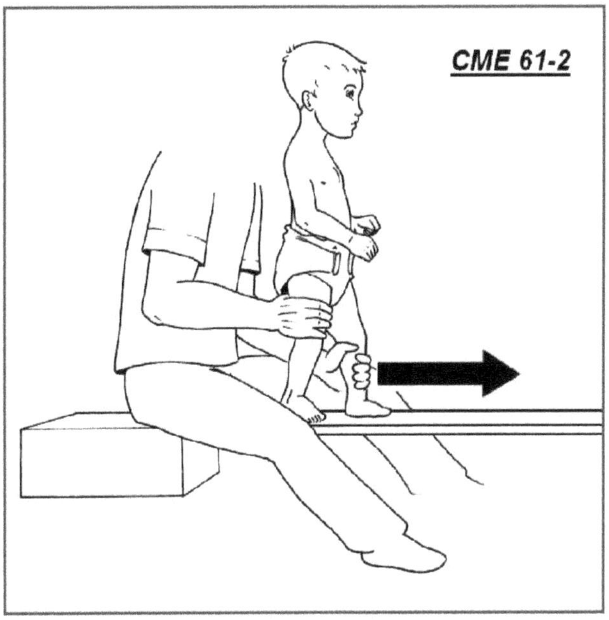

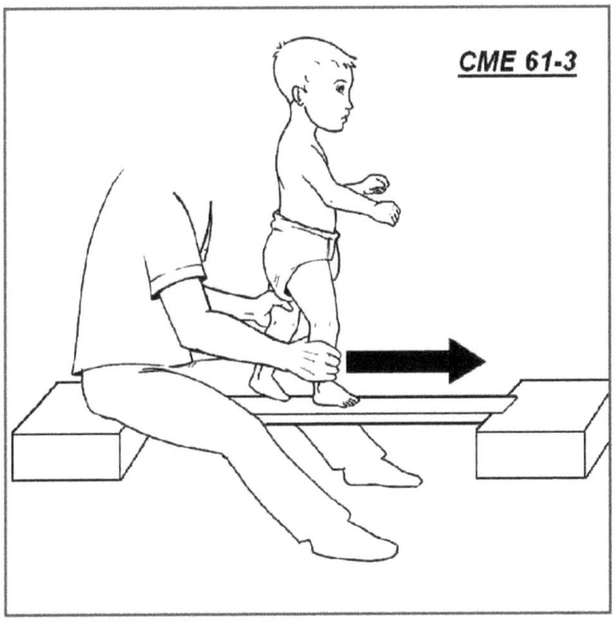

CME 62 STEPS ON BALANCE BOARD BY BELOW KNEES

MAIN GOAL: To provoke balance reactions.

INITIAL POSITION: In a standing position next to the edge of the balance board.

HOLD: By both hands below his/her knees as illustrated in figure 1.

MOVEMENT: Slowly drive the child to take small steps until reaching the end of the balance board as indicated in figures 2 and 3. Repeat the exercise as instructed by your CME therapist.

EXPECTED IMPROVEMENT: The child can take the directed steps easily all along the balance board.

CME 62 STEPS ON BALANCE BOARD BY BELOW KNEES

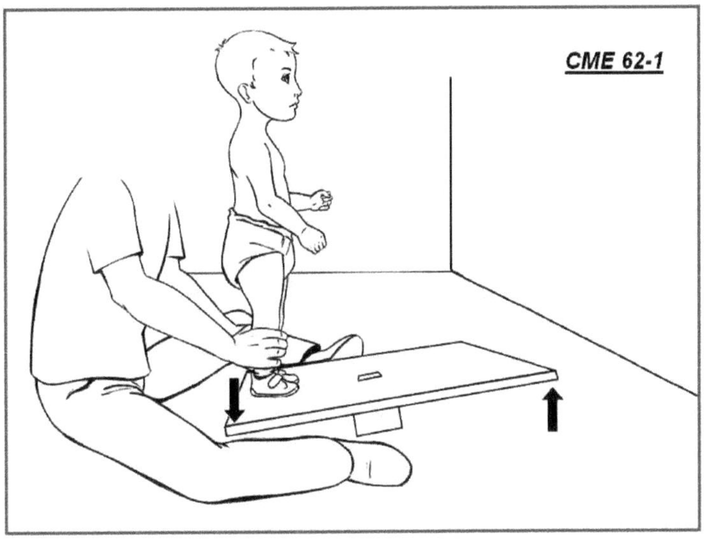

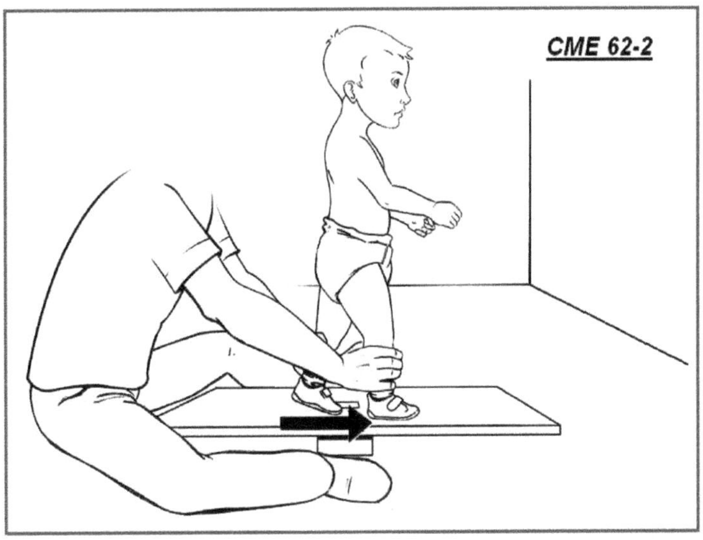

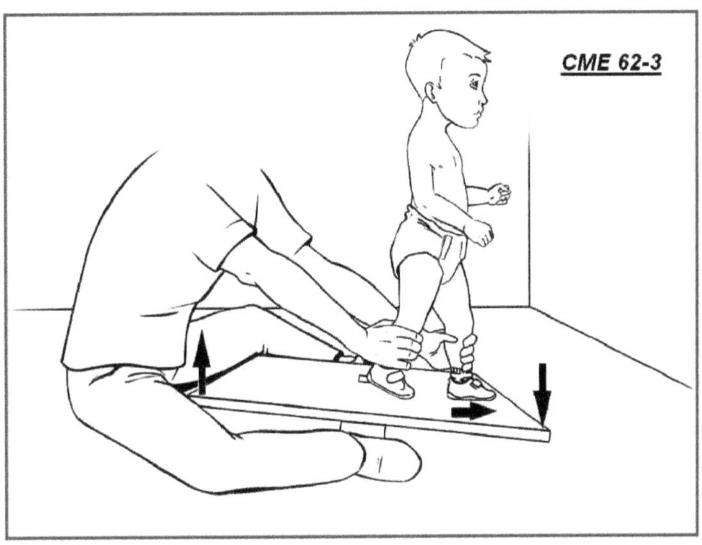

CME 63 STEPS ON BALANCE BAR BY BELOW KNEES

MAIN GOAL: To provoke balance reactions.

INITIAL POSITION: In a standing position on the box in front of the balance bar.

HOLD: By both hands below his/her knees as illustrated in figure 1.

MOVEMENT: Slowly move the child ahead taking small steps as indicated in figure 2. Repeat the sequence until reaching the end of the balance bar as illustrated in figures 3 and 4. Repeat the exercise as instructed by your CME therapist.

EXPECTED IMPROVEMENT: The child is able to take the guided steps easily all along the balance bar.

CME 63 STEPS ON BALANCE BAR BY BELOW KNEES

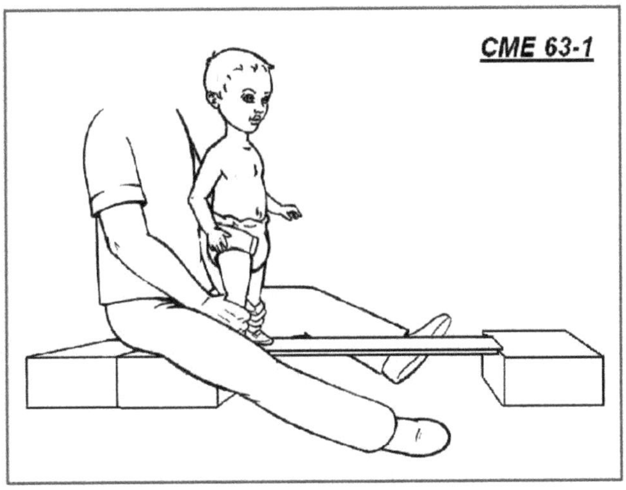
CME 63-1

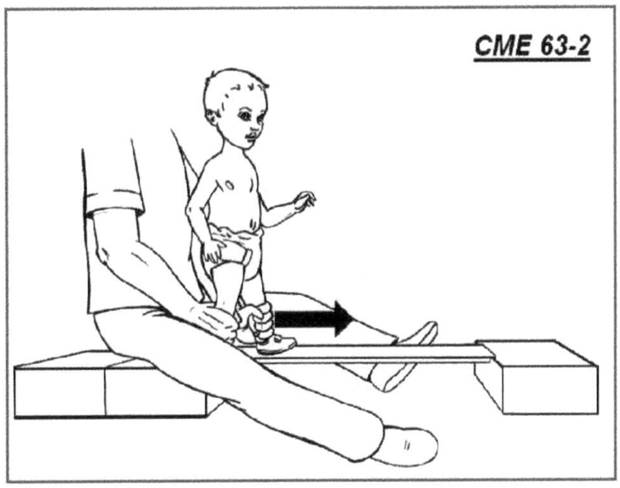
CME 63-2

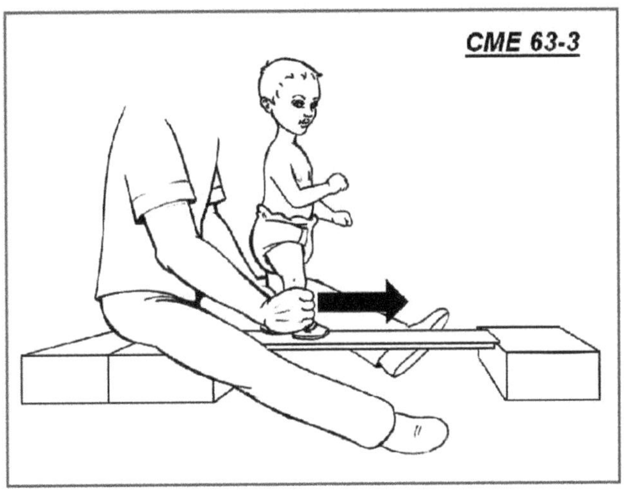
CME 63-3

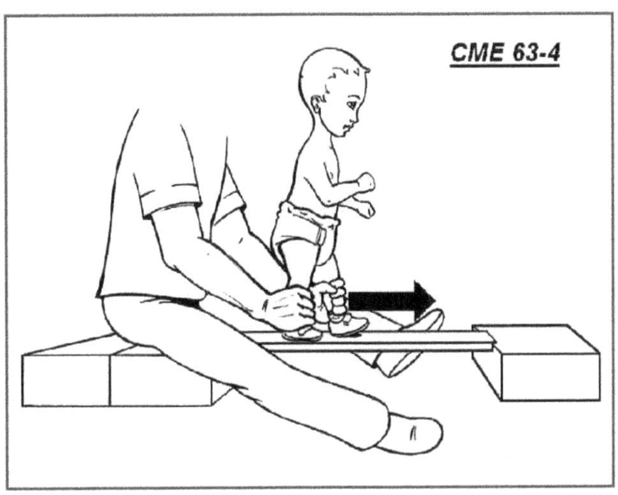
CME 63-4

CME 64 STEPS BY ANKLES

MAIN GOAL: To provoke stability and balance reactions.

INITIAL POSITION: In a standing position on the table as shown in figure 1.

HOLD: Move your right hand to the right ankle (fig. 2), and afterwards move the left hand to the left ankle as illustrated in figure 3.

MOVEMENT: Slowly move the child forward, taking small steps as indicated in figures 4 and 5. Repeat the sequence until reaching the end of the table. Repeat the exercise as instructed by your CME therapist.

EXPECTED IMPROVEMENT: The child takes guided steps easily all along the table.

CME 64 STEPS BY ANKLES

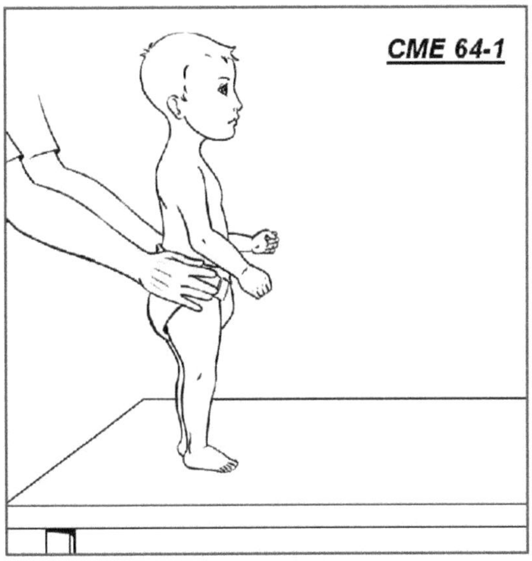

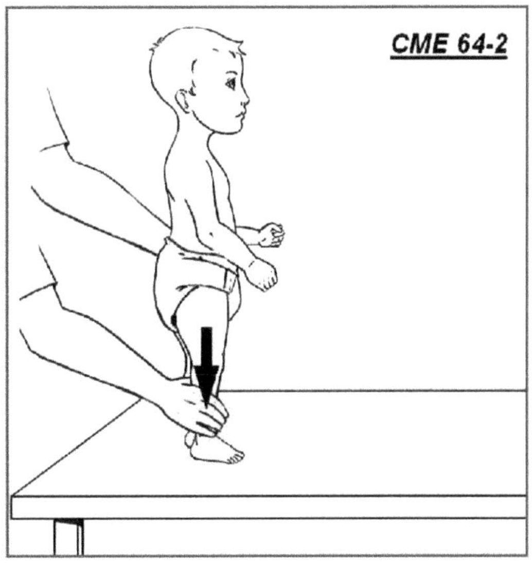

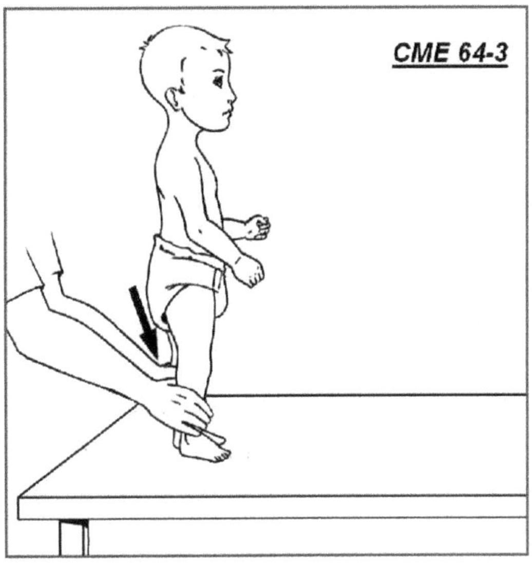

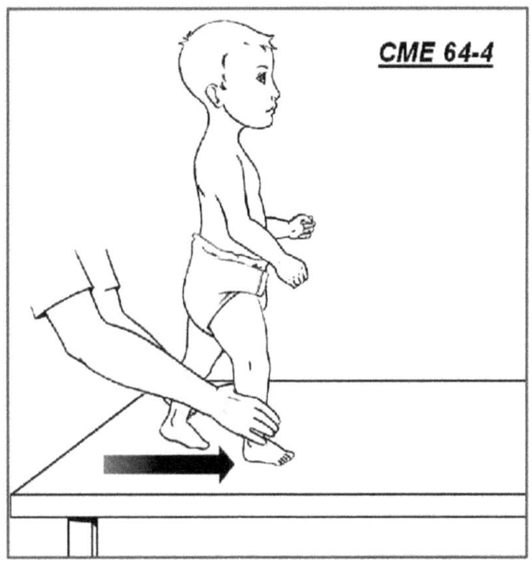

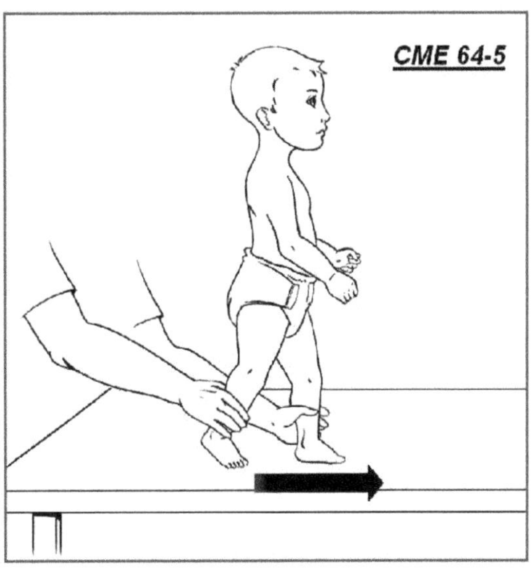

CME 65 STEPS BY STRAPS ON ONE WRIST AND OPPOSITE ANKLE

MAIN GOAL: To provoke stability in walking.

INITIAL POSITION: In standing position on the table.

HOLD: By soft tourniquet straps placed on the left wrist and right ankle, as illustrated in figure 1 and 2.

MOVEMENT: Slowly move the child forward in order to provoke small steps by pushing as indicated in figures 3 and 4. (Multidirectional arrow: stability needed to be compensated with quick and precise movements to any necessary direction). Repeat the sequence until reaching the end of the table. To repeat the exercise, follow the instructions of your CME therapist.

EXPECTED IMPROVEMENT: The child takes guided steps with no difficulty all along the table.

CME 65 STEPS BY STRAPS ON ONE WRIST AND OPPOSITE ANKLE

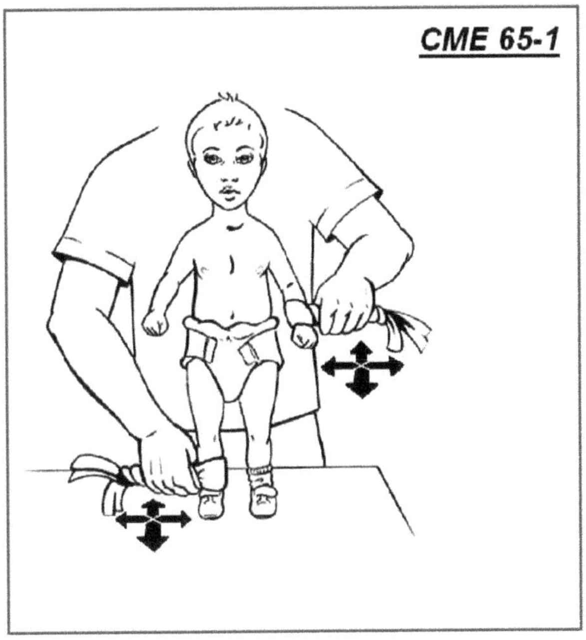

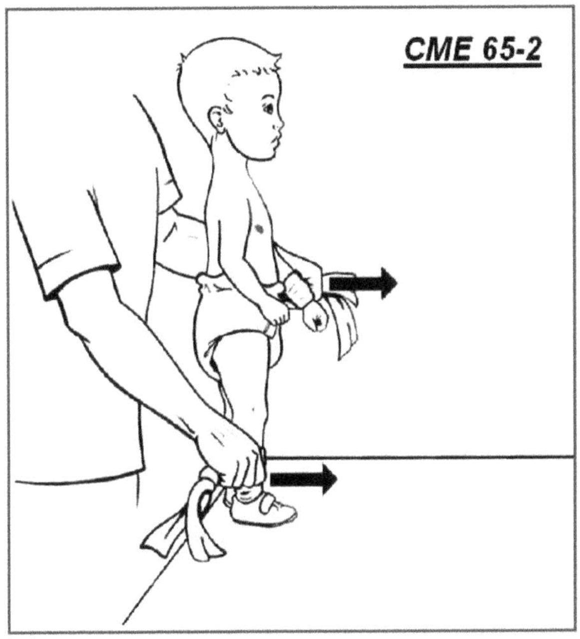

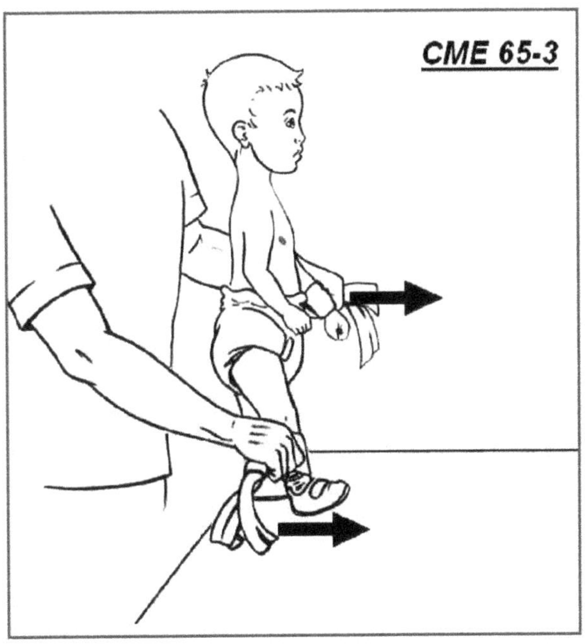

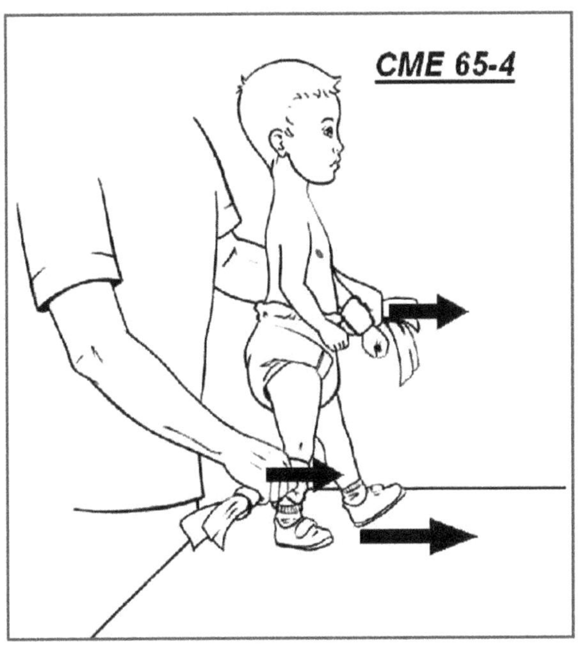

CME 66 STEPS BY LOOP STRAP ON WAIST

MAIN GOAL: To provoke spontaneous steps and standing stability.

INITIAL POSITION: In a standing position on the table.

HOLD: Place the long strap around the waist and pull in the opposite direction at both ends of the long strap, as illustrated in figure 1 and 2.

MOVEMENT: Slowly push the child forward in order to provoke small steps as indicated in figures 3 and 4. Repeat the sequence until reaching the end of the table. Repeat the exercise as instructed by your CME therapist.

EXPECTED IMPROVEMENT: The child takes guided steps easily all along the table.

CME 66 STEPS BY LOOP STRAP ON WAIST

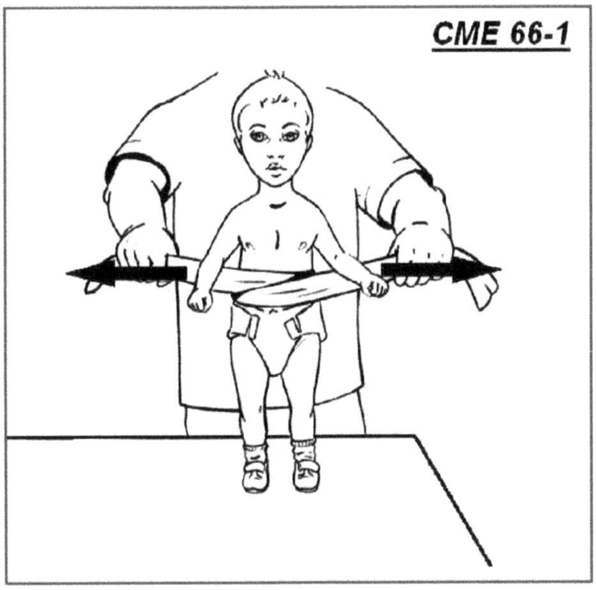

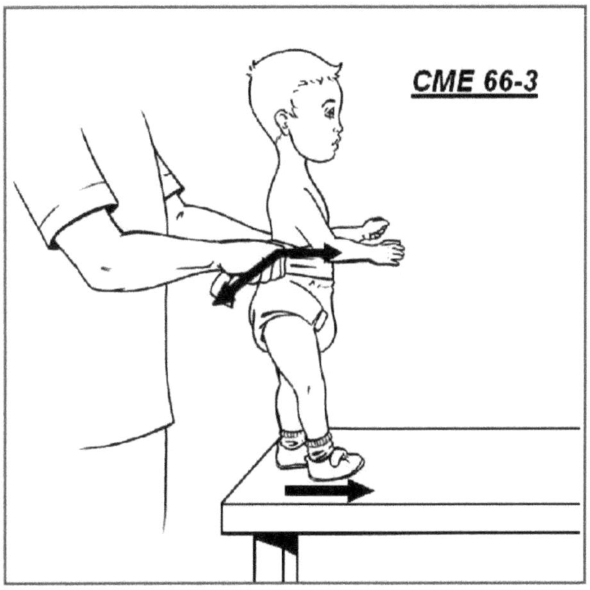

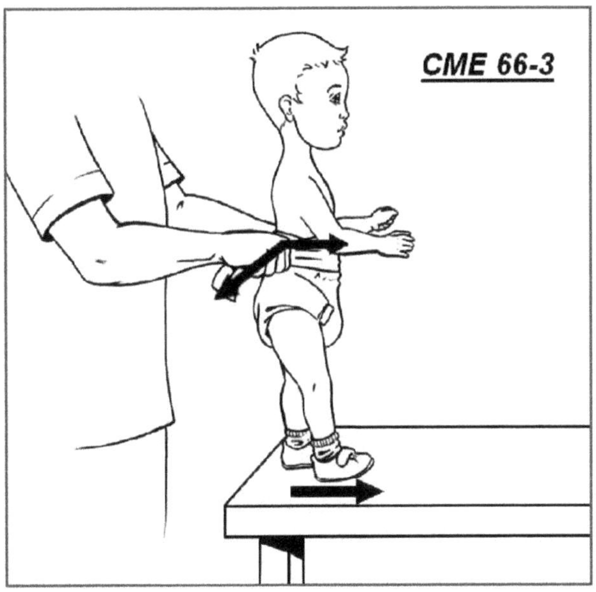

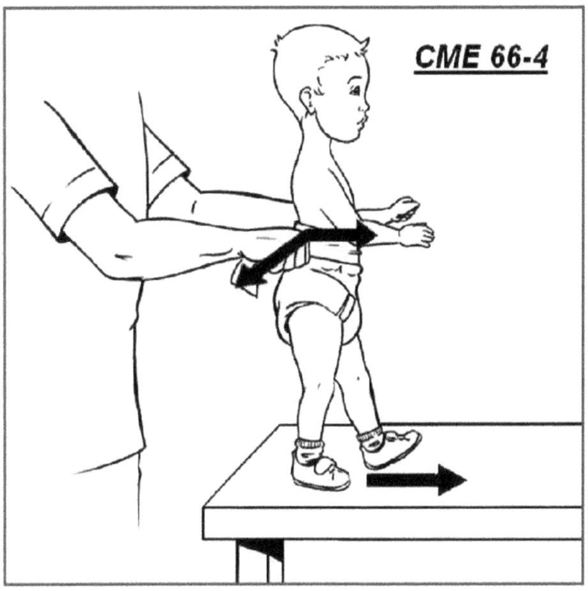

CME 67 STEP OUT OF 10 CM. BOX - SWING LEG FREE

MAIN GOAL: To provoke automatic step.

INITIAL POSITION: In a standing position inside the box.

HOLD: By both hands by below his/her knees as illustrated in the figure 1.

MOVEMENT: Release the left leg and keeping stability wait until the child moves the left leg out of the box as indicated in figure 2. Now switch the support to the left leg as indicated in figure 3 and maintain stability until the child takes the step out of the box as shown in figure 4. Repeat the exercise as instructed by your CME therapist.

EXPECTED IMPROVEMENT: The child is able to take steps quickly out of the box.

CME 67 STEP OUT OF 10 CM. BOX - SWING LEG FREE

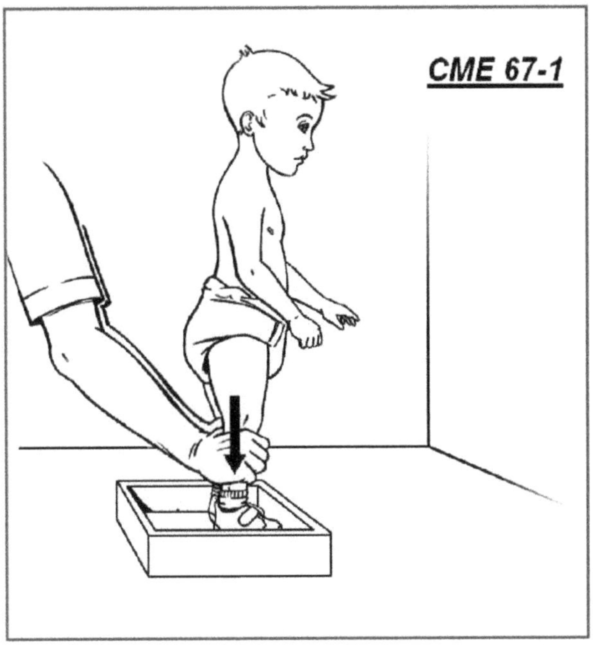

CME 67-1

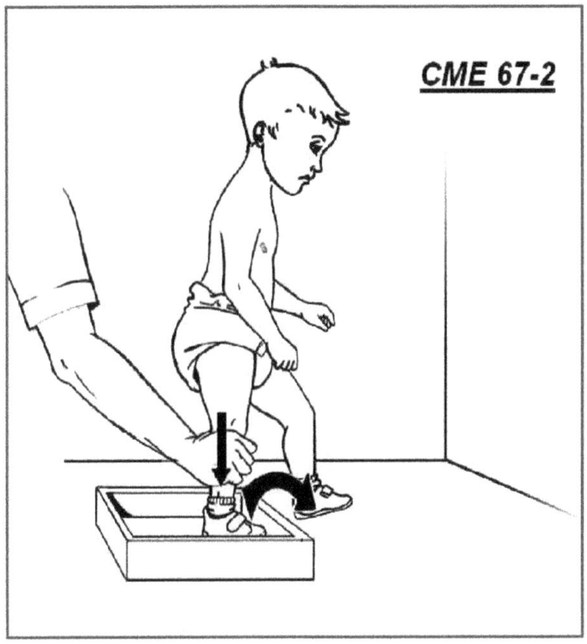

CME 67-2

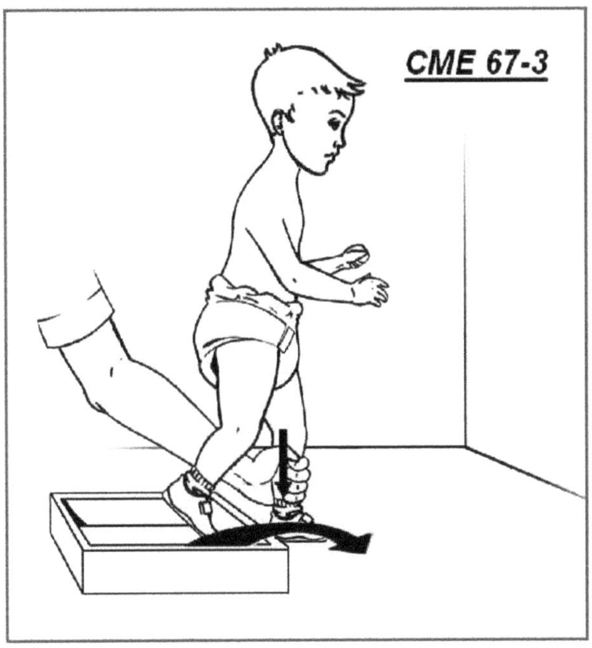

CME 67-3

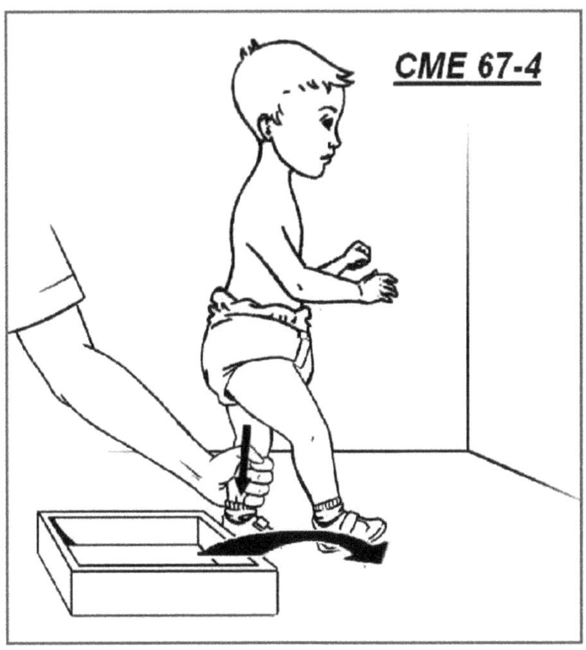

CME 67-4

CME 68 STEPS OUT OF 10 CM. BOX BY CROSSED STRAPS

MAIN GOAL: To provoke spontaneous steps and balance reactions.

INITIAL POSITION: In standing position inside of the 10 cm. box, supported by your chest.

HOLD: By ends of crossed straps (see encircled detail in figure 1), pulling sideways in the opposite direction as indicated in figure 1.

MOVEMENT: Slowly move the child forward by the straps and take off your chest support as illustrated in figure 2. Keep the stability and incite the child by the straps in order to provoke the spontaneous steps out of the box as shown in the figures 3 and 4. Repeat the exercise as instructed by your CME therapist.

EXPECTED IMPROVEMENT: The child is able to take the steps easily, out of the box.

CME 68 STEP OUT OF 10 CM. BOX BY CROSSED STRAPS

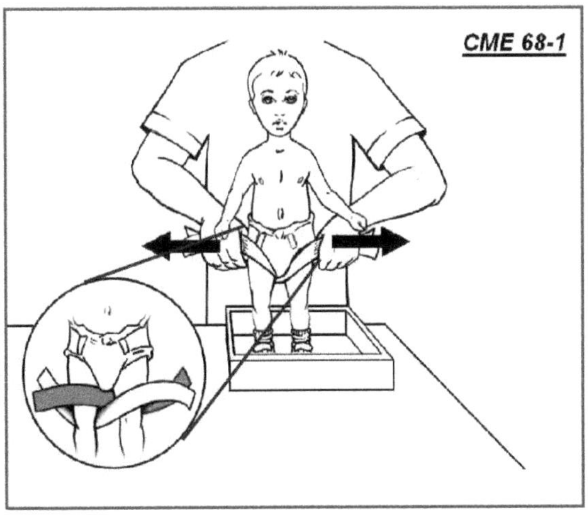

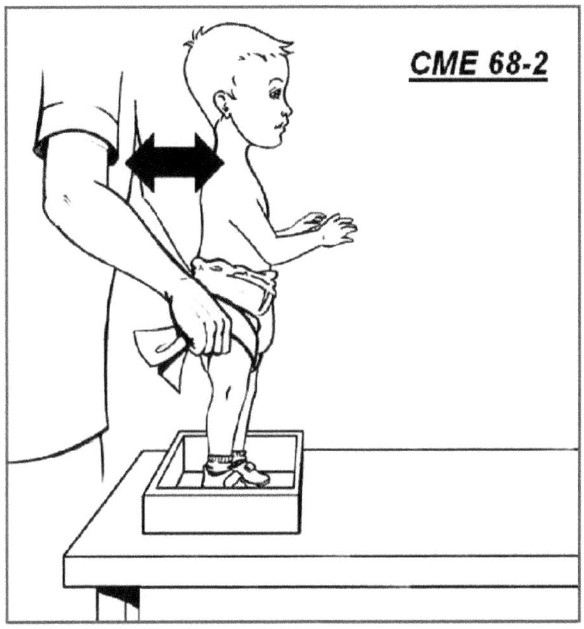

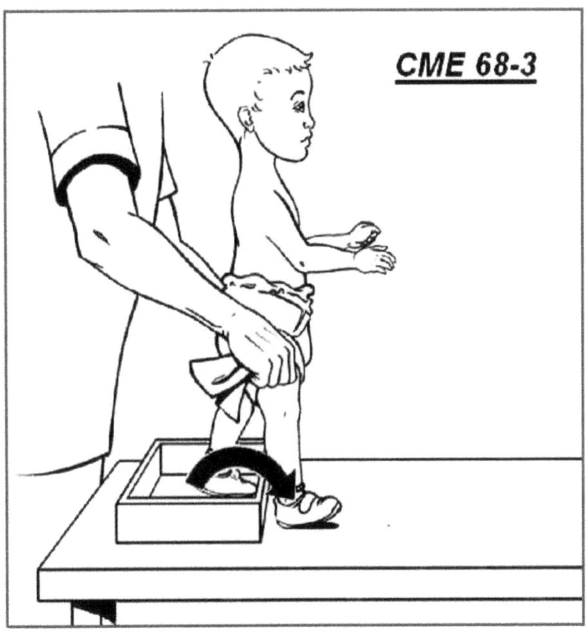

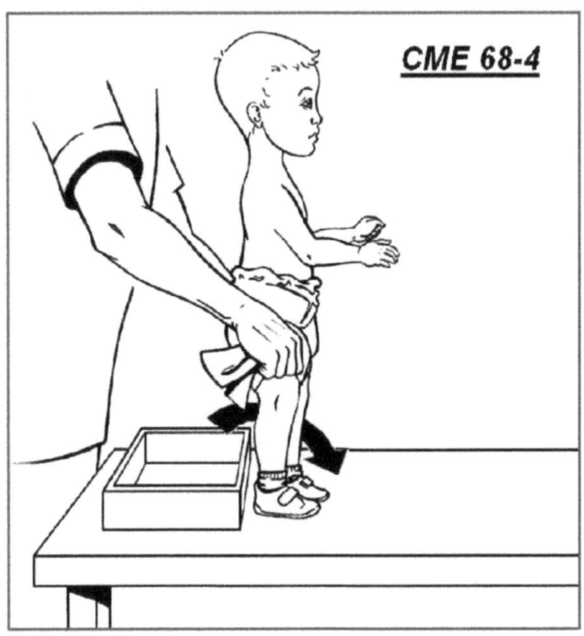

CME 69 STEPS IN/OUT OF 10 CM. BOX BY OPPOSED STRAPS

MAIN GOAL: To provoke spontaneous steps and balance reactions.

INITIAL POSITION: In standing position in front of the two 10 cm. boxes, supported on your chest.

HOLD: With your left hand on the strap on the waist, and the other hand on the straps on hips, pulling in opposite directions as illustrated in figure 1 and 2.

MOVEMENT: Slowly push the child forward in order to provoke the step into the boxes as illustrated in figures 3 and 4. Repeat the sequence until the child steps out of the boxes as indicated in figures 5 and 6. Repeat the exercise as instructed by your CME therapist.

EXPECTED IMPROVEMENT: The child takes guided steps in and out of the boxes easily.

CME 69 STEPS IN/OUT OF 10 CM. BOX BY OPPOSED STRAPS

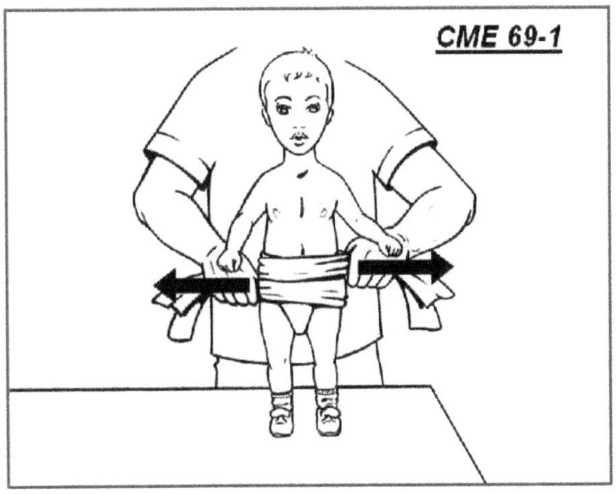

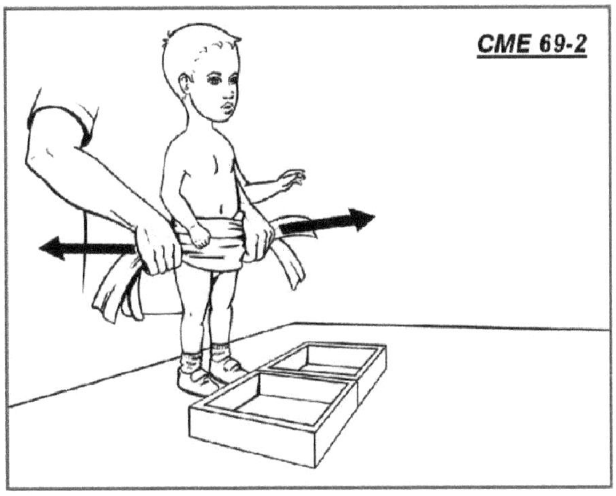

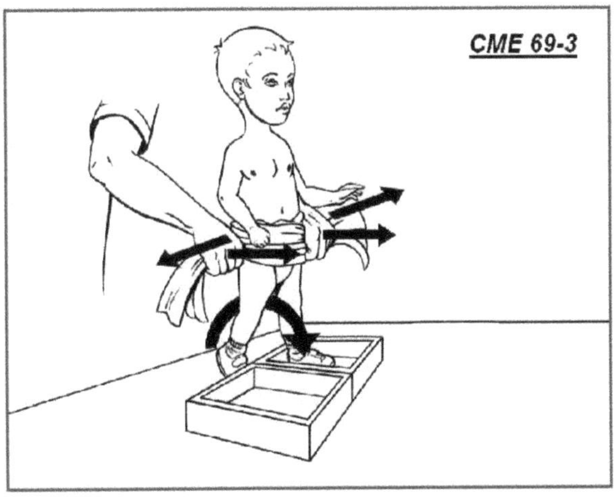

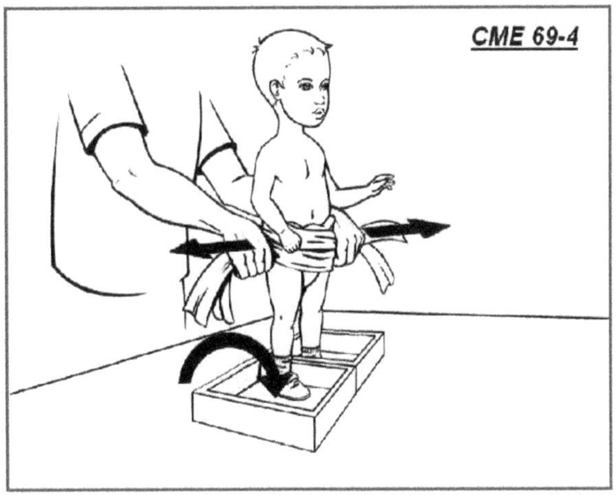

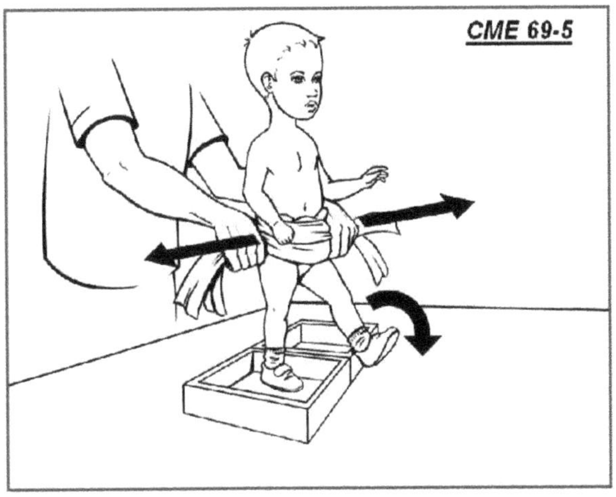

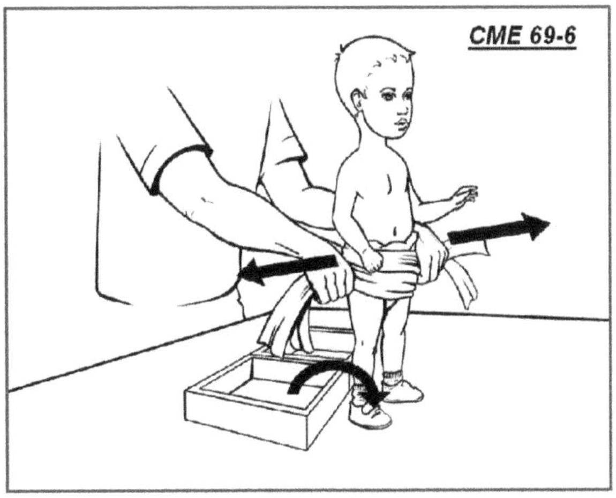

CME 70 STEPS BY LONG STRAP ON NAPE

MAIN GOAL: To provoke a walking reaction.

INITIAL POSITION: In a standing position with the back leaning against the wall.

HOLD: Place the long strap behind the neck (on the nape) holding the strap as illustrated in figure 1.

MOVEMENT: Pull the straps softly towards you, with an upward diagonal direction (black arrow) in order to provoke steps as indicated in figures 3, 4 and 5. Repeat the exercise as instructed by your CME therapist.

EXPECTED IMPROVEMENT: The child is able to take guided steps.

CME 70 STEPS BY LONG STRAP ON NAPE

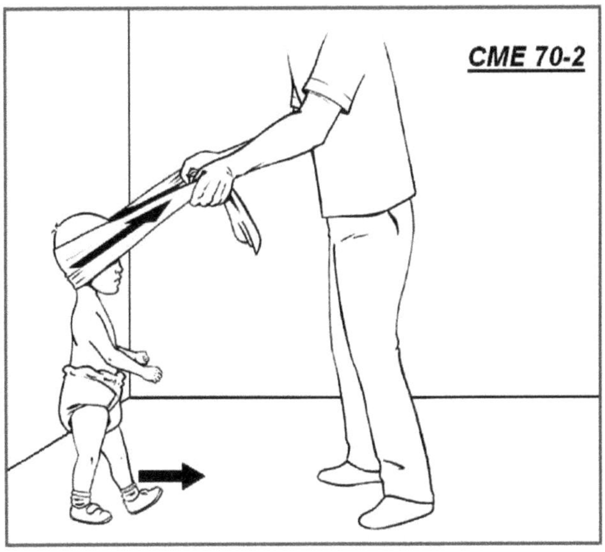

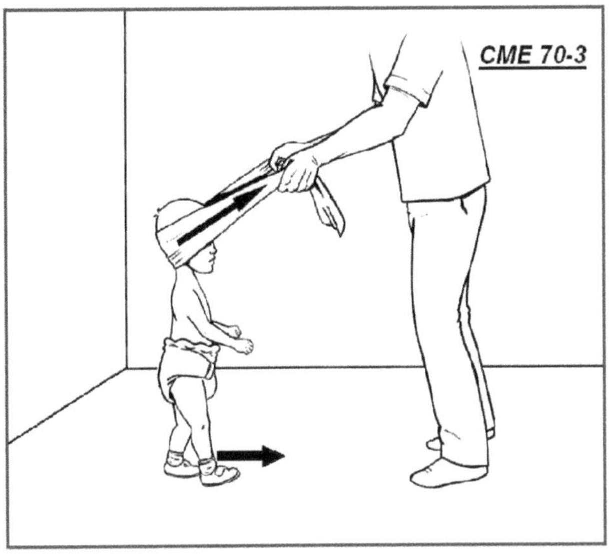

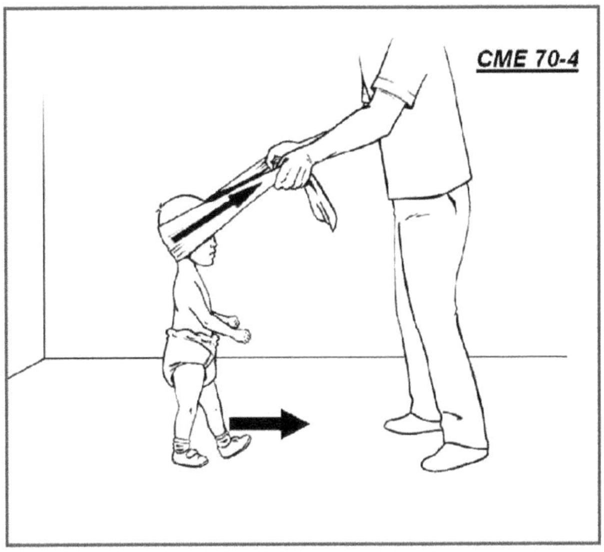

CME 71 STEPS IN/OUT OF 10 CM. BOX BY CROSSED STRAPS

MAIN GOAL: To provoke automatic walking reaction.

INITIAL POSITION: In standing position in front of the 10 cm. boxes as indicated in figure 1.

HOLD: Place the straps under crotch as illustrated in the inserted circle of figure 1.

MOVEMENT: Slowly push the child forward to provoke the first step with one foot inside the box (fig. 3) followed by the other leg (fig. 4). Provoke the steps out of the boxes without stopping as indicated in figure 5 and 6. Repeat the sequence as instructed by your CME therapist.

EXPECTED IMPROVEMENT: The child takes guided steps easily in and out of the 10 cm. boxes.

CME 71 STEPS IN/OUT OF 10 CM. BOX BY CROSSED STRAPS

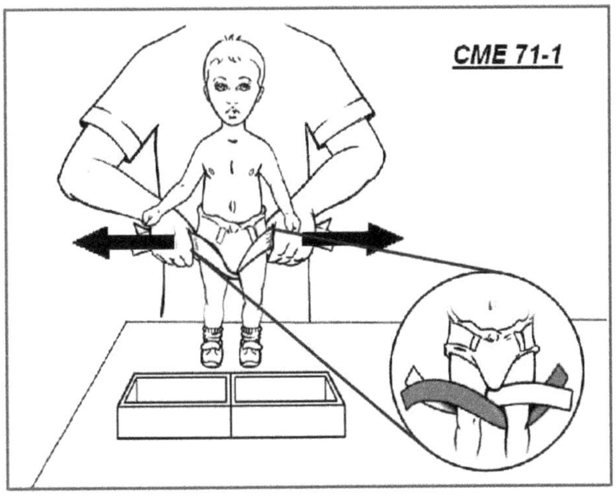

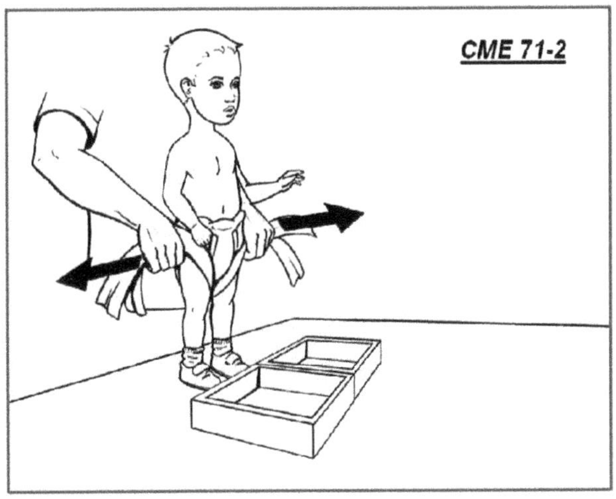

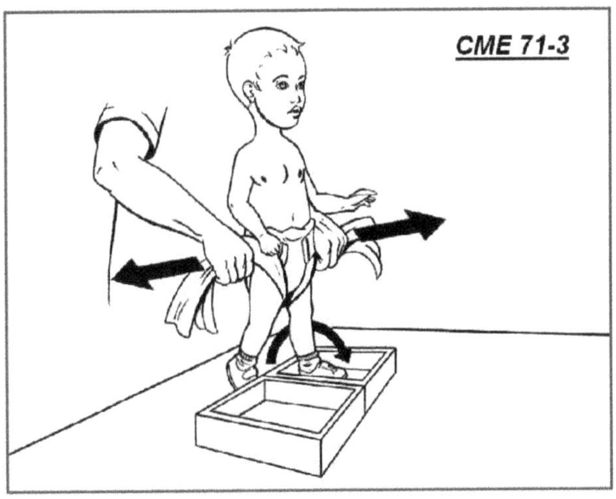

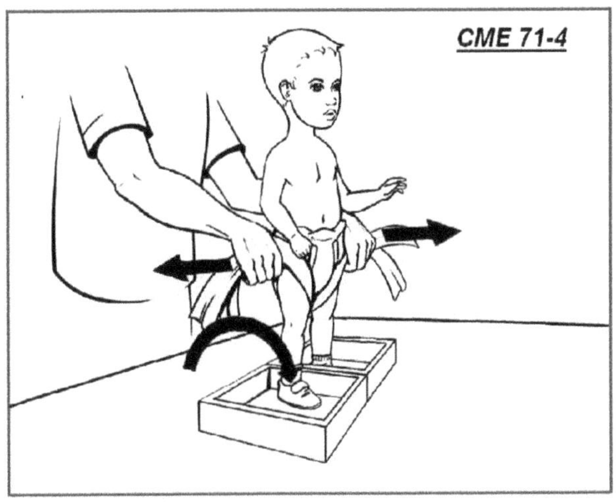

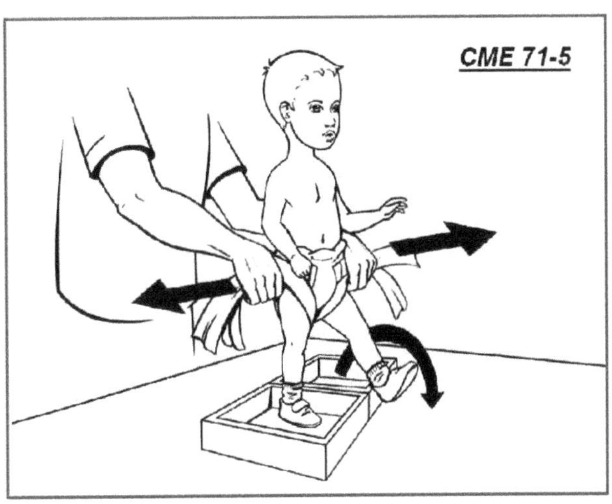

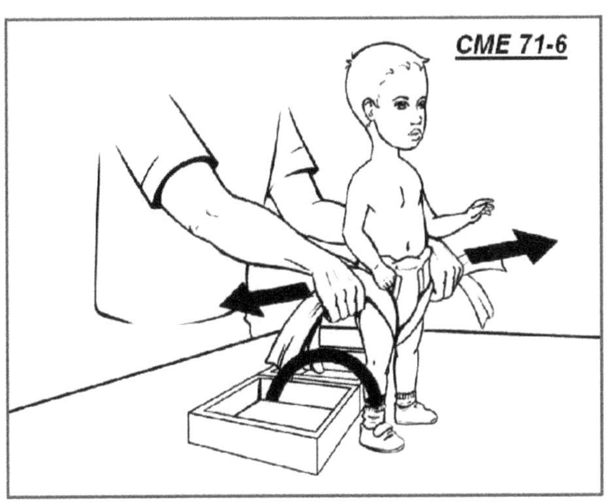

CME 72 STEP UP 15 CM. BOX BY COMBINATION

MAIN GOAL: To provoke stable reactions.

INITIAL POSITION: In standing position in front of the 15 cm. box.

HOLD: By the right thigh and left ankle as illustrated in figure 1.

MOVEMENT: Slowly move the left foot on top of the box as indicated in figure 2, then push up the right leg as shown in figure 3. Hold this position for 5 seconds and move back down the right leg as indicated in figure 4, followed by the left leg down backward as indicated in figure 5. Repeat the sequence as instructed by your CME therapist.

EXPECTED IMPROVEMENT: The child takes guided steps up and down backward with no difficulty.

CME 72 STEP UP 15 CM. BOX BY COMBINATION

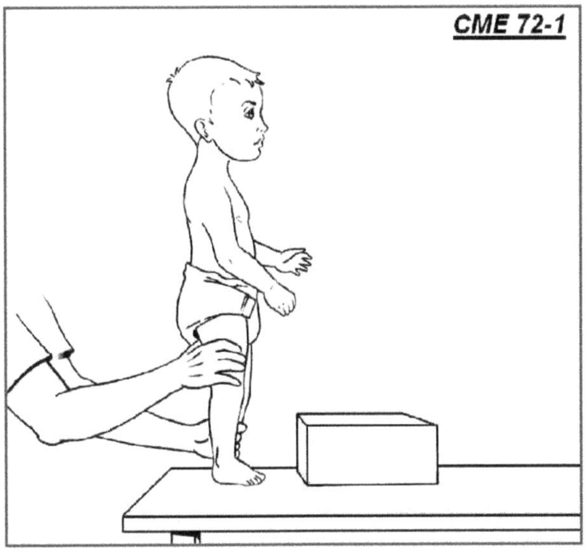

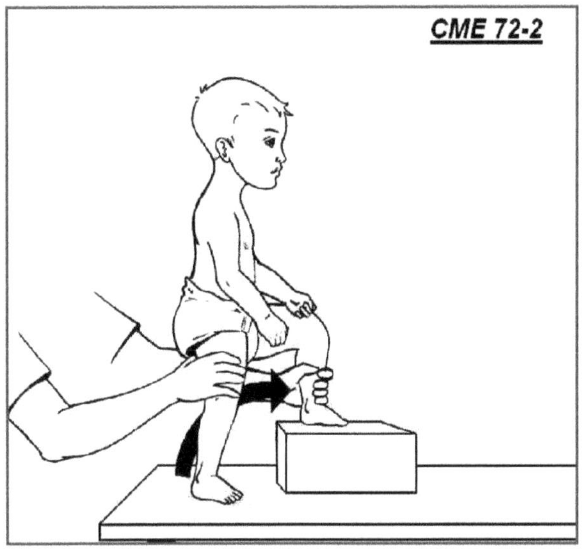

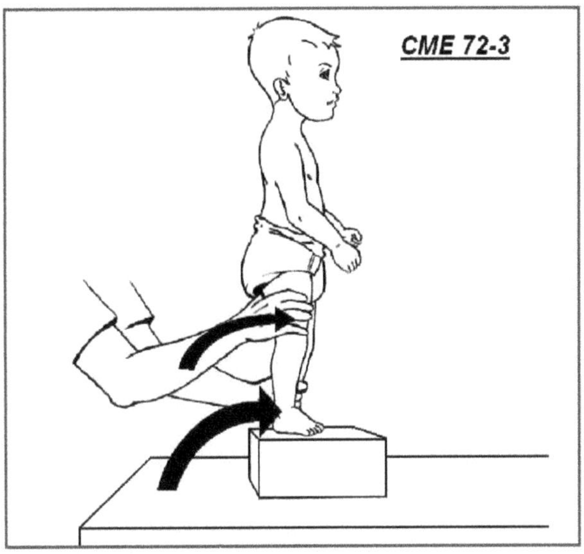

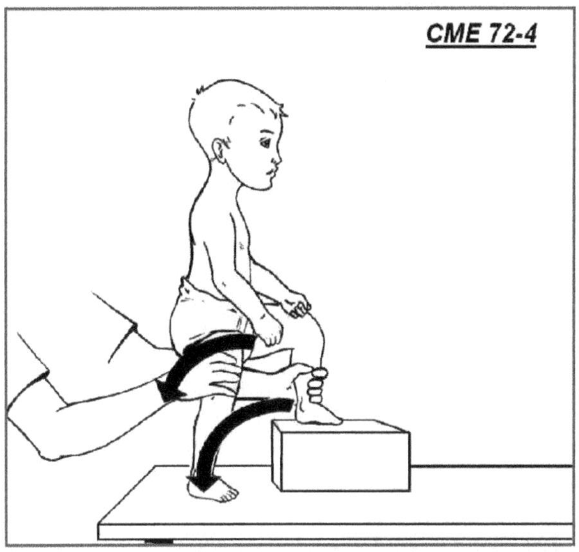

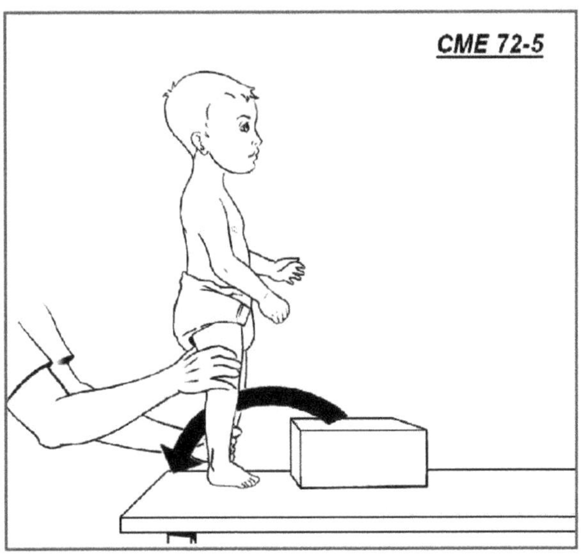

CME 73 STEP UP/DOWN BACKWARD 15 CM. BOX BY BELOW KNEES

MAIN GOAL: To reinforce stable reactions in a standing position

INITIAL POSITION: Ina standing position in front of the box.

HOLD: By both hands below his/her knees as illustrated in figure 1.

MOVEMENT: Slowly place the left foot on top of the 15 cm. box as indicated in figure 2, followed by the right foot as illustrated in figure 3. Keep this position for 5 seconds and now move the right leg back downward as shown in figure 4 followed by the other leg as indicated in figure 5. Repeat the sequence as instructed by your CME therapist.

EXPECTED IMPROVEMENT: The child takes guided steps up and back downward, with no difficulty.

CME 73 STEP UP/DOWN BACKWARD 15 CM. BOX BY BELOW KNEES

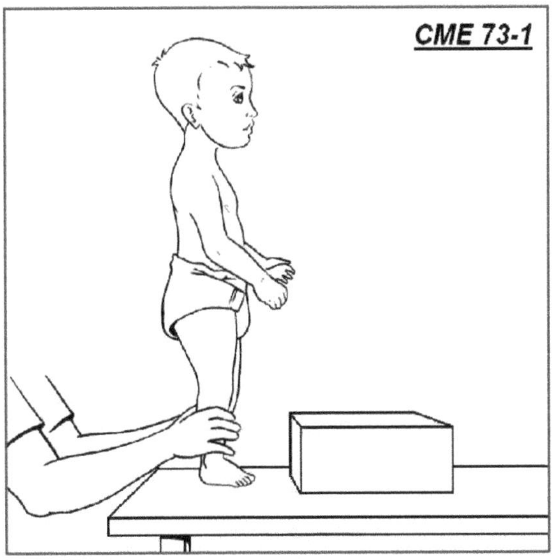

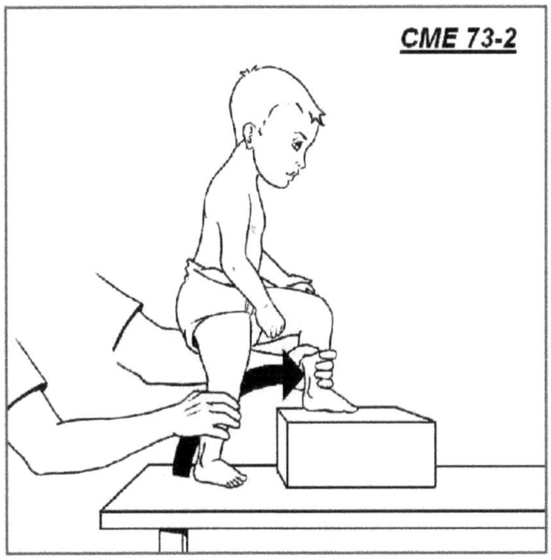

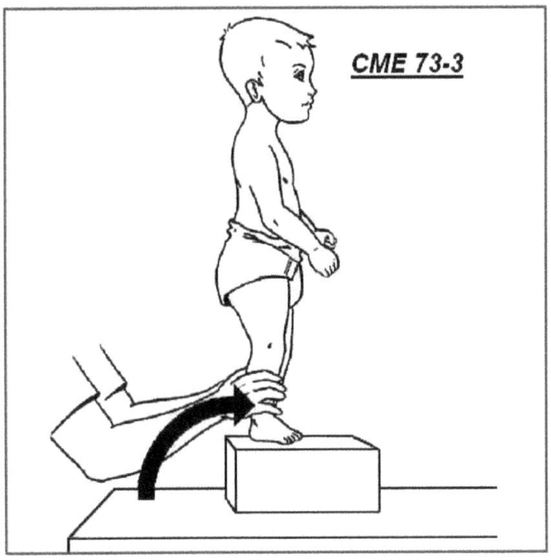

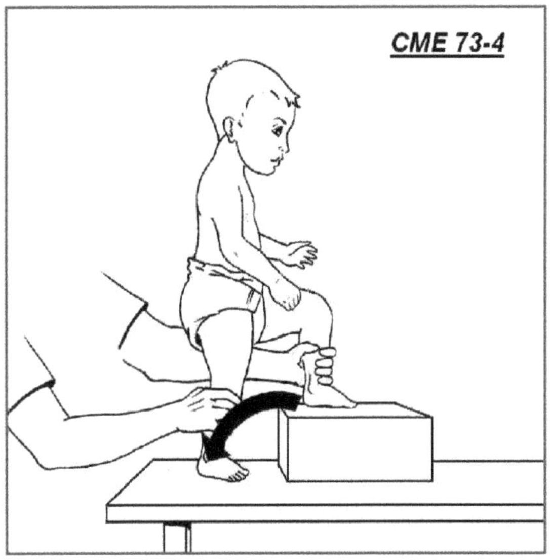

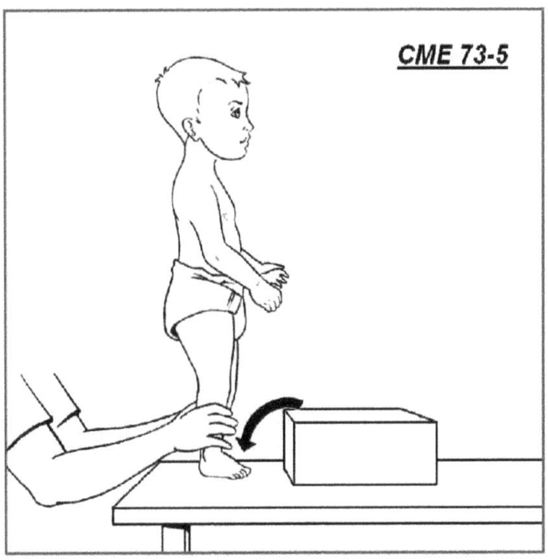

CME 74 STEP UP SMALL RAMP BY CROSSED STRAPS

Exercise Configuration: *Two 15 cm. boxes side by side, plus a balanced board supported on one side. For an even level place one balance bar on the top of the boxes and another on the floor.*

MAIN GOAL: To provoke balance reactions.

INITIAL POSITION: In a standing position in front of the ramp.

HOLD: By both hands by straps close to the thighs as illustrated in figure 2.

MOVEMENT: Slowly move the child forwards (fig. 3) provoking steps as indicated in figure 4 and 5. Repeat the sequence until reaching the end of the ramp. Repeat the exercise as instructed by your CME therapist.

EXPECTED IMPROVEMENT: The child takes guided steps up, easily all along the ramp.

CME 74 STEP UP SMALL RAMP BY CROSSED STRAPS

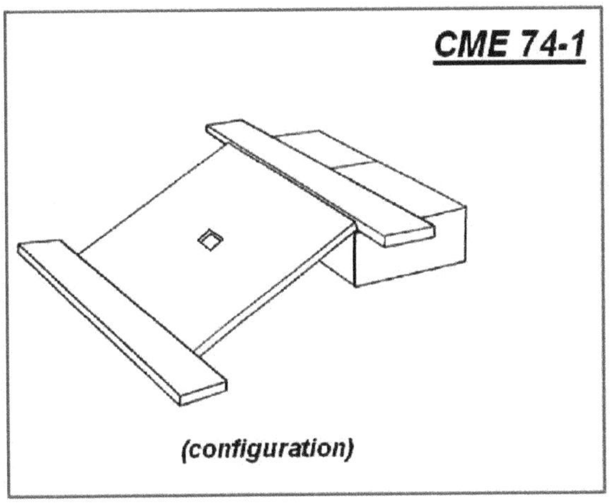

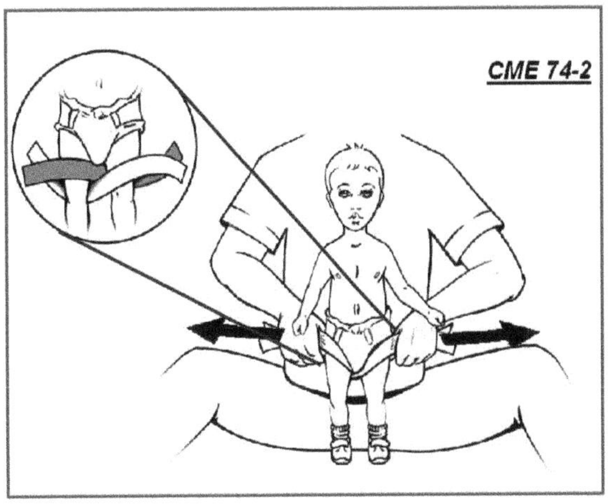

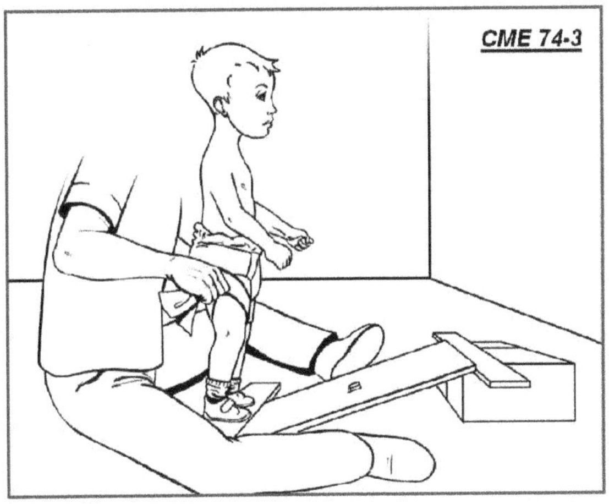

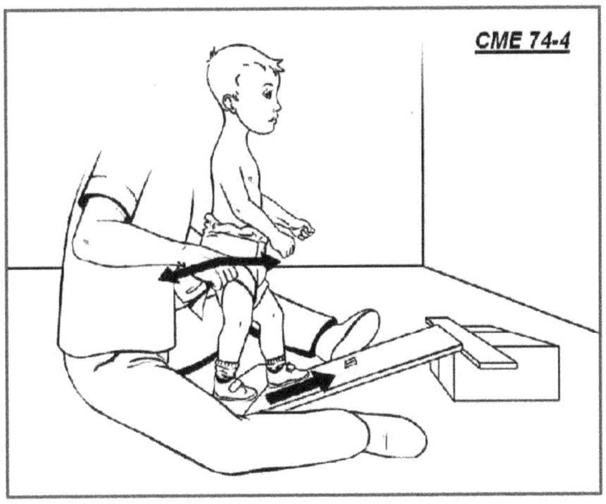

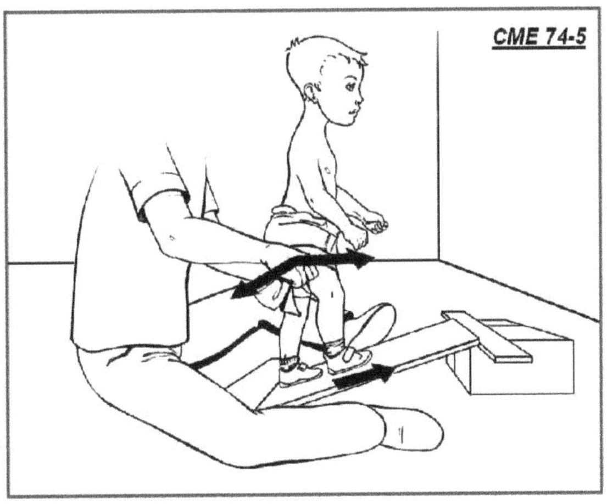

CME 75 STEP DOWN SMALL RAMP BY CROSSED STRAPS

MAIN GOAL: To provoke balance reactions.

INITIAL POSITION: In a standing position on top of the ramp.

HOLD: By both hands by straps, close to the thighs as illustrated in figure 2.

MOVEMENT: Slowly push the child gently forwards (fig. 3) in order to provoke small steps down the ramp as indicated in figure 4 and 5, until reaching the end of the small ramp. Repeat the exercise as instructed by your CME therapist.

EXPECTED IMPROVEMENT: The child takes guided steps downward with no difficulty.

CME 75 STEP DOWN SMALL RAMP BY CROSSED STRAPS

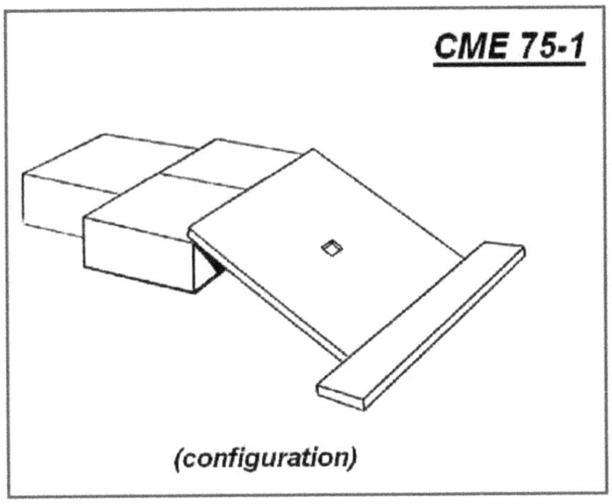

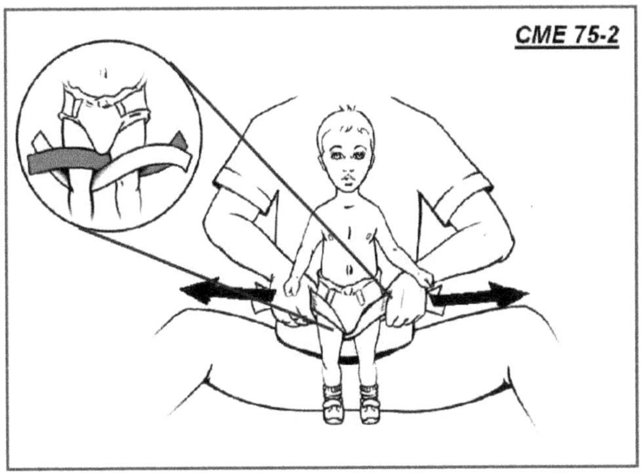

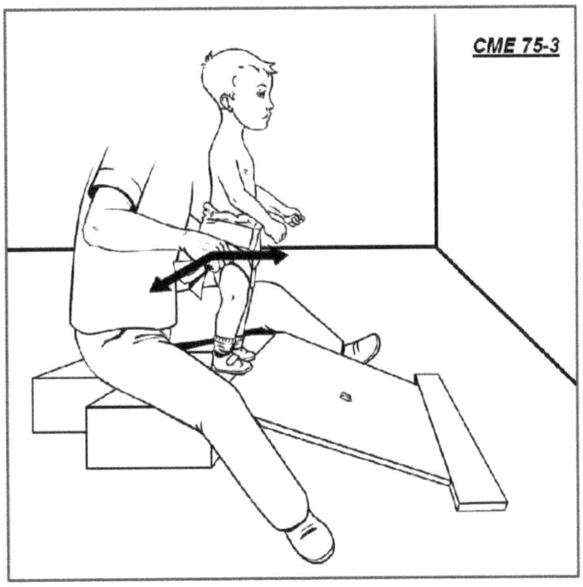

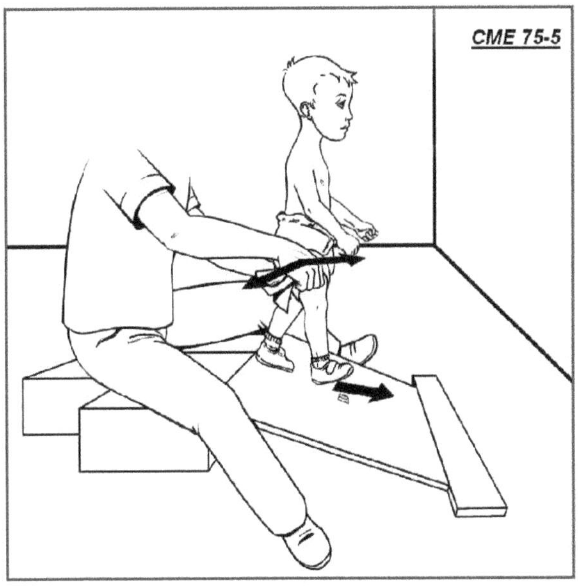

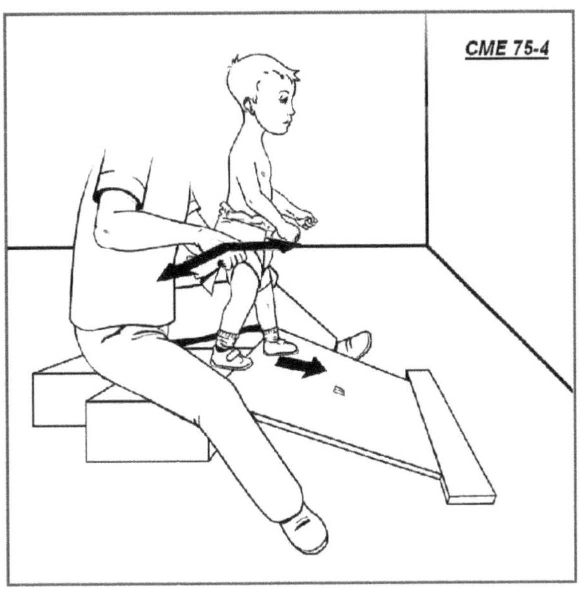

CME 76 STEP UP ON 10 CM. BOX BY STRAPS ON THIGHS

MAIN GOAL: To provoke stable and balance reactions.

INITIAL POSITION: In a standing position in front of the box.

HOLD: By straps around each thigh as illustrated in figure 1.

MOVEMENT: By the strap induce the child to place the right leg on top of the 10 cm. box as illustrated in figures 2 and 3, followed by the left leg as indicated in figure 4. Repeat the exercise as instructed by your CME therapist.

EXPECTED IMPROVEMENT: The child is able step up the box with straps on thighs with no difficulty.

CME 76 STEP UP ON 10 CM. BOX BY STRAPS ON THIGHS

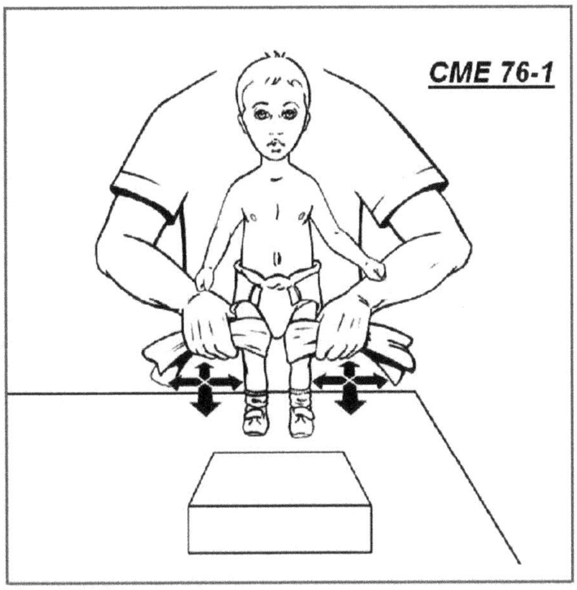

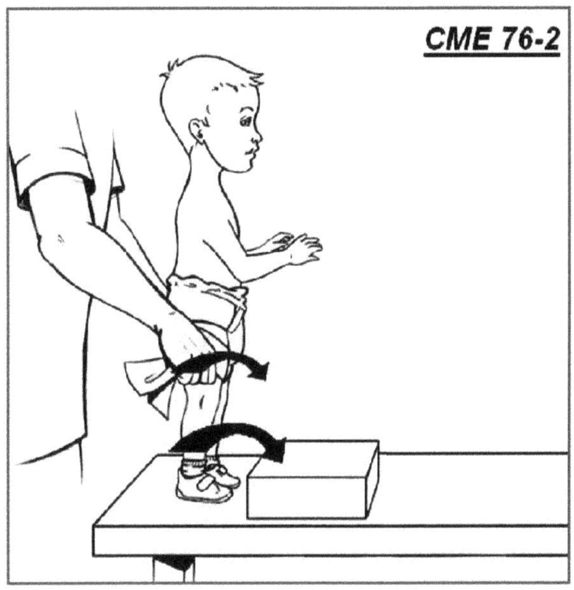

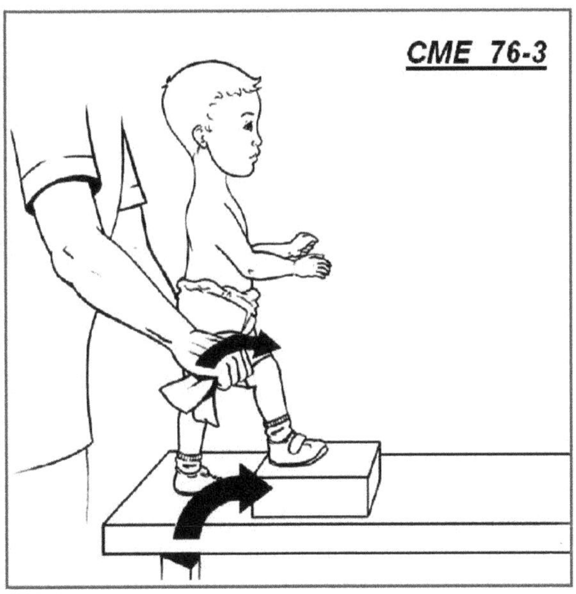

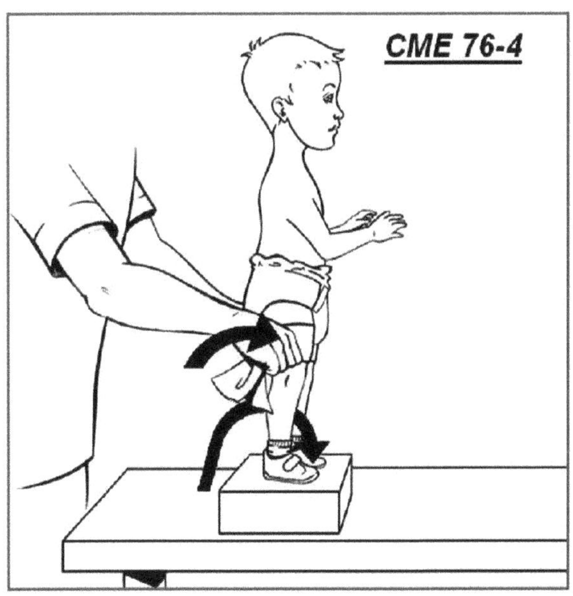

CME 77 STEP UP BIG RAMP BY ANKLES

Exercise Configuration: *Two boxes in a 30 cm. position with 10 cm. box tucked on top, plus a balance board placed diagonally on proximal edge.*

MAIN GOAL: To provoke stable and balance reactions.

INITIAL POSITION: In a standing position in front of the big ramp.

HOLD: By both hands below his/her knees as illustrated in figure 1.

MOVEMENT: Slowly move the child ahead taking small steps as indicated in figure 3. Repeat the sequence until reaching the end of the big ramp. Repeat the exercise as instructed by your CME therapist.

EXPECTED IMPROVEMENT: The child takes guided steps easily all along the balance bar.

CME 77 STEP UP BIG RAMP BY ANKLES

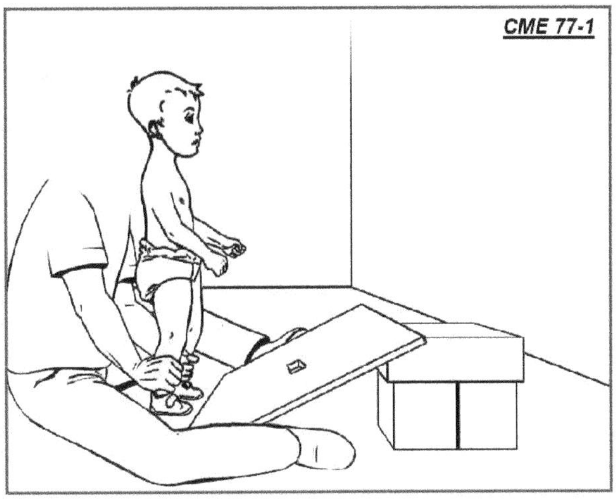

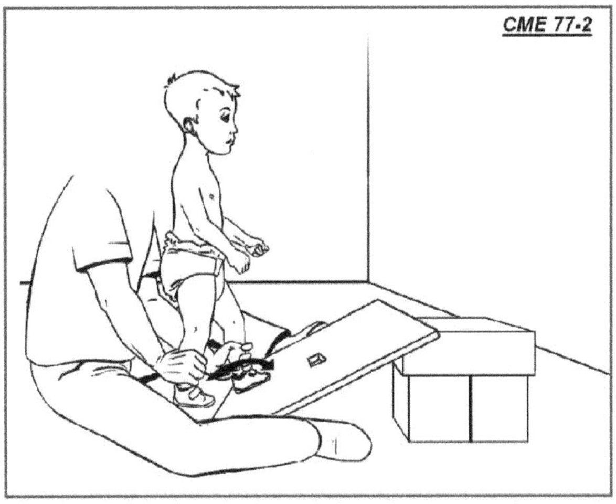

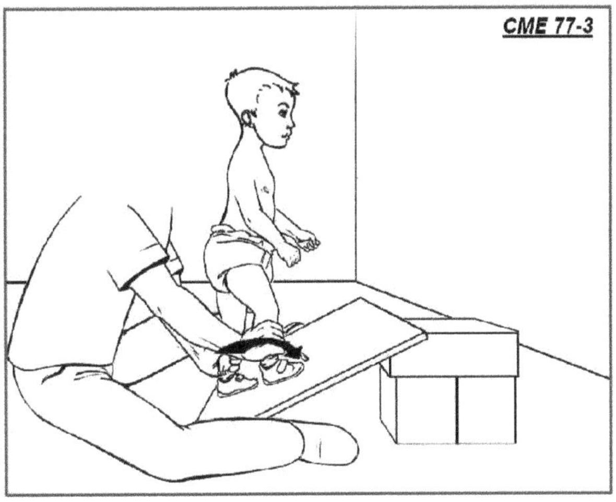

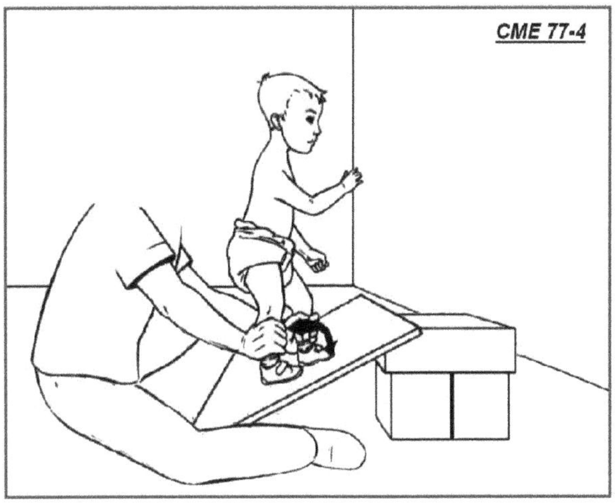

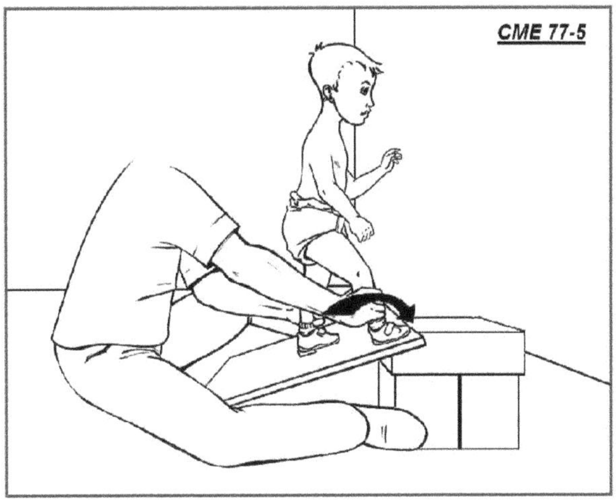

CME 78 STEP UP ON NARROW BALANCE BAR BY ANKLES

Exercise Configuration: *Narrow the balance bar placed under the legs, slightly inclined forward.*

MAIN GOAL: To provoke balance reactions.

INITIAL POSITION: In a standing position on the floor in front of the narrow balance bar, between your legs, as indicated in figures 1 and 2.

HOLD: By both ankles.

MOVEMENT: Place the left foot on the narrow bar as illustrated in figure 3, follow by the right foot as shown in the figure 4. Repeat the exercise as instructed by your CME therapist.

EXPECTED IMPROVEMENT: The child is able to keep the standing position on the narrow bar.

CME 78 STEP UP ON NARROW BALANCE BAR BY ANKLES

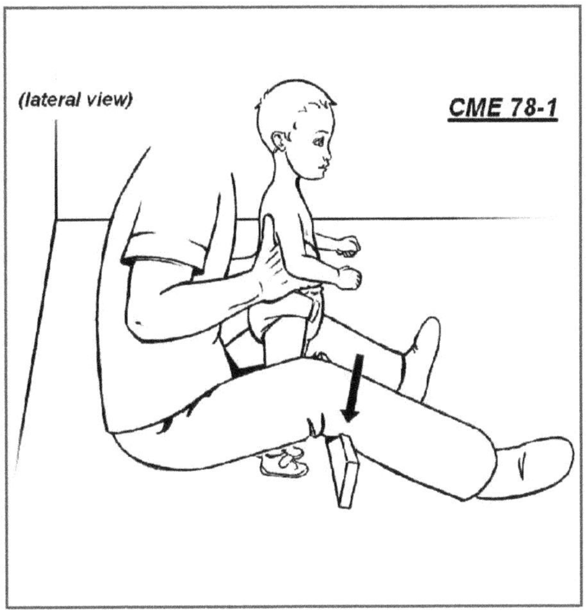

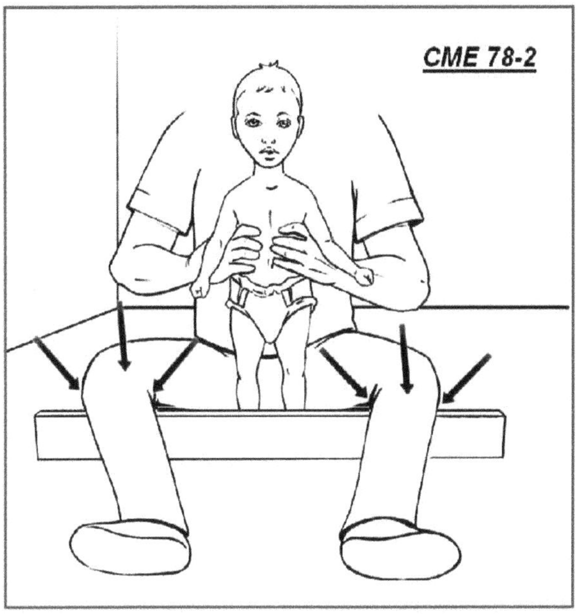

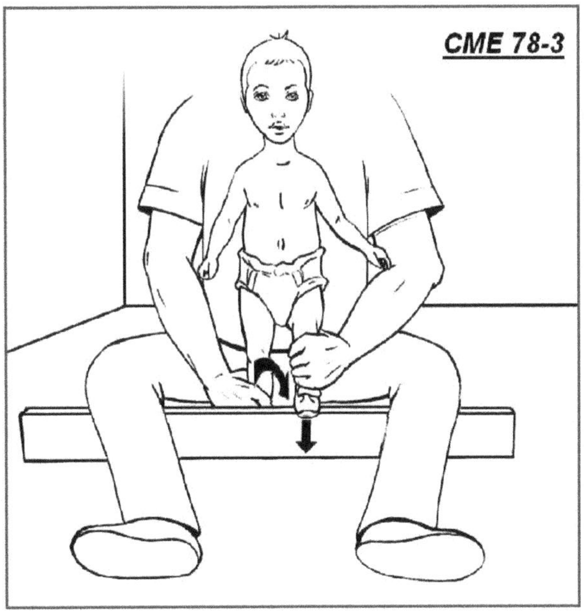

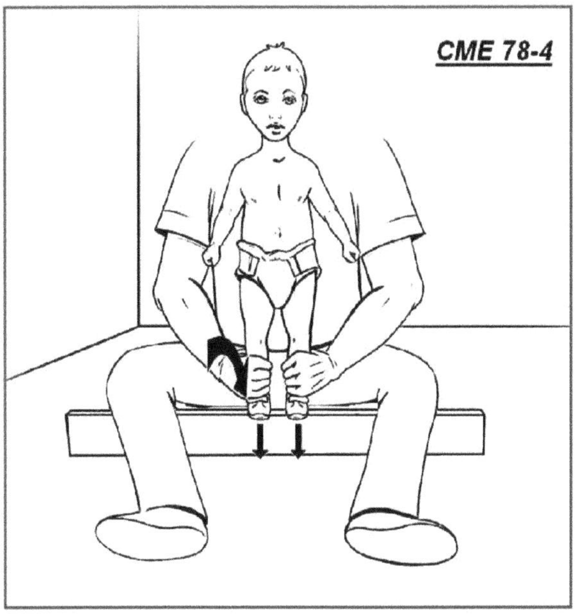

CME 79 ALTERNATE STEPS ON OLYMPIC PODIUM BY ANKLES

Exercise Configuration: *Two 15 cm. boxes side by side, plus one 15 cm. box on top at center.*

MAIN GOAL: To provoke stability and balance reactions.

INITIAL POSITION: In standing a position in front of the Olympic podium.

HOLD: By both ankles as illustrated in the figure 1.

MOVEMENT: Place the left foot on top of the first box as indicated in figure 2, now place the right foot on top of the second box (fig. 3), and reach the top placing the left foot (fig. 4). Subsequently, step down to the first box by the right foot (fig, 5), followed by the left foot to the floor (fig. 6), finish the sequence with the right foot on the floor (fig. 7). Repeat the exercise as instructed by your CME therapist.

EXPECTED IMPROVEMENT: The child takes guided steps easily, up and down Olympic podium.

CME 79 ALTERNATE STEPS ON OLYMPIC PODIUM BY ANKLES

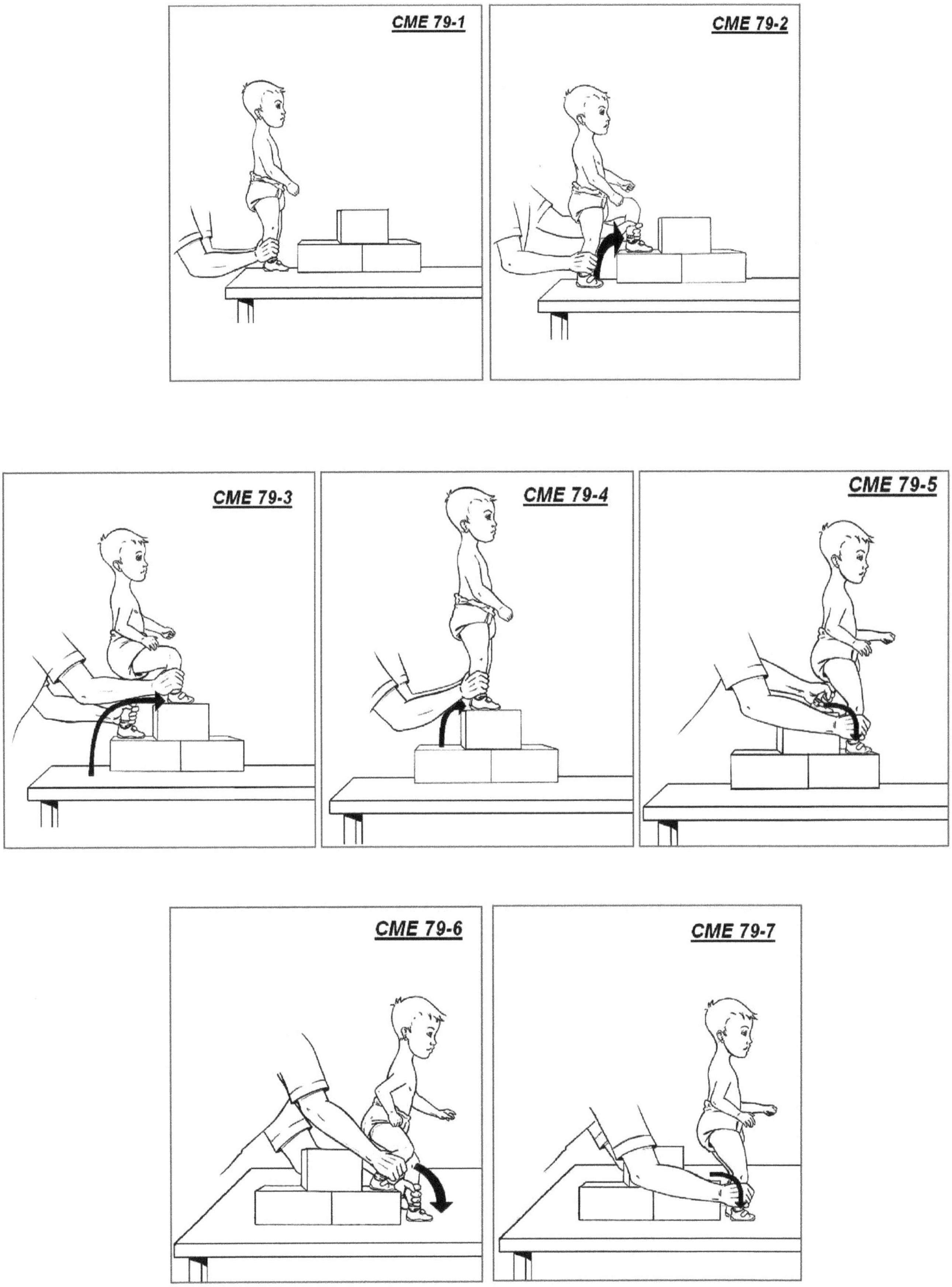

CME 80 STEPS ON PARALLEL BALANCE BARS BY ANKLES

Exercise Configuration: *To parallel sets of balance bar set up as illustrated in figure 1.*

MAIN GOAL: To provoke balance reactions.

INITIAL POSITION: In standing position on the box in front of the parallel balance bars.

HOLD: With both hands by his/her ankles as illustrated in figure 2.

MOVEMENT: Slowly move the child ahead taking small steps as indicated in figures 3 and 4. Repeat the sequence until reaching the end of the balance bars. Repeat the exercise as instructed by your CME therapist.

EXPECTED IMPROVEMENT: The child takes guided steps easily all along the balance bars.

CME 80 STEPS ON PARALLEL BALANCE BARS BY ANKLES

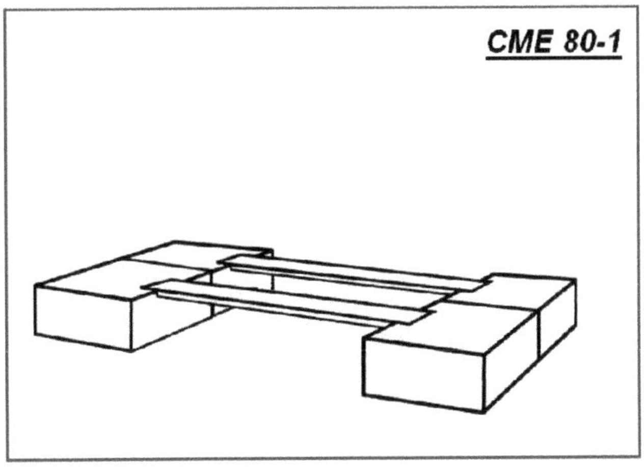

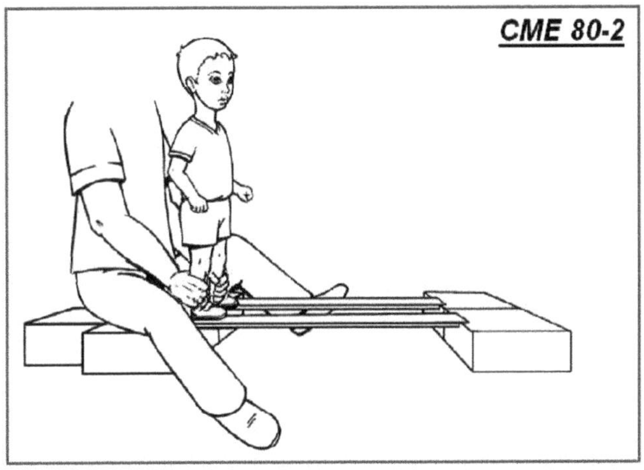

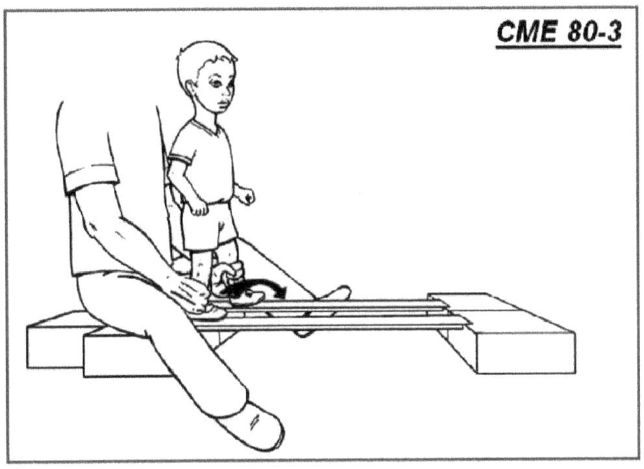

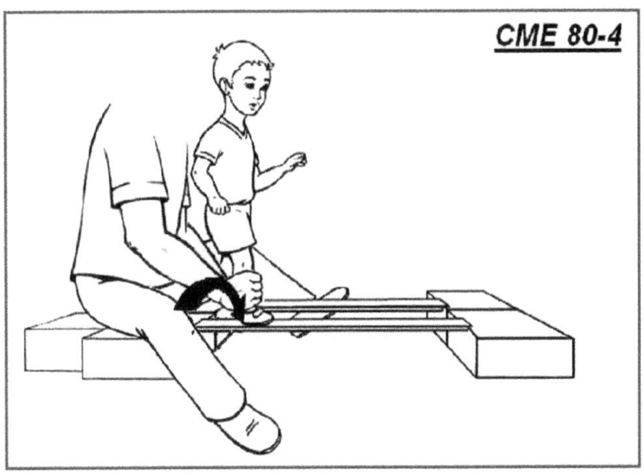

CME 81 ZIG-ZAG STEPS BY ANKLES

Exercise Configuration: *Two 15 cm. boxes with slots face to face plus 2 balance bars in a narrow position on the slots.*

MAIN GOAL: To provoke balance reactions.

INITIAL POSITION: In a standing position on the floor in front of the narrow parallel balance bars.

HOLD: By both ankles as illustrated in figure 1.

MOVEMENT: Place the child in the middle of the proximal narrow bar as indicated in figure 1. Take the step to the distal bar with the left foot (fig. 2), followed by the right foot (fig. 3). Proceed by taking the left foot back and to the right side as indicated in figure 4, followed by the right foot also back and to the right side (fig. 5). Repeat the sequence with a zig-zag shape, until reaching the right end of the narrow parallel balance bars. Repeat the exercise as instructed by your CME therapist.

EXPECTED IMPROVEMENT: The child is able to take zig-zag steps easily all along the narrow parallel balance bar.

CME 81 ZIG-ZAG STEPS BY ANKLES

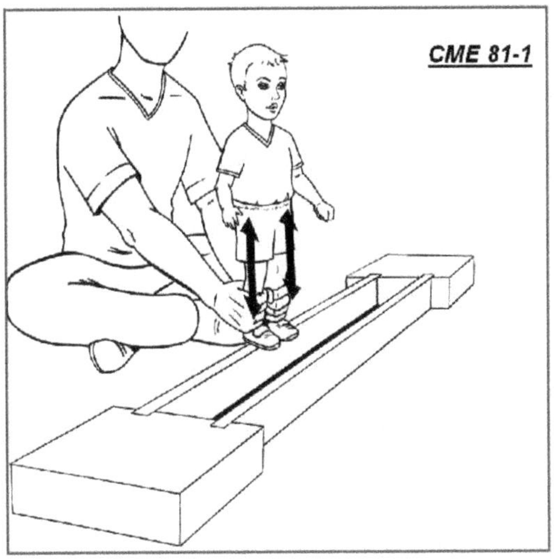

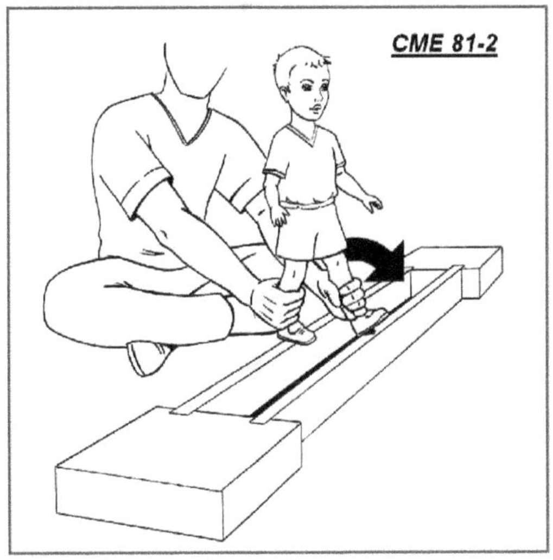

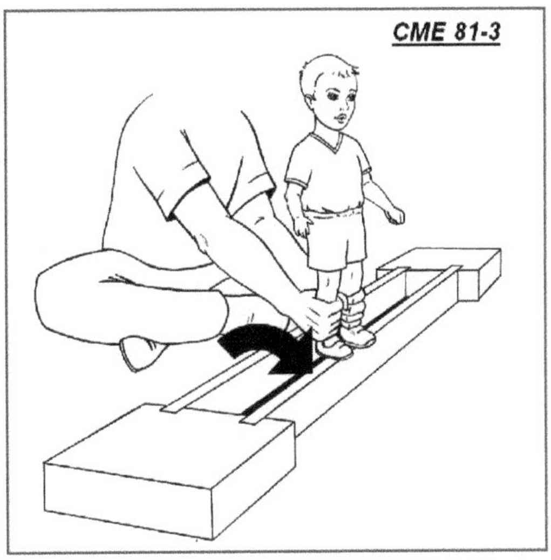

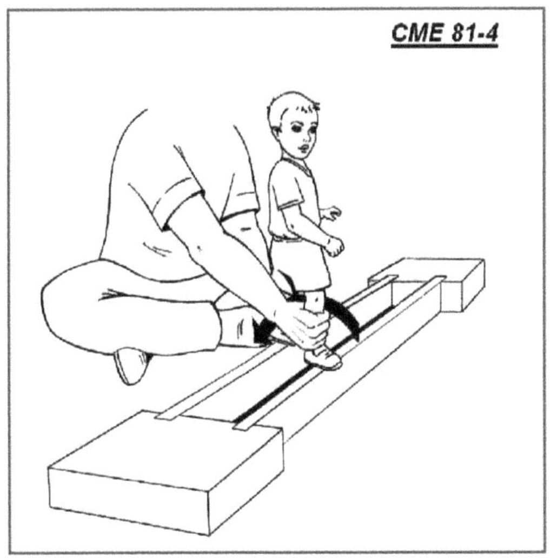

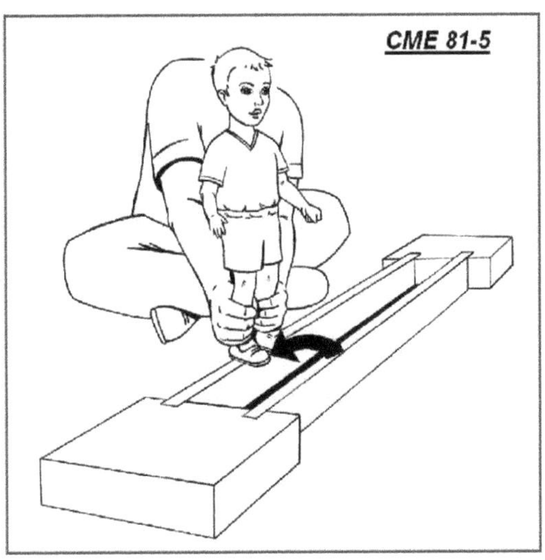

CME 82 DRAGGING FORWARD STEPS ON FREE NARROW BALANCE BARS BY ANKLES

Exercise Configuration: *Two bars resting on the narrow edge, 20 centimeters apart, touching the wall.*

MAIN GOAL: To provoke balance reactions.

INITIAL POSITION: In a standing position on the floor in front of the free narrow balance bars.

HOLD: By both ankles, as illustrated in figure 1.

MOVEMENT: Place the right foot on top of the right narrow balance (fig. 2), and then place the other foot on the other bar (fig. 3). Slowly move the child forward, taking small dragging steps as indicated in figures 4, 5, 6 and 7, until reaching the end of the narrow balance bars. Repeat the exercise as instructed by your CME therapist.

EXPECTED IMPROVEMENT: The child takes dragging steps all along the free narrow balance bars with no difficulty.

CME 82 DRAGGING FORWARD STEPS ON FREE NARROW BALANCE BARS BY ANKLES

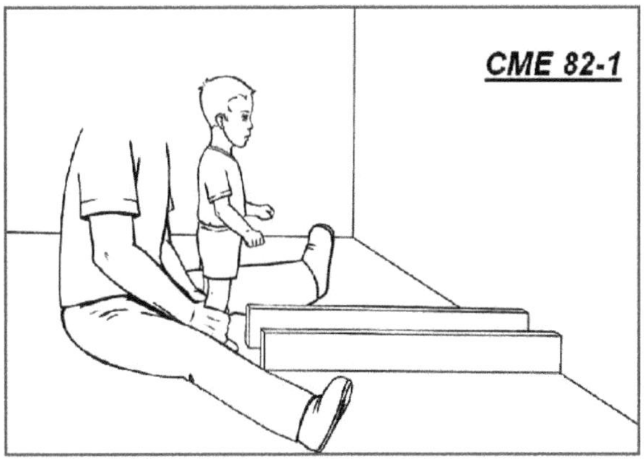

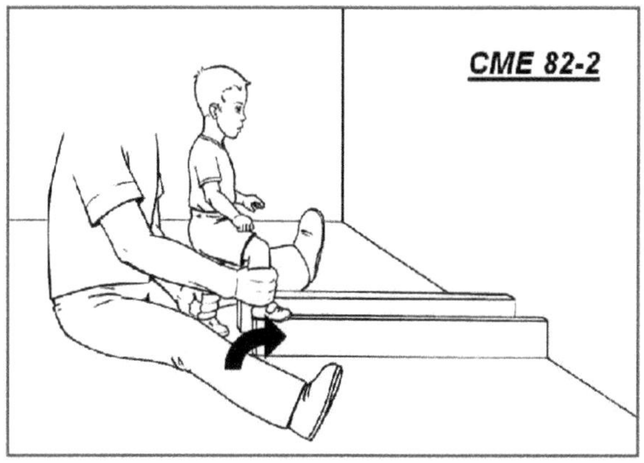

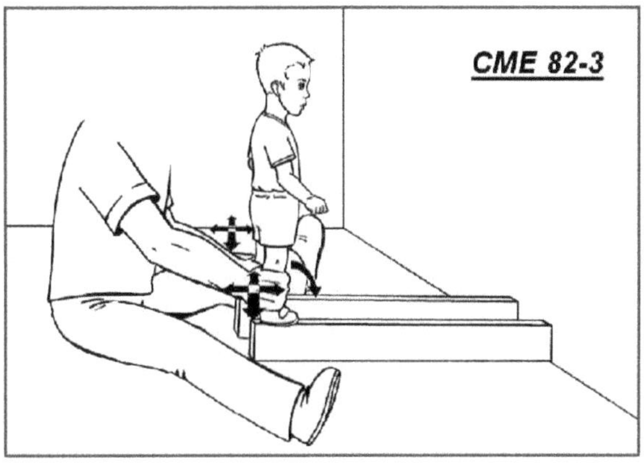

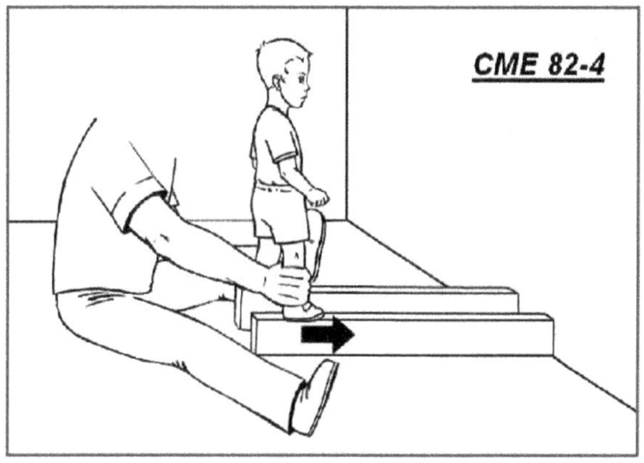

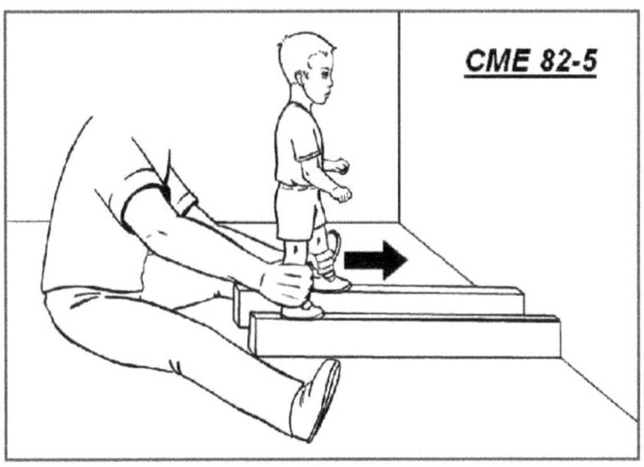

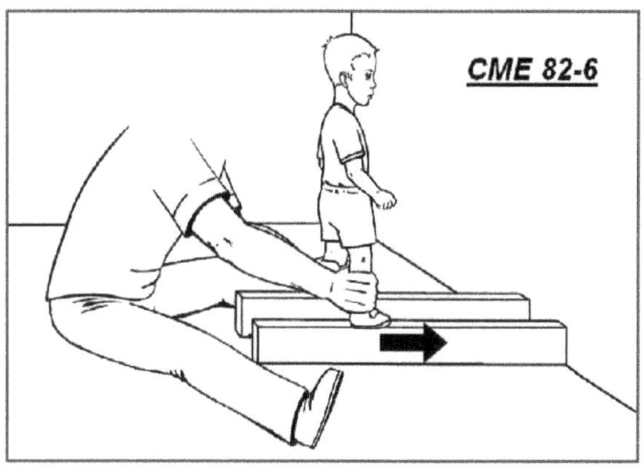

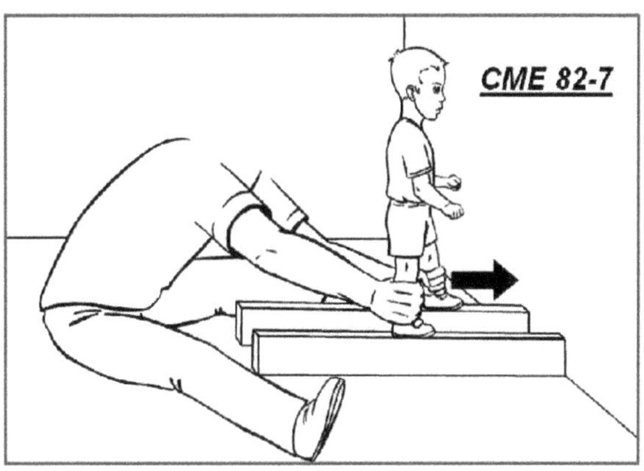

CME 83 SIDEWAY STEPS ON FREE NARROW BALANCE BAR BY ANKLES

MAIN GOAL: To provoke balance reactions.

INITIAL POSITION: In a standing position in front of the balance bar.

HOLD: By both ankles as illustrated in figure 1.

MOVEMENT: Slowly and carefully place the left foot on top of the narrow bar as indicated in figure 2, and take the step on top of the narrow bar with the right foot as shown in figure 3. Now maintaining balance at all times, make sliding sideways steps to the right until reaching the right end of the narrow balance bar as illustrated in figures 4, 5, 6 and 7. Repeat the exercise as instructed by your CME therapist.

EXPECTED IMPROVEMENT: The child is able to maintain balance all along the sliding steps.

CME 83 SIDEWAY STEPS ON FREE NARROW BALANCE BAR BY ANKLES

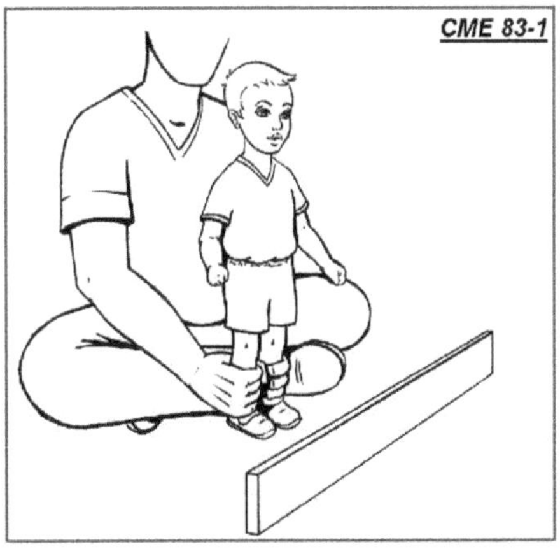

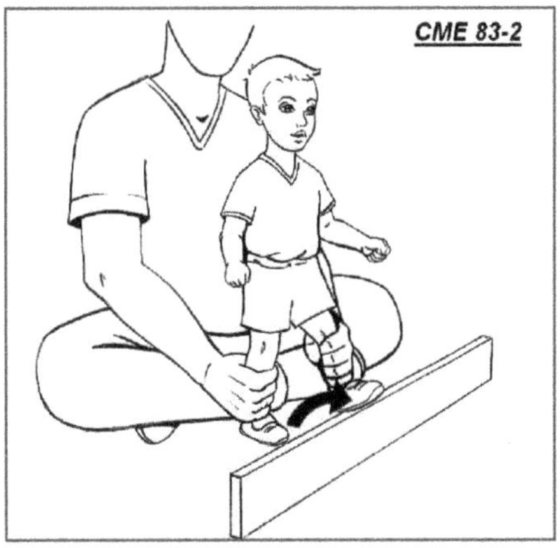

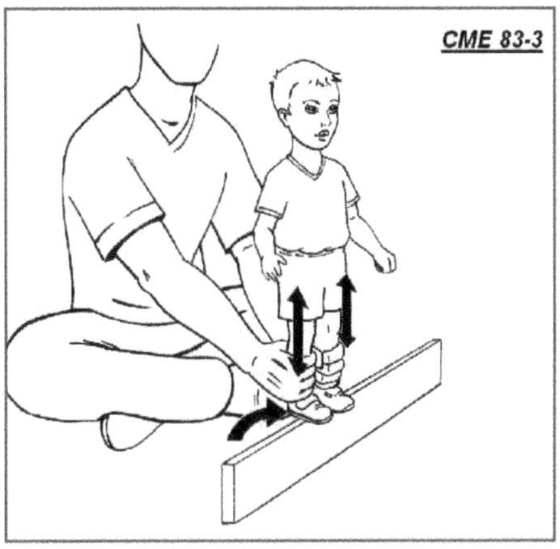

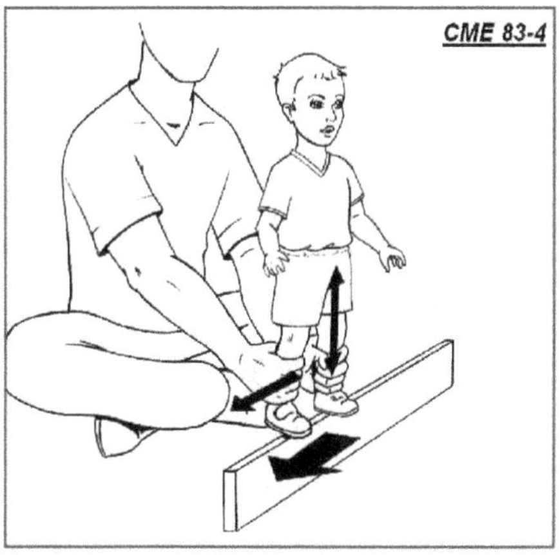

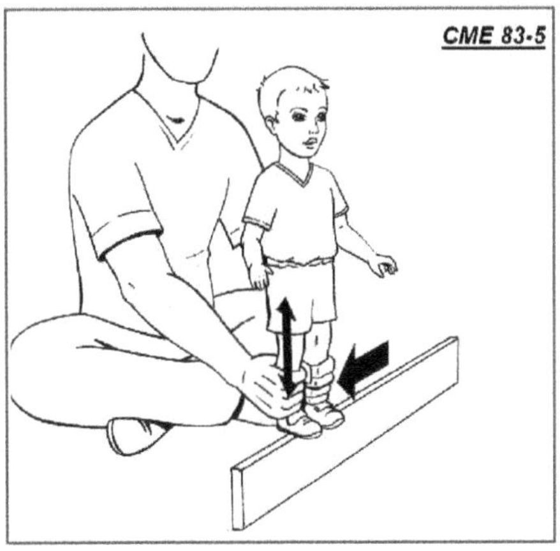

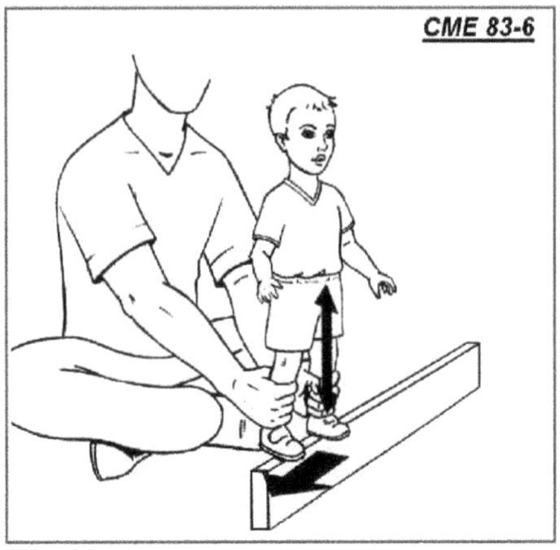

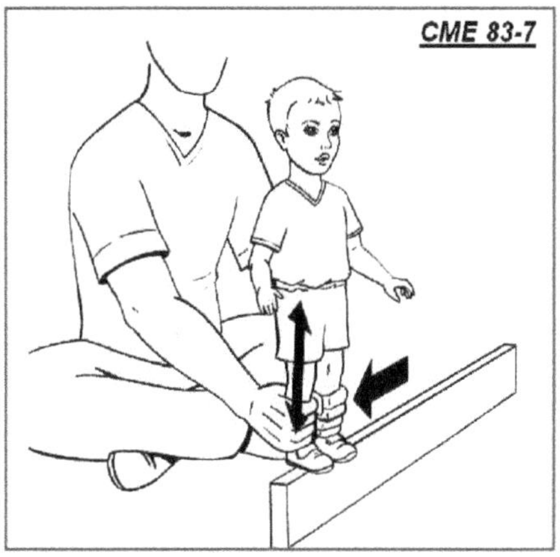

CME 84 STEPS ON LADDER BY ANKLES

Exercise Configuration: *2 boxes of 30 cm; one on top of the other with a strap in between to avoid sliding.*

MAIN GOAL: To provoke balance reactions.

INITIAL POSITION: In a standing position in front of the ladder.

HOLD: By ankles as illustrated in figure 1.

MOVEMENT: Slowly place the right foot first in the first step (fig. 2), and after the left foot (fig. 3). Now move the left foot on top of the ladder (fig. 4), followed by the right one (fig. 5). Lift the child carefully and take him/her back to the floor as indicated in figures 6 and 7. Repeat the exercise as instructed by your CME therapist.

EXPECTED IMPROVEMENT: The child is able to step up the ladder and float down, without difficulty.

CME 84 STEPS ON LADDER BY ANKLES

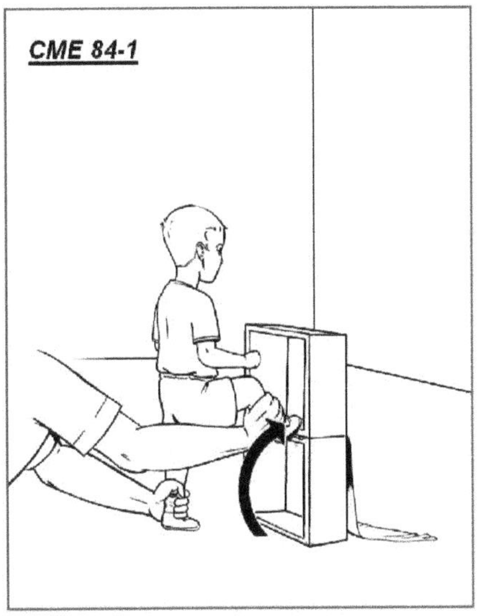

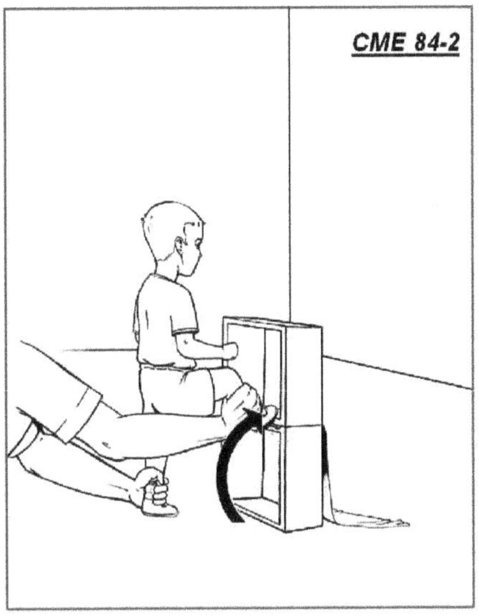

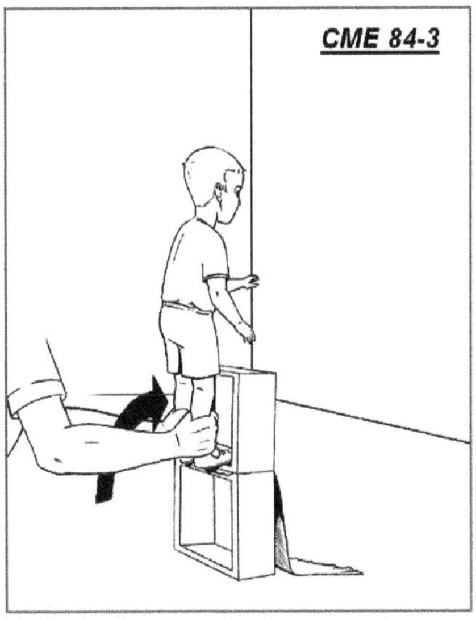

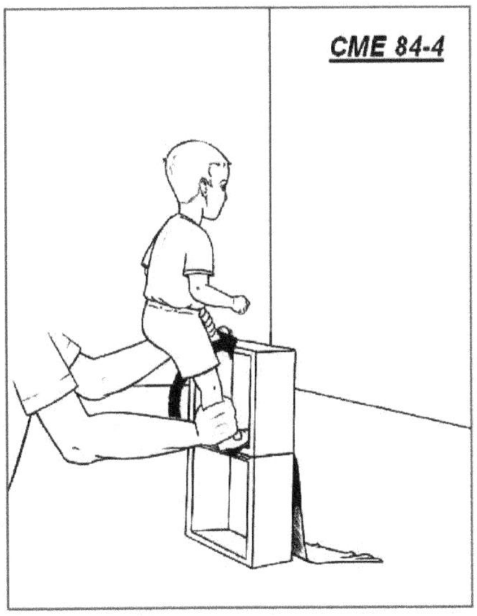

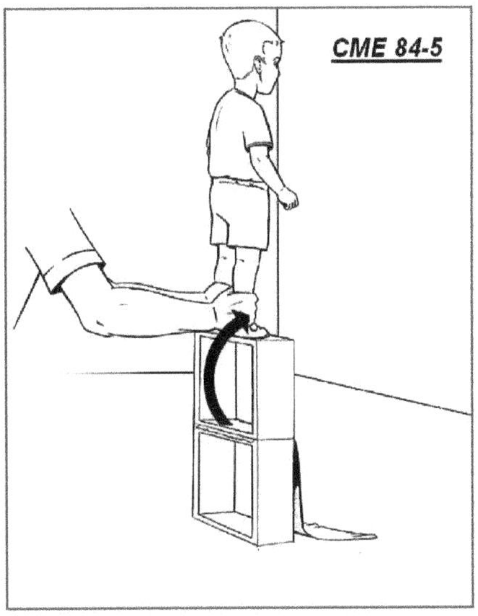

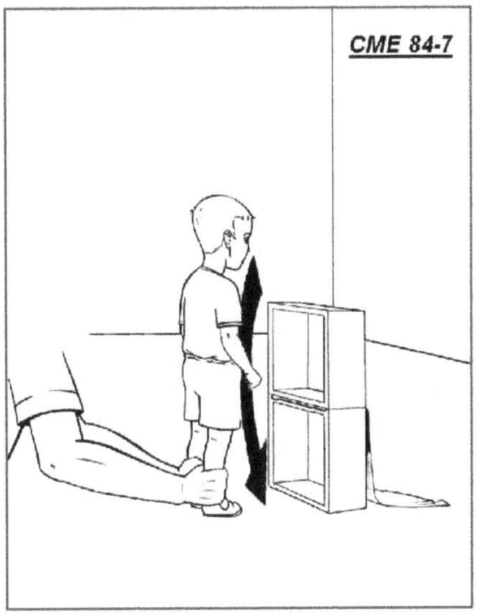

CME 85 STEP ON CUBE TO 15 CM. BOX BY ONE ANKLE

Exercise Configuration: *One balance board with a cube tucked in the middle, one balance bar along one edge, and one 15 cm. box on top of the bar and the board, as illustrated in figure 1.*

MAIN GOAL: To provoke balance reactions.

INITIAL POSITION: In a standing position in front of the configuration.

HOLD: Only by the right ankle as illustrated in figure 2.

MOVEMENT: Slowly place the right foot on top of the small cube as indicate in figure 3, keep stability and wait until the child moves the left foot on top of the 15 cm. box by himself/herself as shown in figure 4, at this time you move the right foot on top of the box to complete the sequence as shown in figure 5. Repeat the exercise as instructed by your CME therapist.

EXPECTED IMPROVEMENT: The child is able to take guided steps without difficulty from the floor to the box.

CME 85 STEP ON CUBE TO 15 CM. BOX BY ONE ANKLE

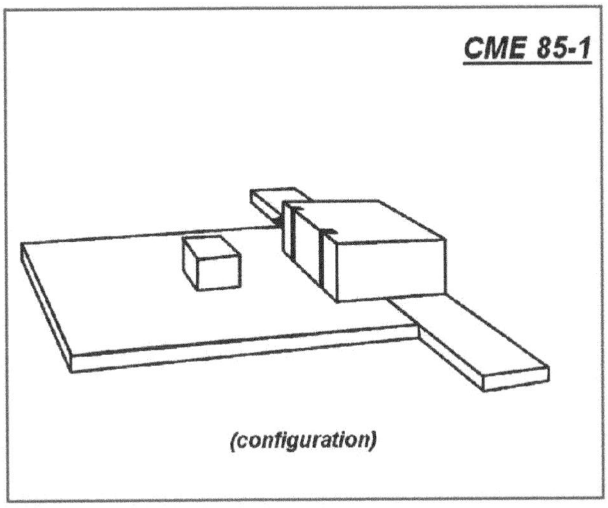

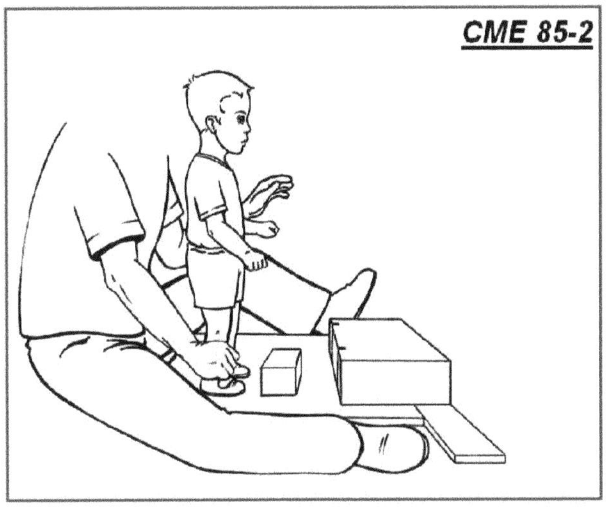

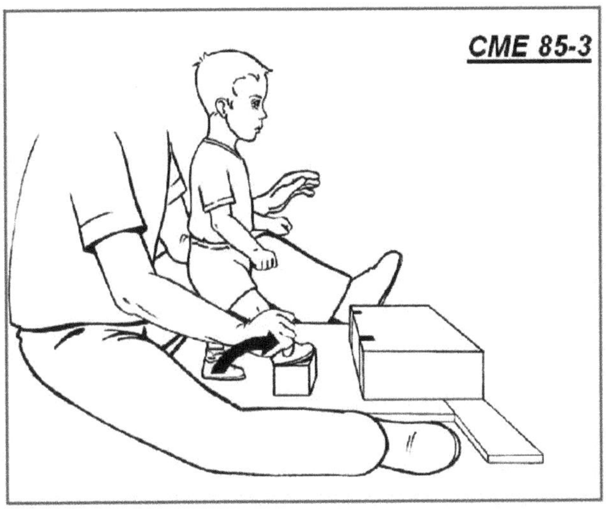

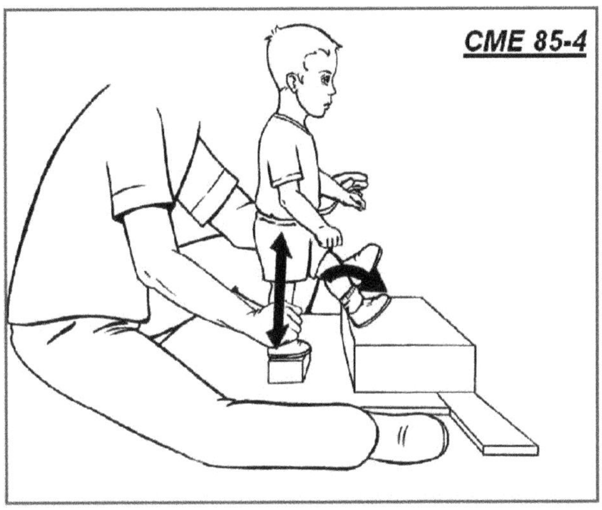

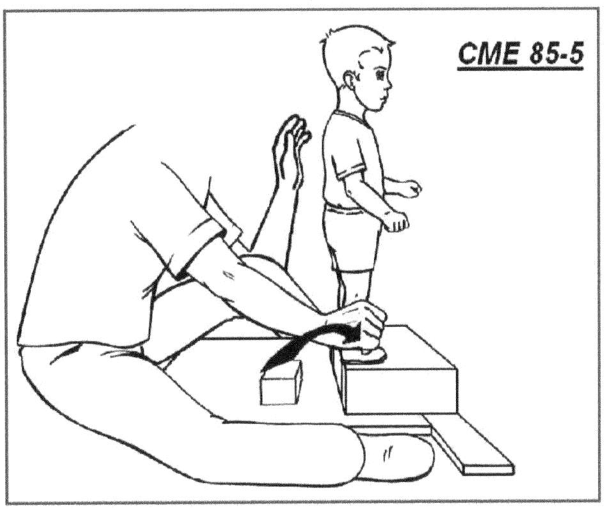

CME 86 MIMIC STEPS ON 10 CM. BOX BY ANKLES

MAIN GOAL: To provoke stability in a standing position

INITIAL POSITION: In a standing position at the right side of the 10cm. box.

HOLD: By both ankles as illustrated in figure 1.

MOVEMENT: Slowly place the left foot at the same level of the surface of the box and firmly contain the position as indicated in figure 2, now do the same movement with the right leg in order to achieve a perfect floating standing posture as shown in figure 3. Move back the left and right foot to the floor (fig. 4), (fig. 5). Repeat the exercise as instructed by your CME therapist.

EXPECTED IMPROVEMENT: The child is able to take the mimic steps without difficulty.

CME 86 MIMIC STEPS ON 10 CM. BOX BY ANKLES

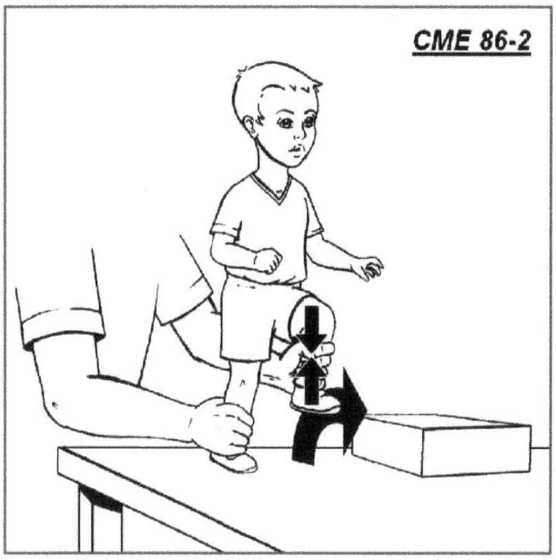

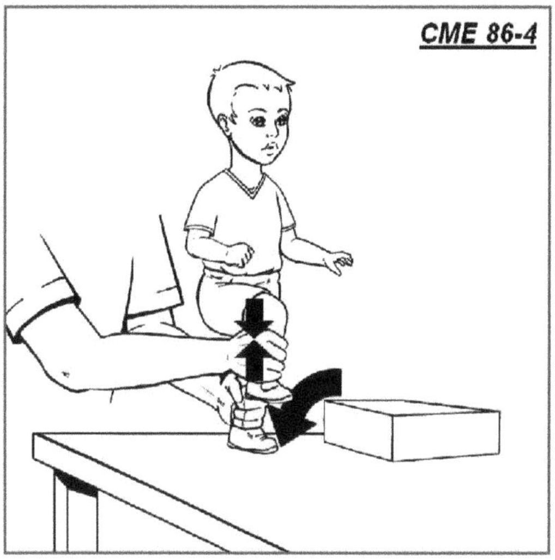

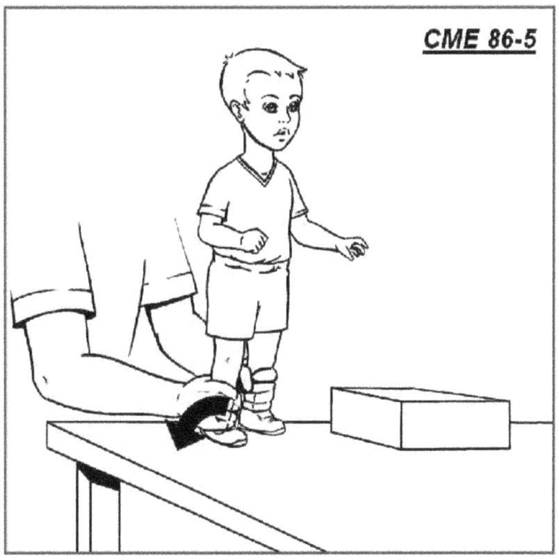

CME 87 DOUBLE PLACING BY ANKLES

Exercise Configuration: *2 boxes in a 30 cm. position, plus one 10 cm. box tucked on top.*

MAIN GOAL: To provoke standing stability and balance reactions.

INITIAL POSITION: In a standing position on top of the boxes with heels at the proximal edge as indicated in figure 1.

HOLD: Firmly by both ankles (fig. 1).

MOVEMENT: Slowly bend the left leg backwards as illustrated in figure 2 and sustaining weight on your hands, simultaneously bend the right leg backwards and the left leg forward to the boxes as illustrated in the sequence of figures 2, 3 and 4, and repeat with a non stop pace, the synchronized sequence moving the right foot on the box as indicated in figures 5 and 6. Repeat the exercise as instructed by your CME therapist.

EXPECTED IMPROVEMENT: The child is able to resist the guided steps on top of the boxes easily.

CME 87 DOUBLE PLACING BY ANKLES

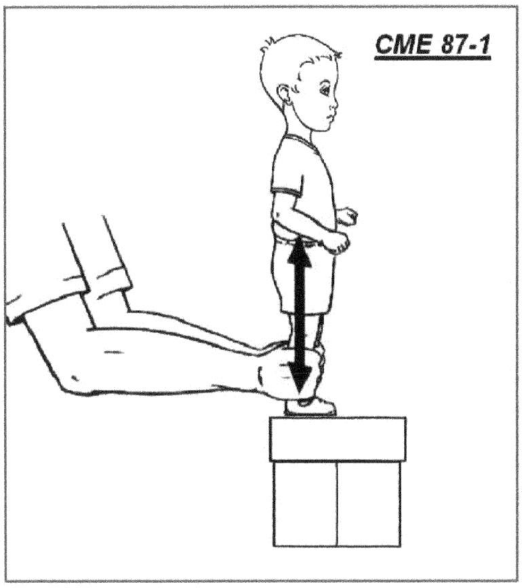

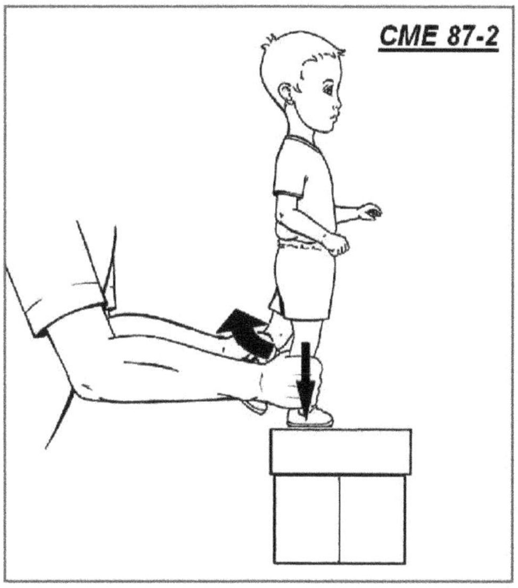

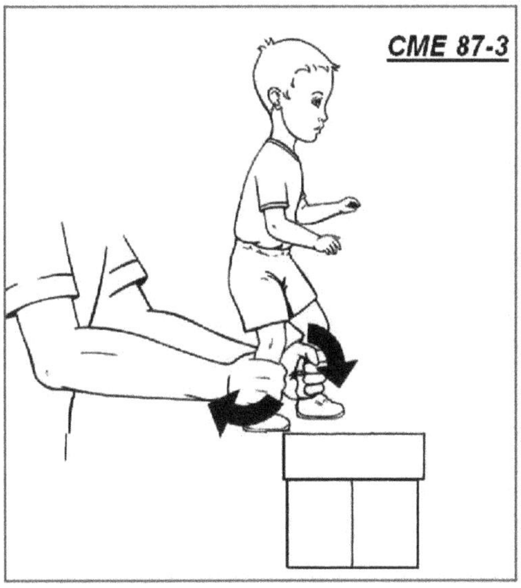

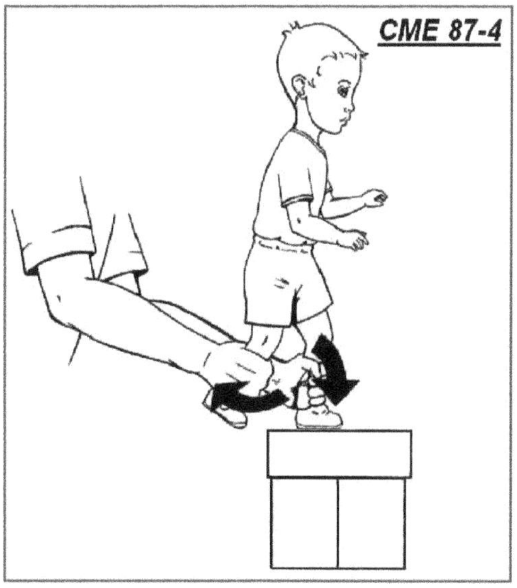

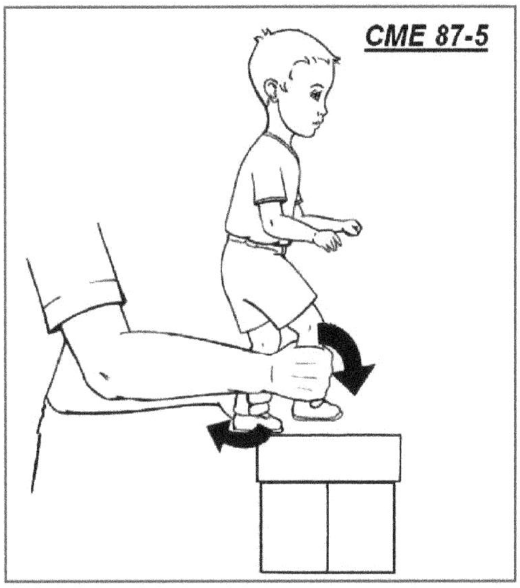

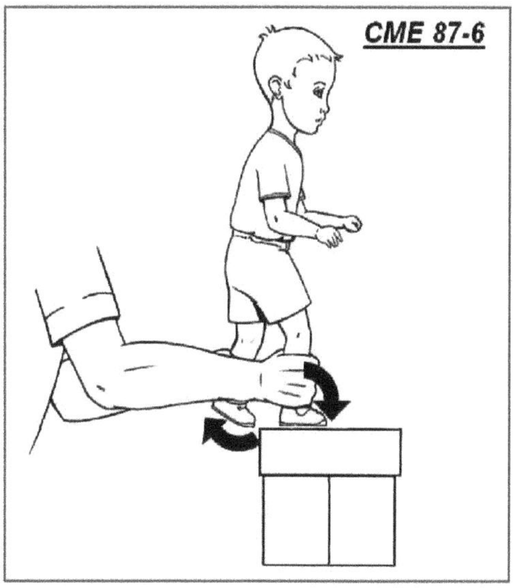

CME 88 STEPS BY STRAPS ON ANKLES

MAIN GOAL: To provoke stability in standing and walking, and also enhance balance reactions.

INITIAL POSITION: In a standing position on the table.

HOLD: By straps around each ankle as tourniquet as illustrated in figures 1 and 2.

MOVEMENT: Slowly by the straps, move the child forward taking small steps as indicated in figures 3, 4 and 5. Repeat the sequence until reaching the end of the table. Repeat the exercise as instructed by your CME therapist.

EXPECTED IMPROVEMENT: The child is able to take guided steps easily, all along the table.

CME 88 STEPS BY STRAPS ON ANKLES

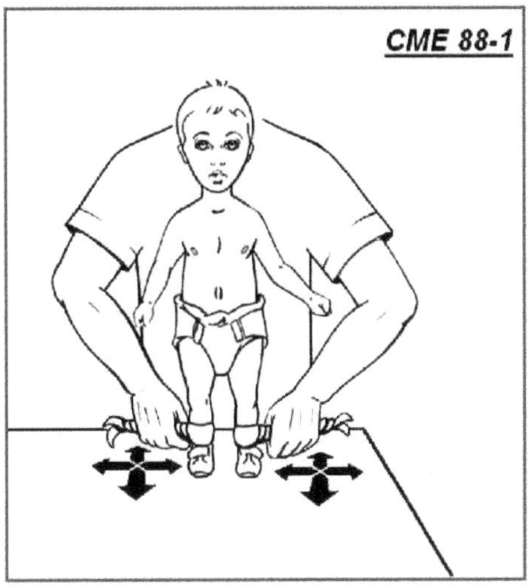

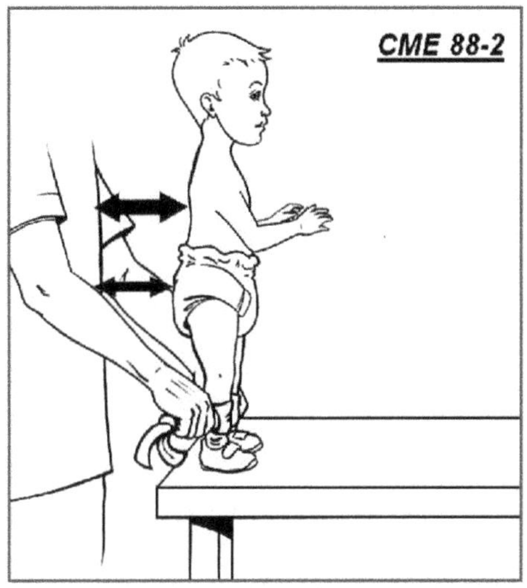

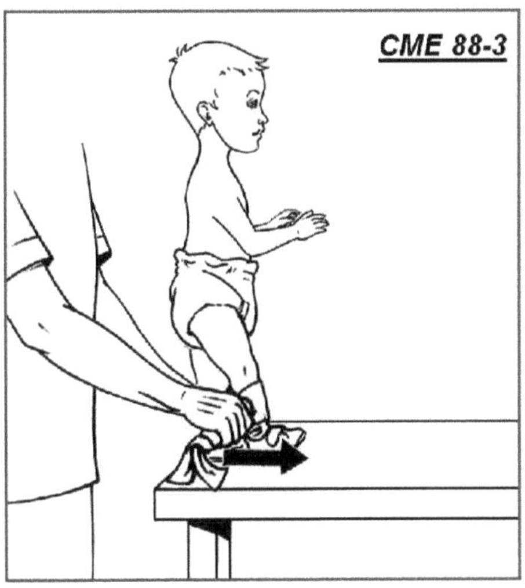

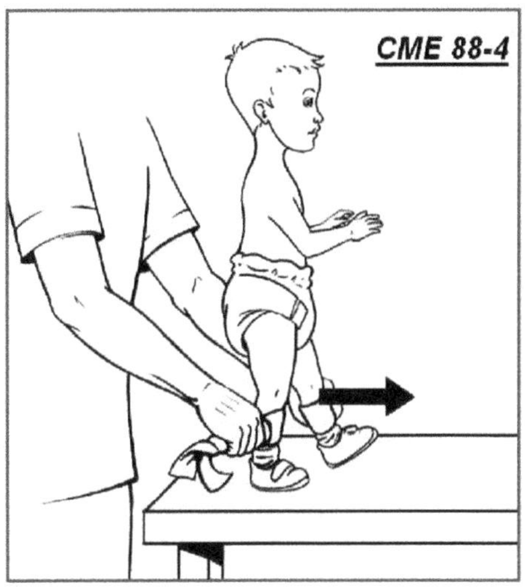

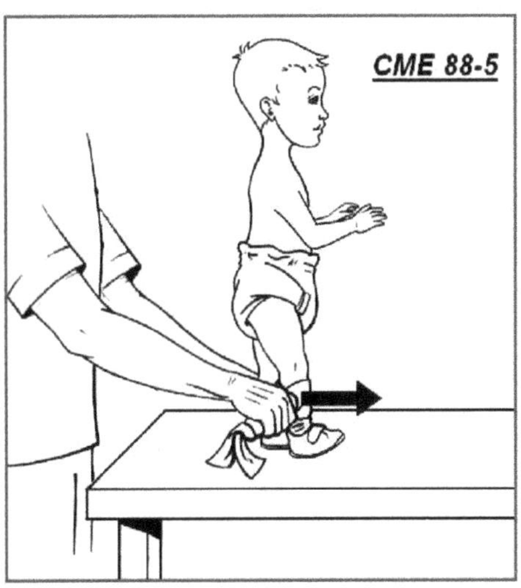

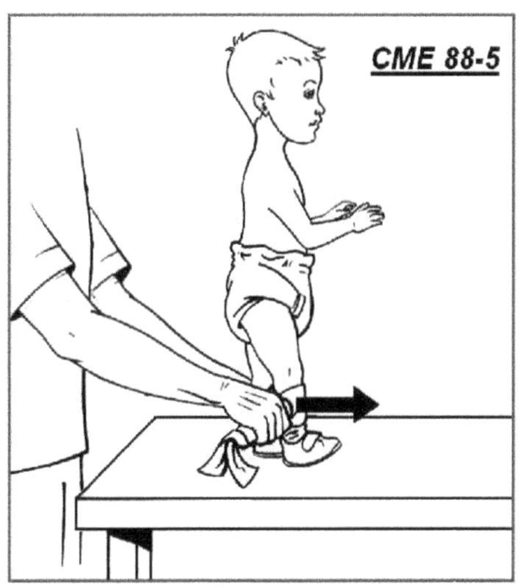

CME 89 STEP ON BALANCE BOARD BY STRAPS ON ANKLES

MAIN GOAL: To provoke balance reactions.

INITIAL POSITION: In a standing position on the proximal edge of the balance board.

HOLD: By straps around each ankle as illustrated in figure 1, and 2.

MOVEMENT: By the straps, slowly move the child forward, taking small steps until reaching the distal edge of the balance board, as indicated in figures 3, 4, 5 and 6. Repeat the exercise as instructed by your CME therapist.

EXPECTED IMPROVEMENT: The child is able to take guided steps easily, all along the balance board.

CME 89 STEP ON BALANCE BOARD BY STRAPS ON ANKLES

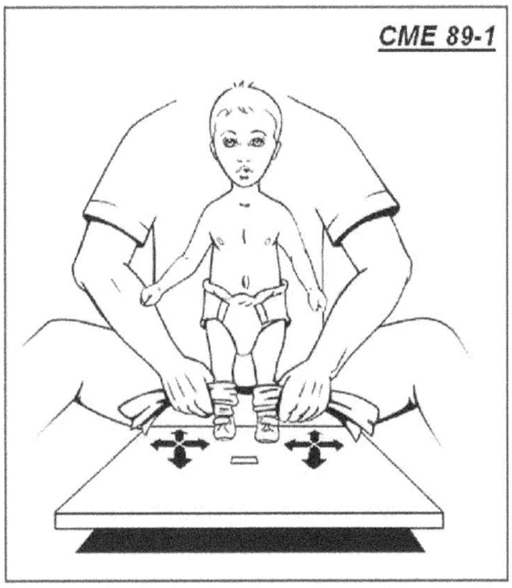
CME 89-1

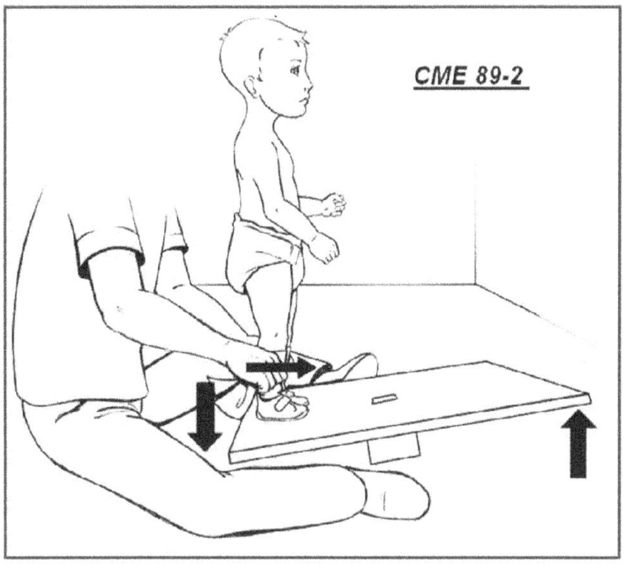
CME 89-2

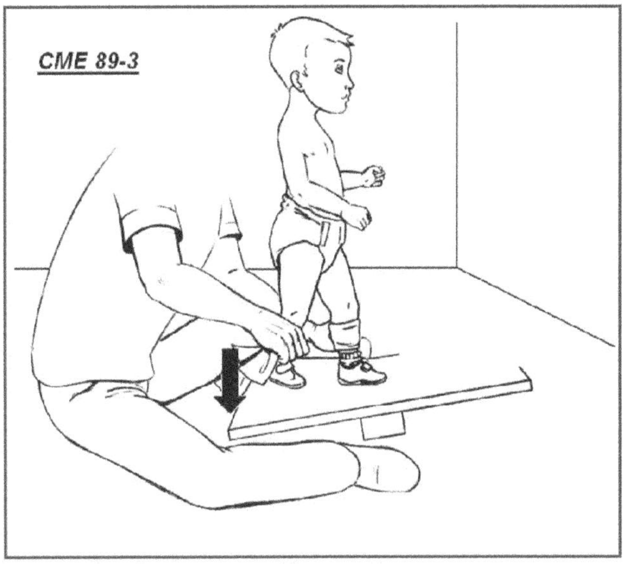
CME 89-3

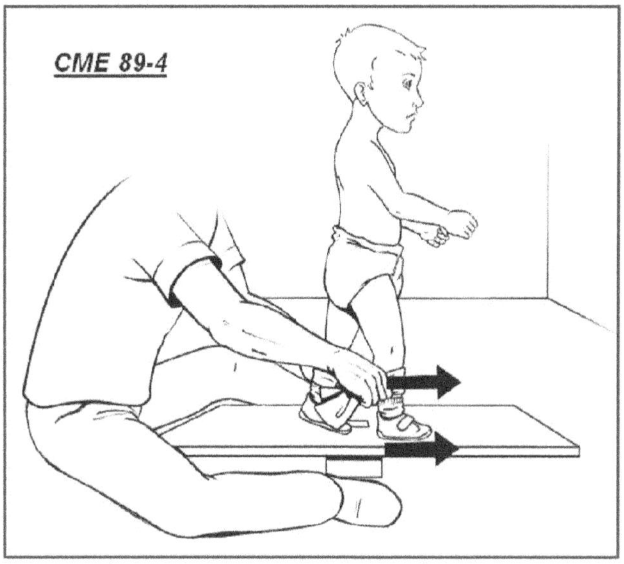
CME 89-4

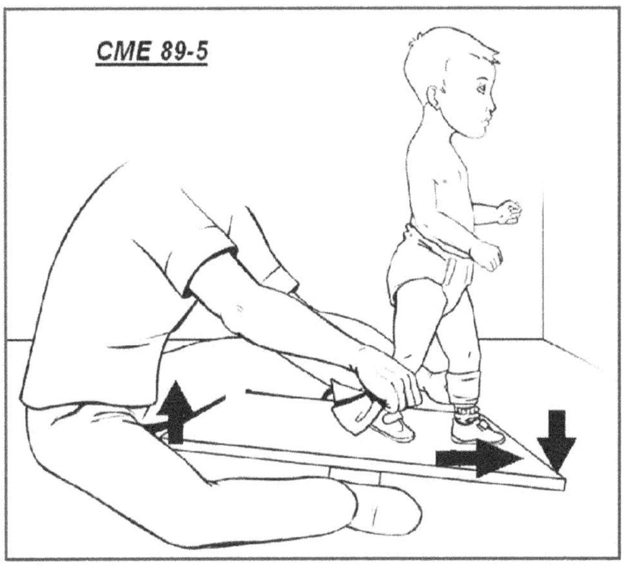
CME 89-5

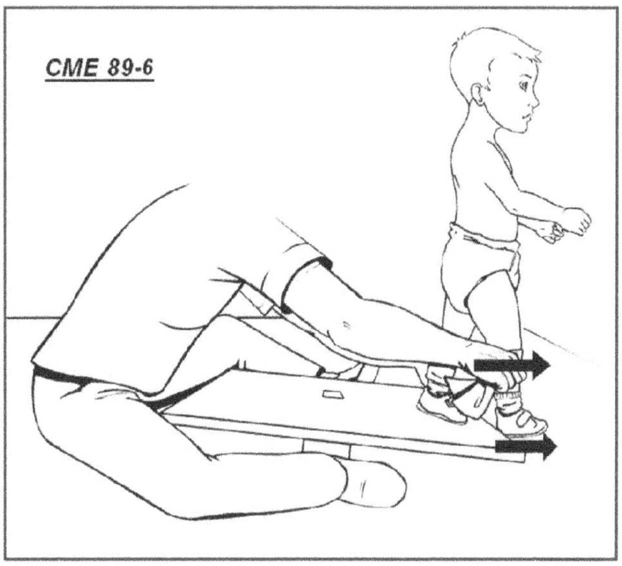
CME 89-6

CME 90 STEP ON CUBE TO 15 CM. BOX BY STRAPS ON ANKLES

Exercise Configuration: *One balance board with a cube tucked in the middle, one balance bar along one edge, and one 15 cm. box on top of the bar and the board, as illustrated in figure 1.*

MAIN GOAL: To provoke balance reactions.

INITIAL POSITION: In a standing position in front of the small cube.

HOLD: By straps around each ankle as illustrated in figures 2 and 3.

MOVEMENT: Place the right foot on top of the small cube as illustrated in figure 4, and immediately move the left foot on top of the box as indicated in figure 5, and finish the sequence with the right foot on top of the box as shown in figure 6. Take the child back to the initial position and repeat the sequence, beginning the movement with the left foot. Repeat the exercise as instructed by your CME therapist.

EXPECTED IMPROVEMENT: The child is able to take guided steps from the cube to the box, with no difficulty.

CME 90 STEP ON CUBE TO 15 CM BOX BY STRAPS ON ANKLES

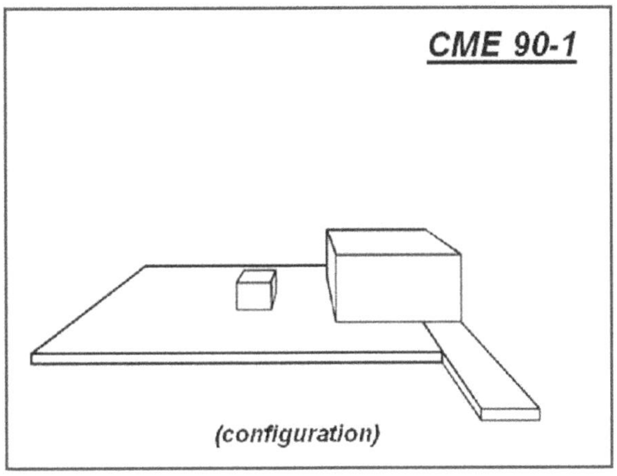

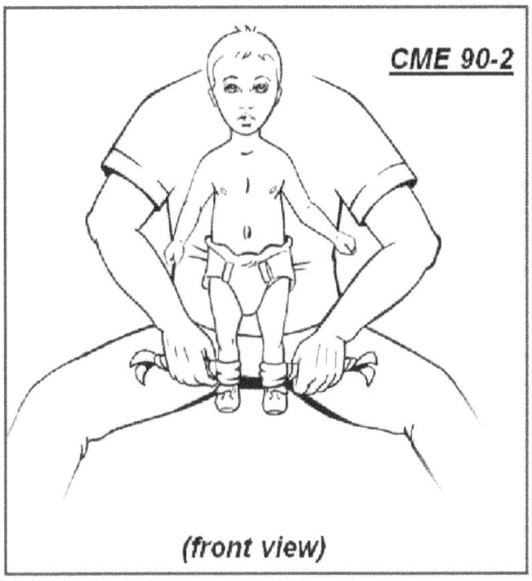

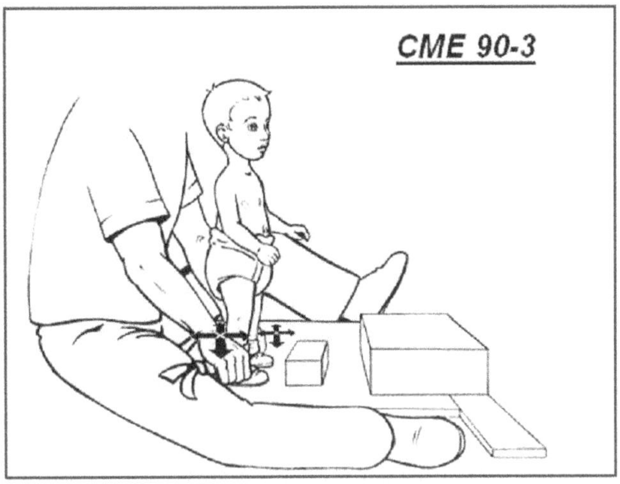

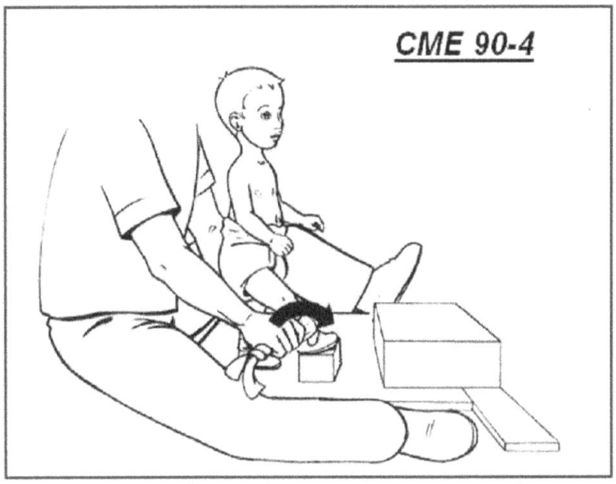

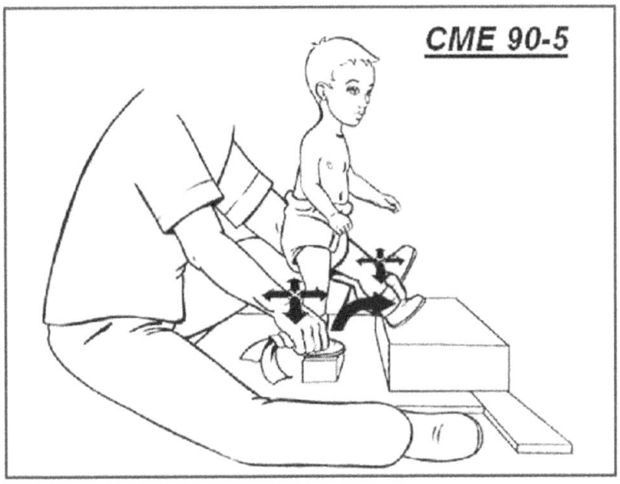

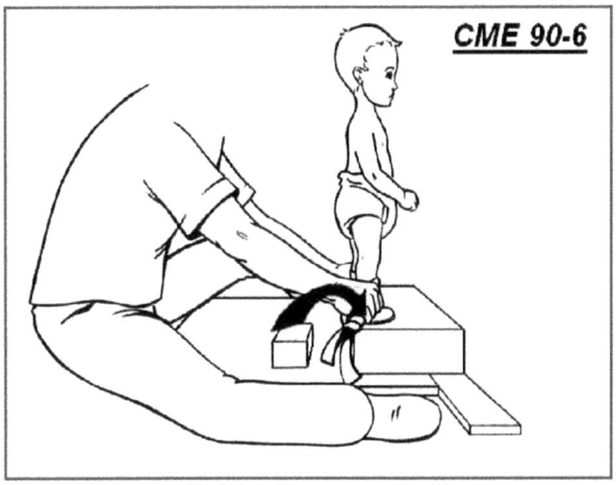

CME 91 STEPS ON BALANCE BAR BY STRAPS ON ANKLES

MAIN GOAL: To provoke balance reactions.

INITIAL POSITION: In a standing position on the box in front of the balance bar.

HOLD: By straps around each ankle as illustrated in figure 1. (The multidirectional arrows indicated the direction of any necessary pulses in order to compensate balance).

MOVEMENT: Slowly move the child ahead by the straps in order to take small steps all along the balance bar as indicated in figure 3, 4 and 5. Repeat the steps until reaching the box at the end of the balance bar. Repeat the exercise as instructed by your CME therapist.

EXPECTED IMPROVEMENT: The child is able to take guided steps easily, all along the balance bar.

CME 91 STEPS ON BALANCE BAR BY STRAPS ON ANKLES

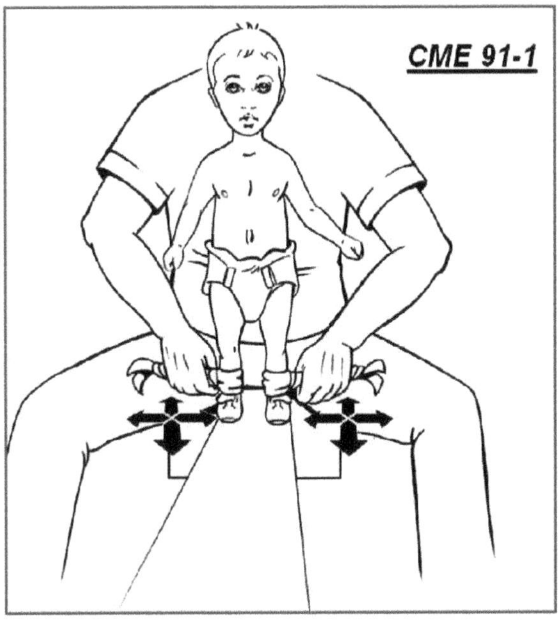

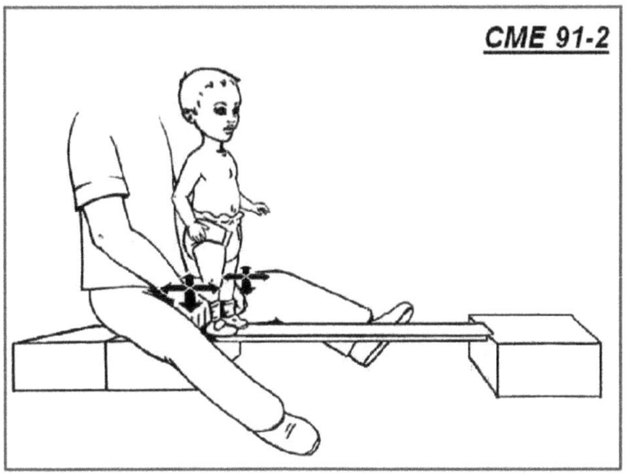

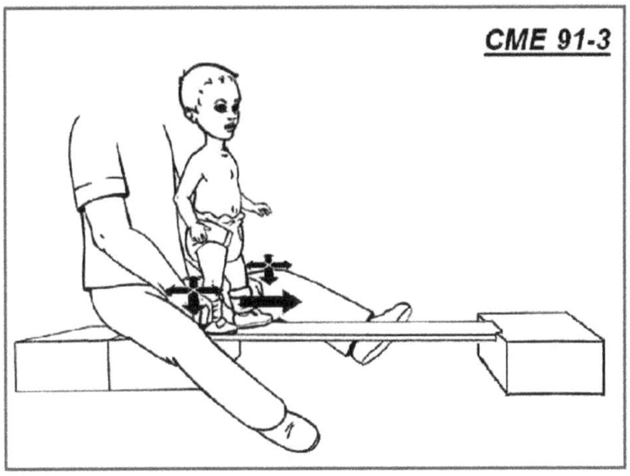

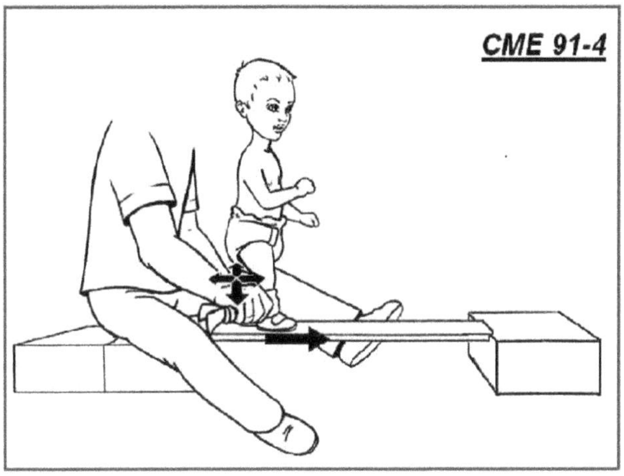

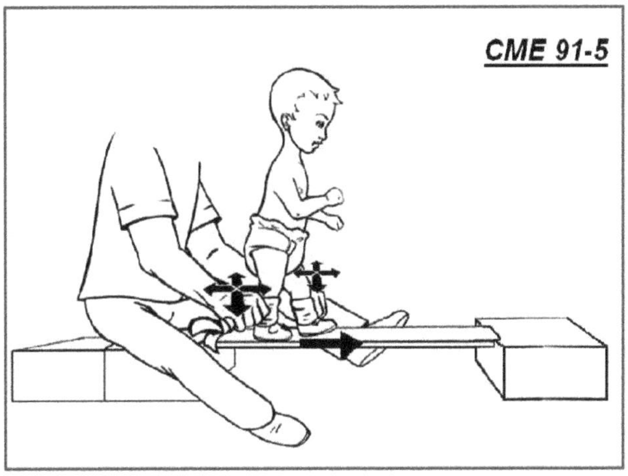

CME 92 STEPS ON 30 CM. BOXES BY STRAPS ON ANKLES

Exercise Configuration: *Five boxes in position 30 cm.one after the other, as illustrated in figure 92.*

MAIN GOAL: To provoke balance reactions.

INITIAL POSITION: In a standing position on top of the first the box.

HOLD: By straps around ankles as illustrated in figure 1. (The multidirectional arrows indicate that it is necessary to do balanced compensation pulses at all times).

MOVEMENT: By the straps provoke the child to take small steps on the narrow surface as indicated in figures 2, 3 and 4. Repeat the sequence until reaching the end of the narrow boxes. Repeat the exercise as instructed by your CME therapist.

EXPECTED IMPROVEMENT: The child takes guided steps easily all along the line of narrow boxes.

CME 92 STEPS ON 30 CM. BOXES BY STRAPS ON ANKLES

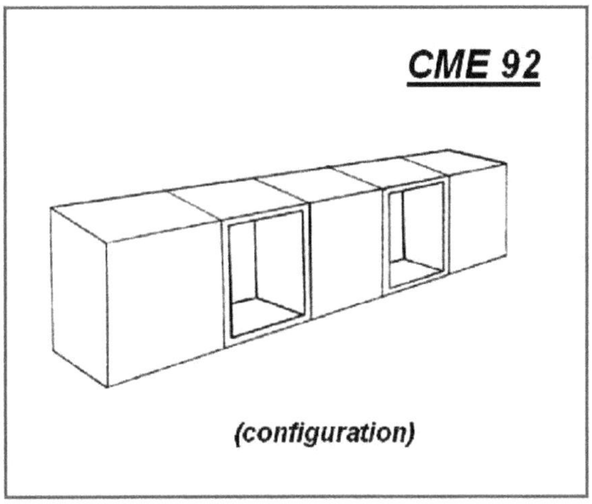

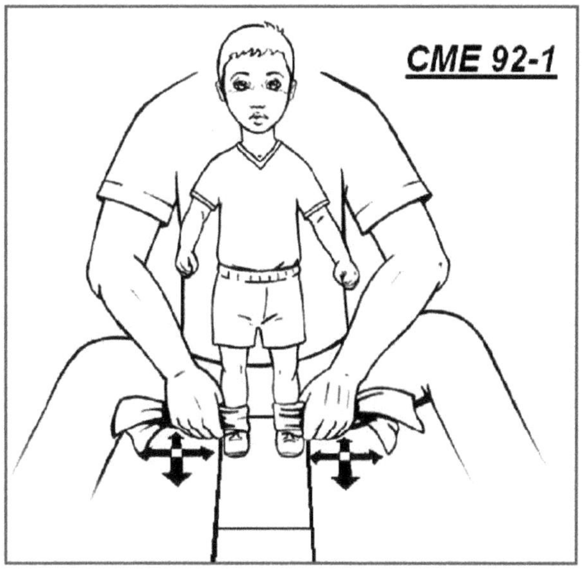

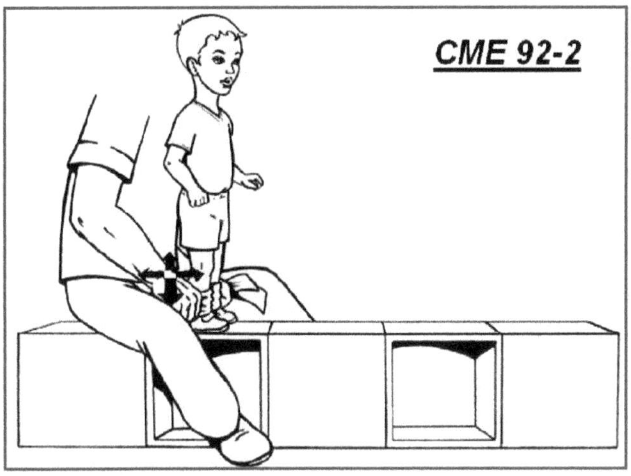

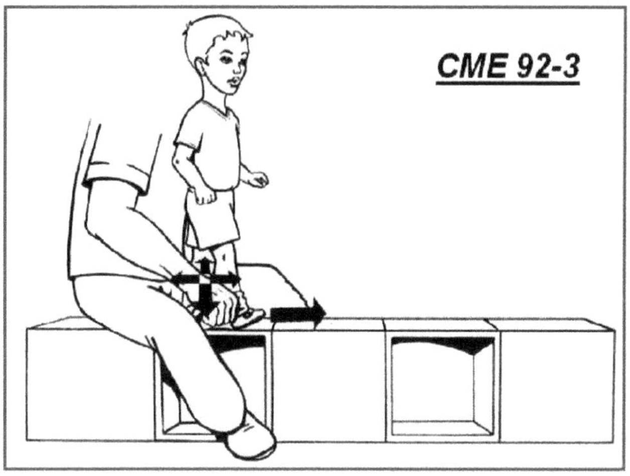

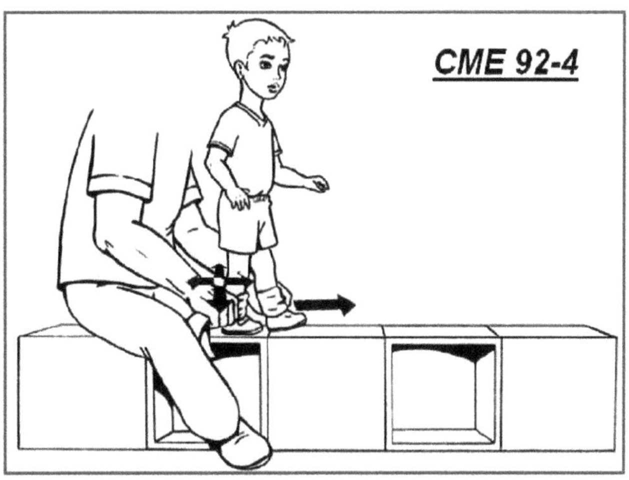

CME 93 RECIPROCAL HOPPING ON 10 CM. BOX BY ANKLES

MAIN GOAL: To achieve stability in a standing position.

INITIAL POSITION: In a standing position in front of the box.

HOLD: By both hands by ankles (fig. 1).

MOVEMENT: To begin, place the right foot on top of the box (fig. 1), lift the child up (fig. 2), and simultaneously switch the legs as indicated in figure 3, in order to place the left foot on top of the box and the right foot on the floor (fig. 4). Repeat the sequence as indicated in figures 5, 6 and 7 to complete a full cycle. Repeat the exercise as instructed by your CME therapist.

EXPECTED IMPROVEMENT: The child is able to do the guided reciprocal hopping, with full stability.

CME 93 RECIPROCAL HOPPING ON 10 CM. BOX BY ANKLES

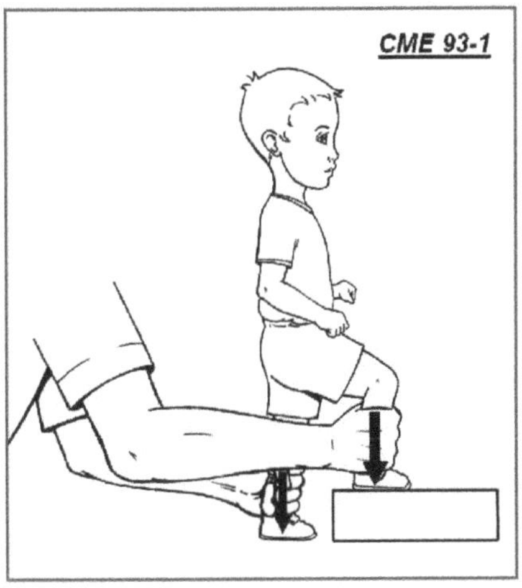
CME 93-1

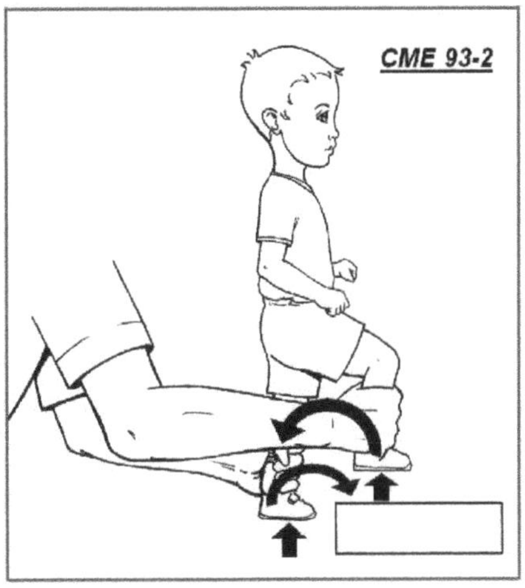
CME 93-2

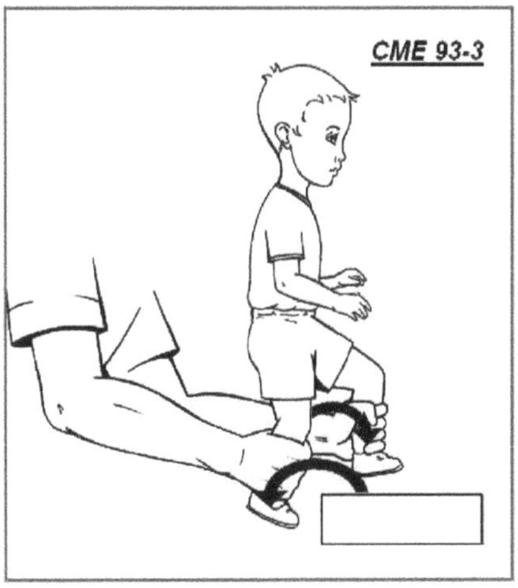
CME 93-3

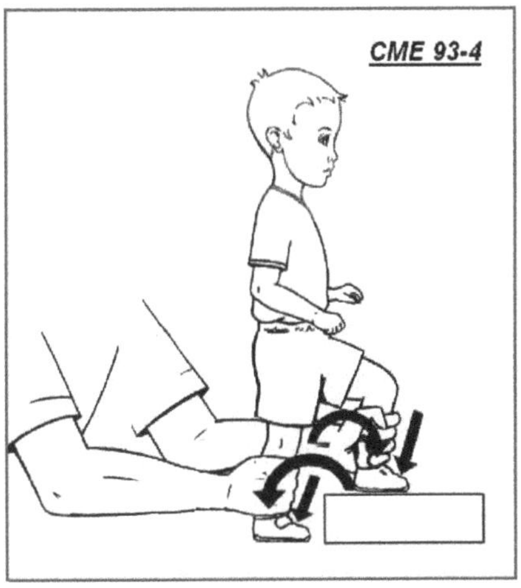
CME 93-4

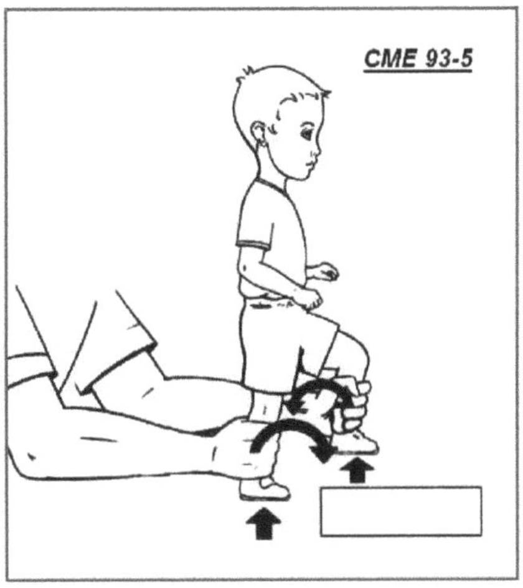
CME 93-5

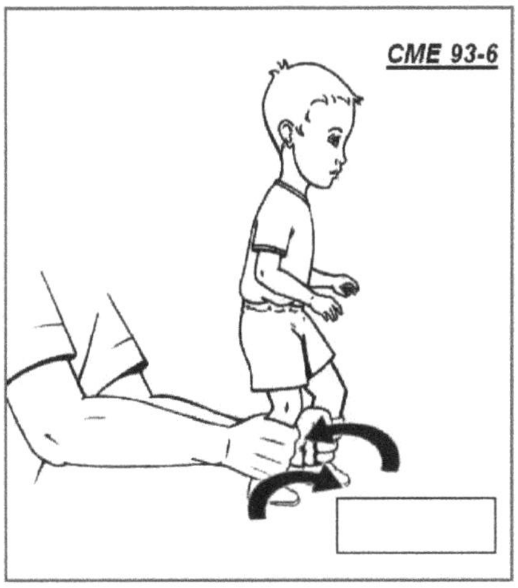
CME 93-6

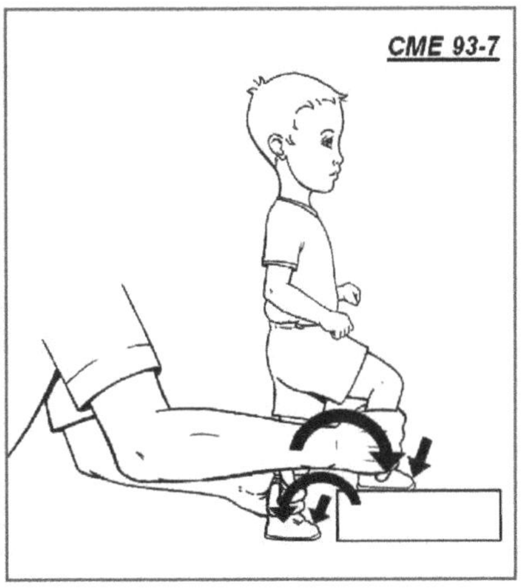
CME 93-7

CME 94 STEPS ON FREE NARROW BALANCE BAR BY ANKLES

MAIN GOAL: To provoke balance reactions.

INITIAL POSITION: In a standing position in front of the free narrow balanced bar (fig. 1).

HOLD: Both hands by the ankles as illustrated in fig.1.

MOVEMENT: Slowly place the right foot on top of the narrow balance bar as illustrated in figure 2. (The multidirectional arrow means that it is mandatory to apply the necessary movements to any convenient direction to maintain the child's stability on top of the narrow bar). Then place the left foot on the bar as indicated in figure 3. Now take small steps as shown in figures 4 and 5. Repeat the sequence until reaching the end of the balance bar. Repeat the exercise as instructed by your CME therapist.

EXPECTED IMPROVEMENT: The child takes guided steps easily all along the balance bar.

CME 94 STEPS ON FREE NARROW BALANCE BAR BY ANKLES

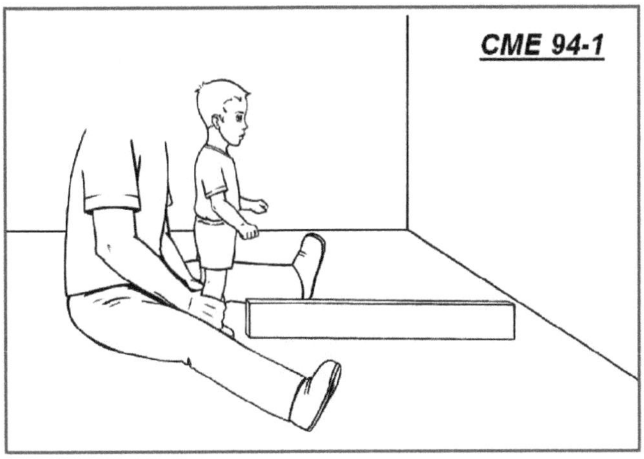

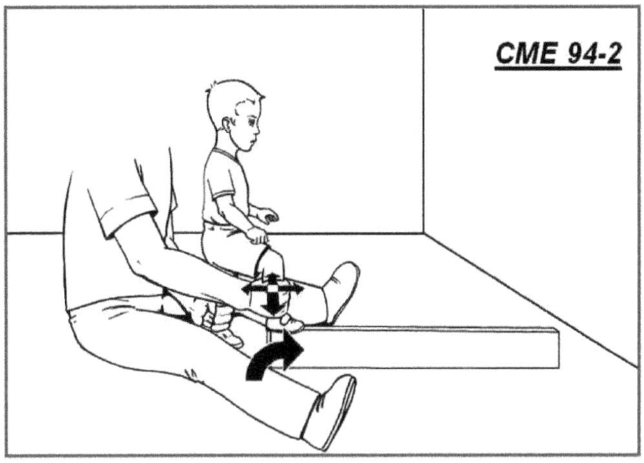

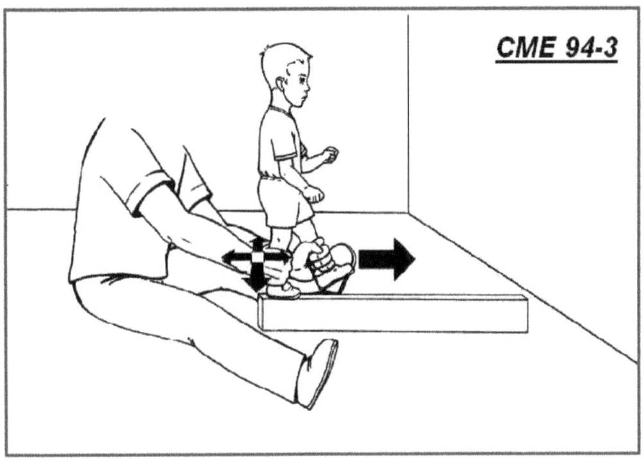

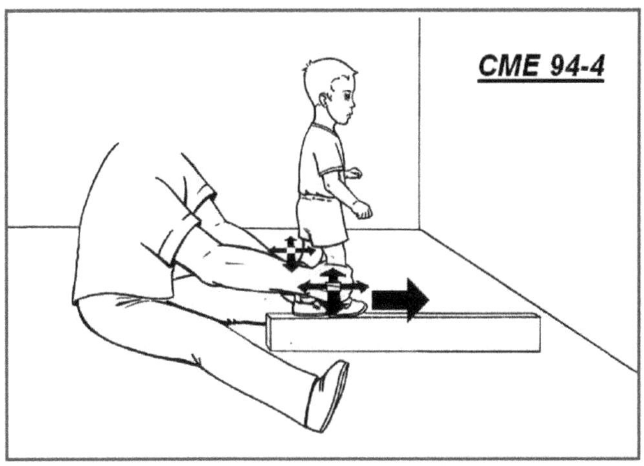

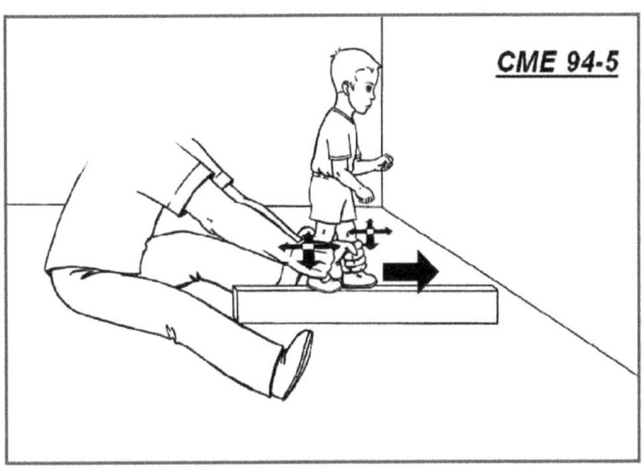

CME 95 ALTERNATE STEPS ON LADDER BY ANKLES

MAIN GOAL: To provoke stability in a standing position.

INITIAL POSITION: In a standing position in front of the ladder made of 2 boxes (fig.1).

HOLD: By both ankles as illustrated in figure 1.

MOVEMENT: Slowly move up the right foot on the first step as indicated in figure 2; subsequently move the left foot all the way to the second box as shown in figures 3 and 4. Complete the upward sequence placing the right foot on top of the boxes (fig. 5). Now, move the child down with left foot to first step (fig 6), and right foot to the floor (fig. 7) followed by the left foot on floor (fig. 8). Repeat the exercise as instructed by your CME therapist.

EXPECTED IMPROVEMENT: The child takes guided steps up and down the ladder with no difficulty.

CME 95 ALTERNATE STEPS ON LADDER BY ANKLES

CME 95-1

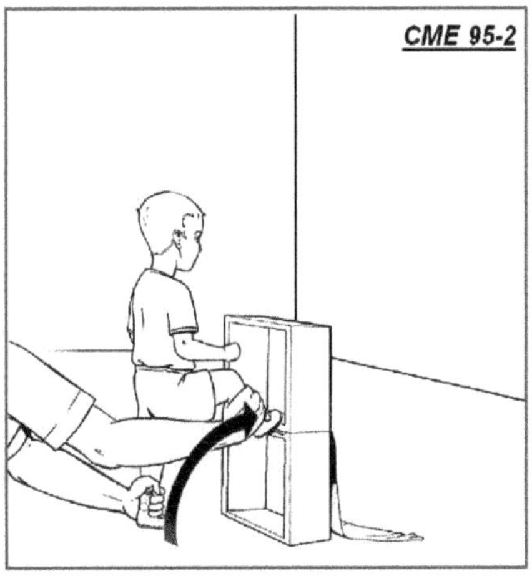
CME 95-2

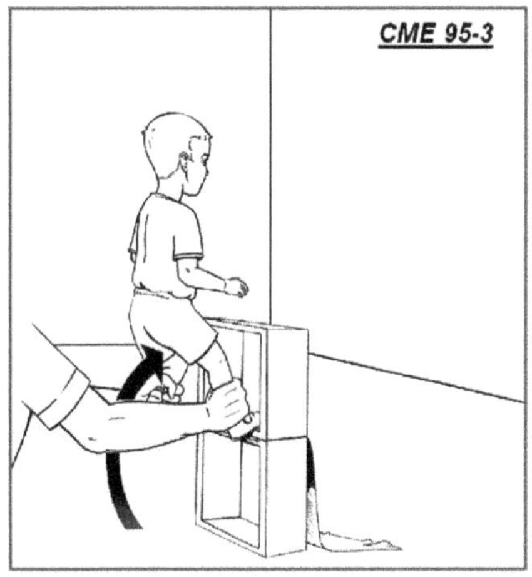
CME 95-3

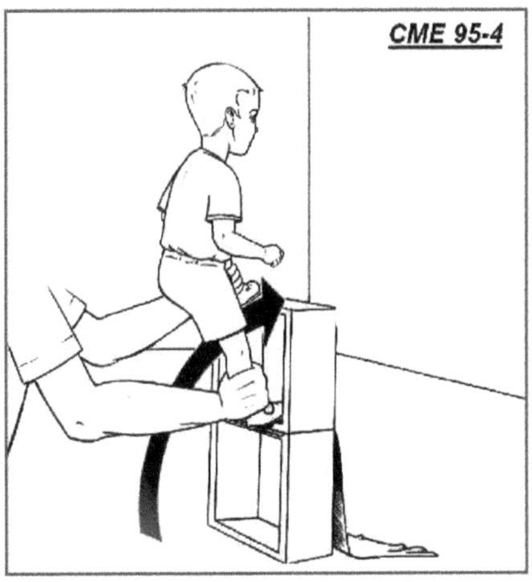
CME 95-4

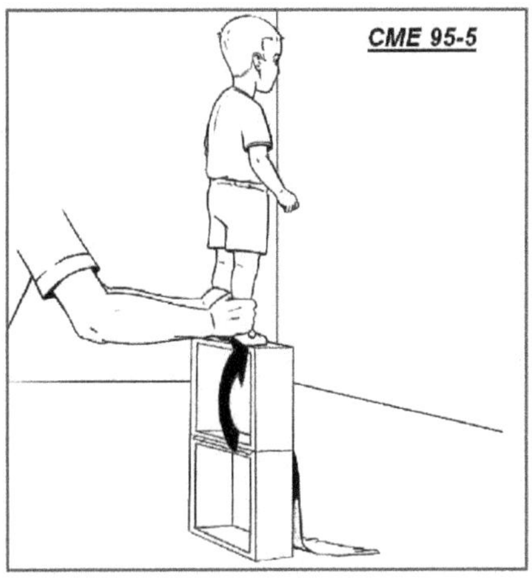
CME 95-5

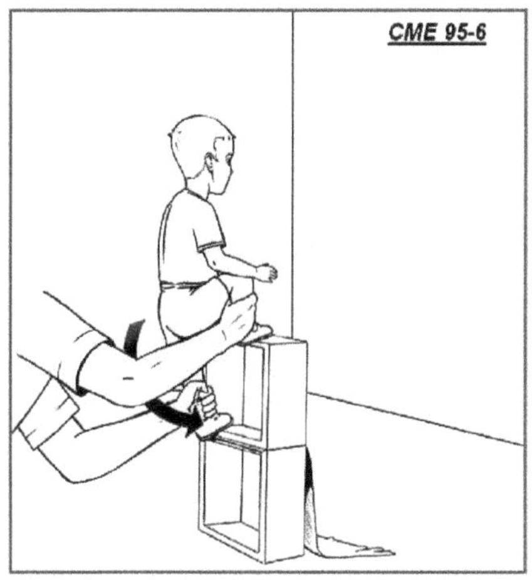
CME 95-6

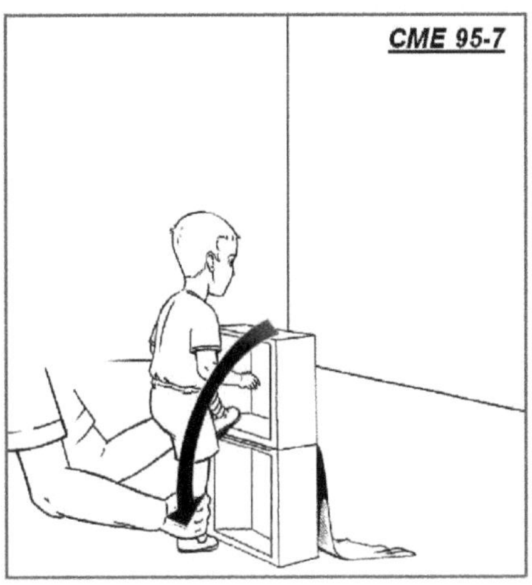
CME 95-7

CME 95-8

CME 96 MIMIC STEPS ON 30 CM. BOX BY ANKLES

MAIN GOAL: To provoke stability in a standing position.

INITIAL POSITION: In a standing position at the right side of the 30 centimeters box (fig. 1).

HOLD: With both hands by the ankles as illustrated in figure 1.

MOVEMENT: Slowly move the left leg up to the height of the box without touching the box as indicated in figure 2. Hold this position firmly by taking all the weight onto your left hand (fig. 2) and then move up the right leg to the invisible box as illustrated in figure 3. Stay in this position for 5 seconds and step down the left and right leg (fig. 4 and 5). Repeat the exercise as instructed by your CME therapist.

EXPECTED IMPROVEMENT: The child takes the mimic steps up and down without difficulty.

CME 96 MIMIC STEPS ON 30 CM. BOX BY ANKLES

CME 97 STANDING ON A DYNAMIC SLOPE 30 CM. BOXES BY ANKLES

MAIN GOAL: To enhance stability and balance reactions in a standing position

INITIAL POSITION: In a standing position on top of the boxes.

HOLD: By both ankles as illustrated in figure 1.

MOVEMENT: Slowly open the left leg sideways to tilt the box around 10 cm. from the middle line as illustrated in figure 2, then return to the center and move the right leg sideways (fig. 3) and come back to the center. Now carefully tilt forward the 2 boxes together (fig. 4), moving back to the center and tilting the boxes backward as shown in figure 5. Move to the center again and now do a reciprocal tilt with the right leg forward and the left leg backwards (fig. 6), taking center and repeating reciprocally with left leg forward and right leg backwards (figure 7). Repeat the sequence as instructed by your CME therapist.

EXPECTED IMPROVEMENT: The child stays standing with full stability during the sloped movements.

CME 97 STANDING ON DYNAMIC SLOPE 30 CM. BOXES BY ANKLES

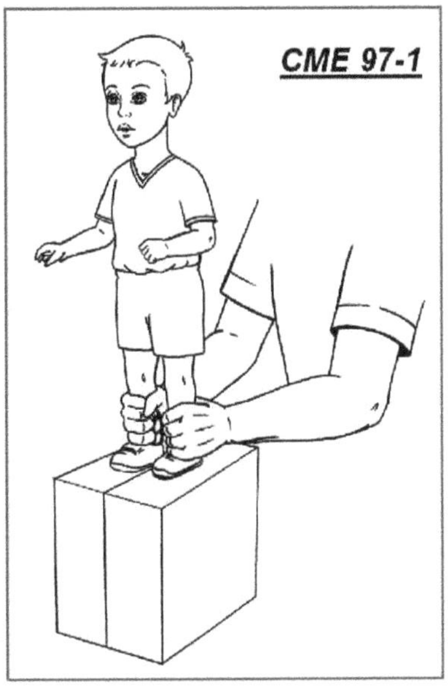

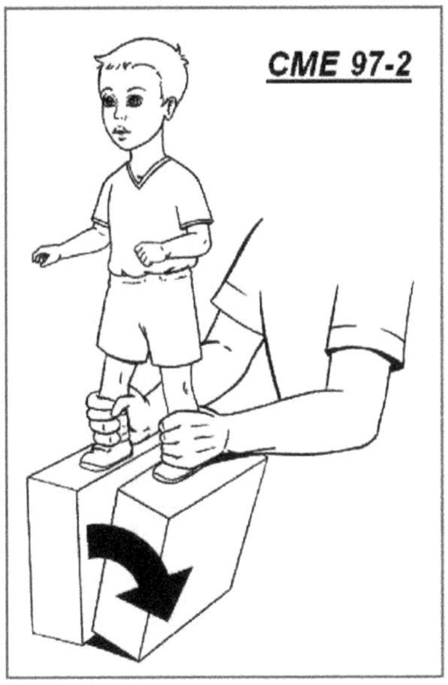

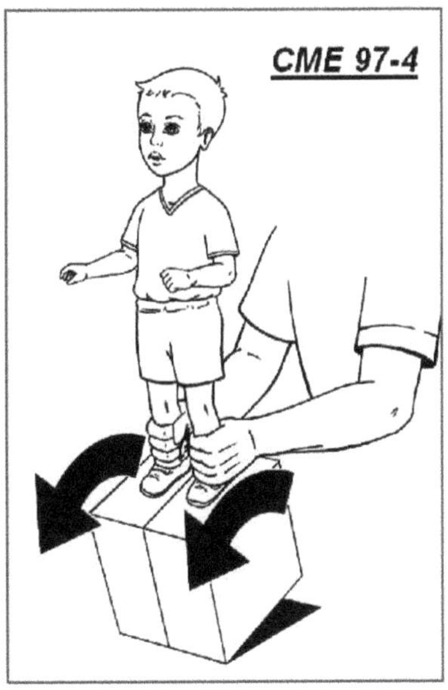

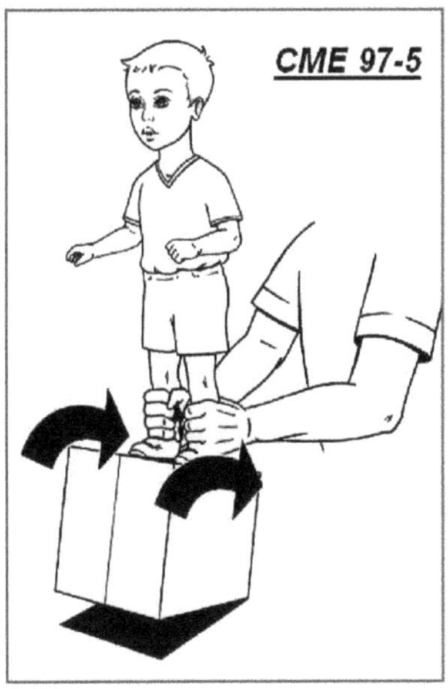

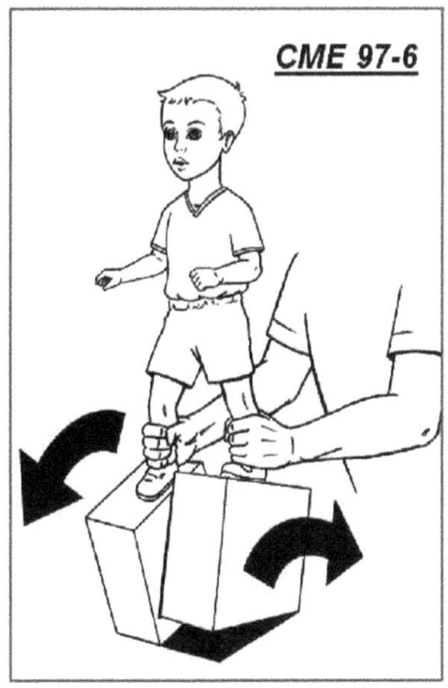

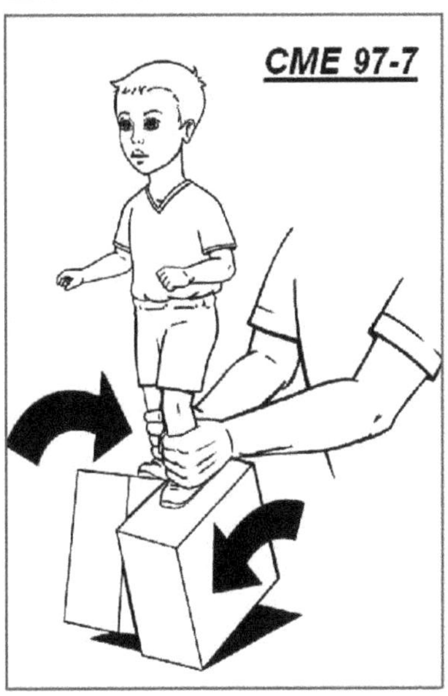

CME 98 STEPS ON UP-SLOPE BOXES BY ANKLES

Exercise Configuration: *Four boxes placed as illustrated in figure 1.*

MAIN GOAL: To provoke balance reactions and reciprocal coordination and stability.

INITIAL POSITION: In a standing position on the floor in front of the slope up the boxes.

HOLD: By both hands by the ankles.

MOVEMENT: Place the right foot on the first box (fig. 2), and then the other leg as illustrated in figure 3. Slowly place the child down on the second box (figs. 4 and 5), and proceed with the sequence until reaching the last box as illustrated in figures 6, 7, 8 and 9. Repeat the exercise as instructed by your CME therapist.

EXPECTED IMPROVEMENT: The child takes guided steps easily all along the up-slope boxes.

CME 98 STEPS ON UP-SLOPE BOXES BY ANKLES

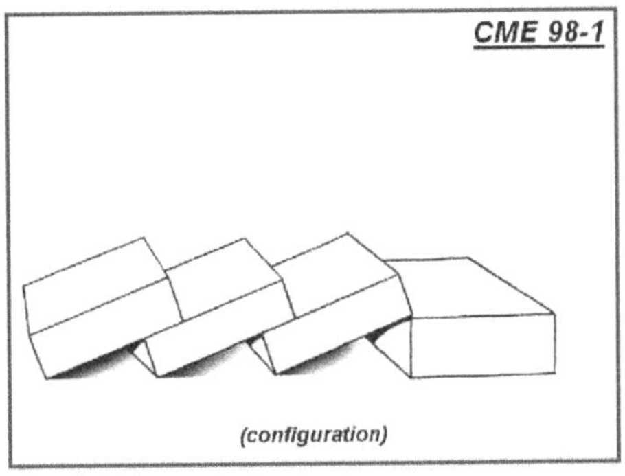

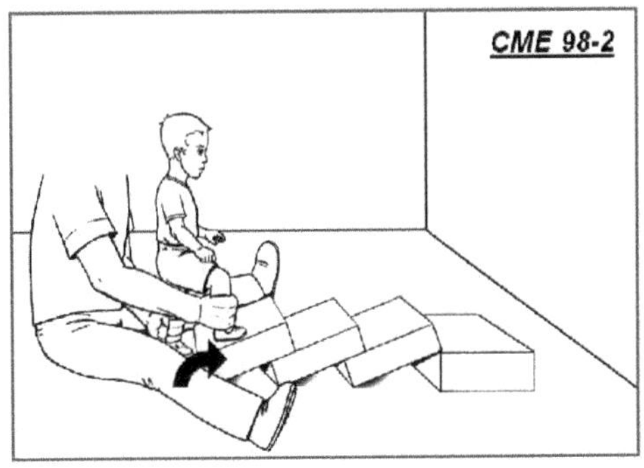

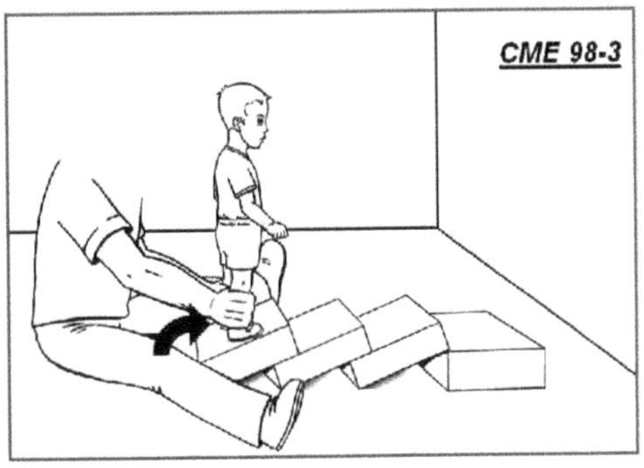

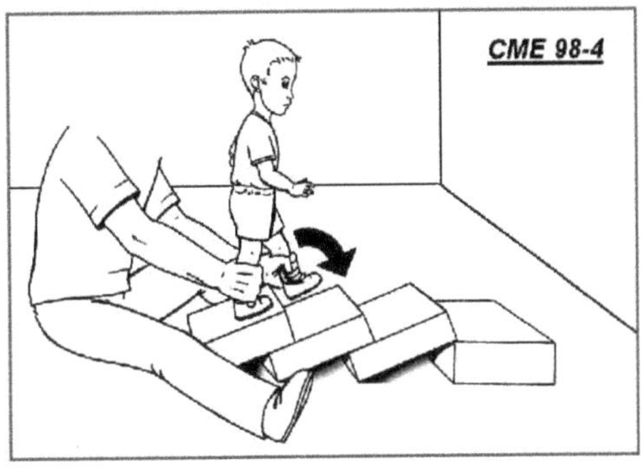

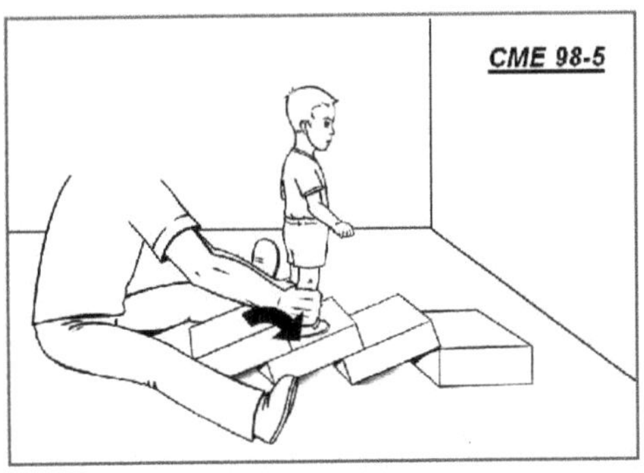

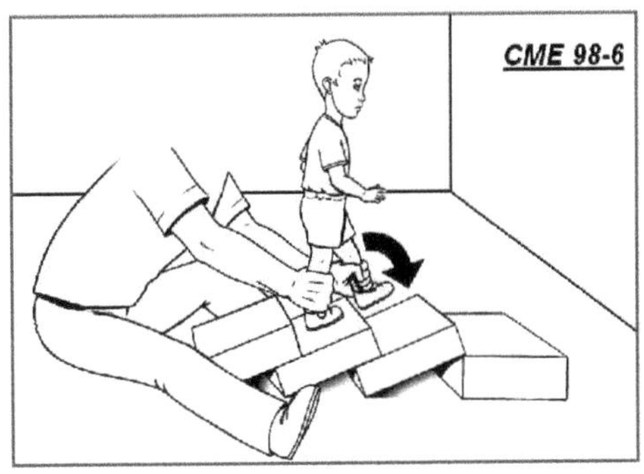

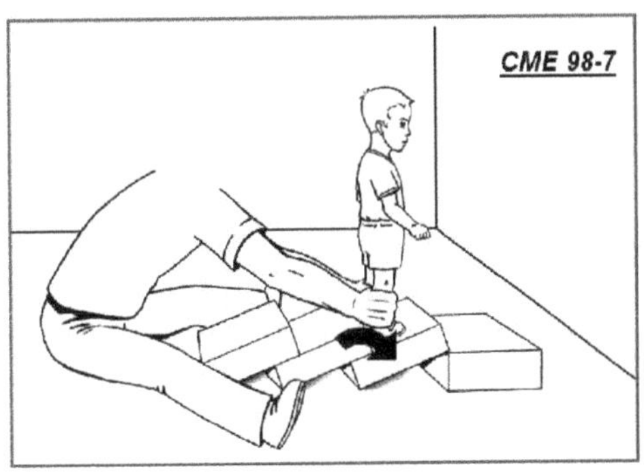

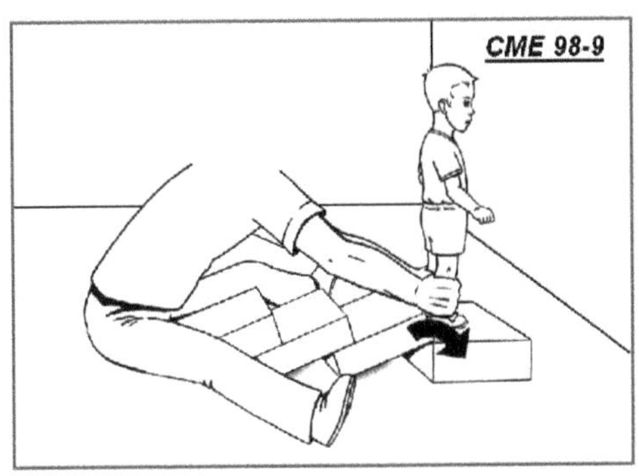

CME 99 SIDEWAY STEPS ON SEE-SAW BALANCE BAR BY ONE ANKLE

Exercise Configuration: *One cube covered by a grip linner and one balance bar on the top as illustrated in figures 1 and 2.*

MAIN GOAL: To provoke balance reactions.

INITIAL POSITION: In a standing position at center of the balance bar.

HOLD: By both hands by his/her ankles as illustrated in figure 3.

MOVEMENT: Release the left ankle as indicated in figure 4, and slowly slide the right foot sideways to the right as indicated in figure 5. Hold the left ankle (fig. 6), releasing the right ankle (fig. 7), and take a step sideways with the left foot as illustrated in figure 8. Repeat the sequence until reaching the end of the see-saw balance bar. Repeat the exercise as instructed by your CME therapist.

EXPECTED IMPROVEMENT: The child is able to take guided steps easily all along the see-saw balance bar.

CME 99 SIDEWAY STEPS ON SEE-SAW BALANCE BAR BY ONE ANKLE

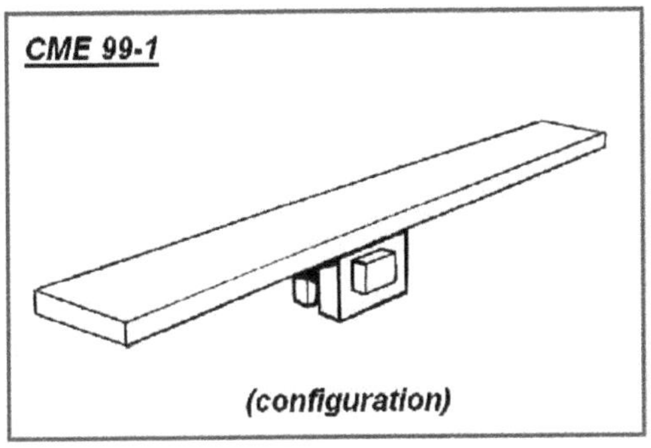

CME 99-1 (configuration)

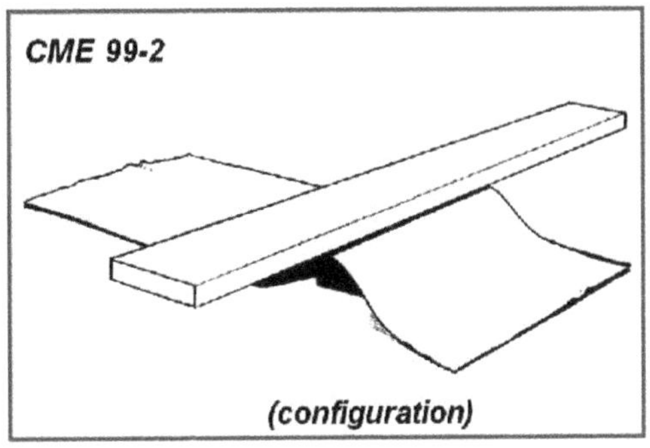

CME 99-2 (configuration)

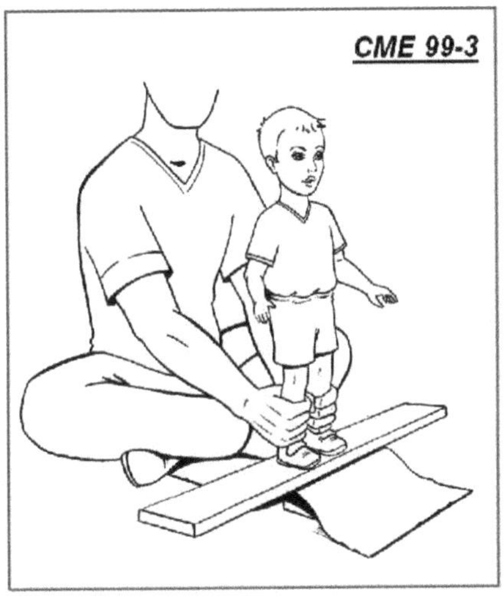

CME 99-3

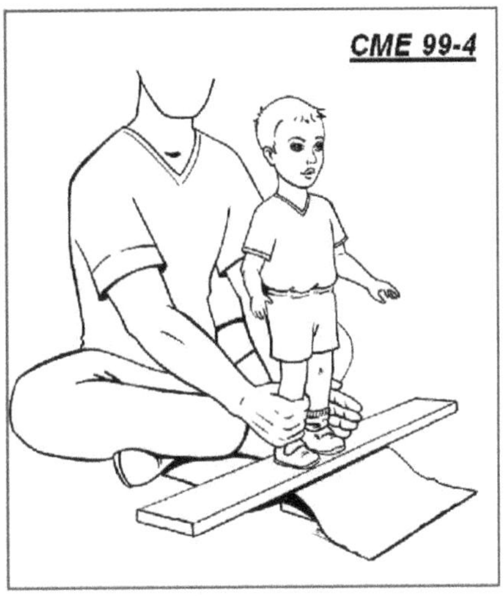

CME 99-4

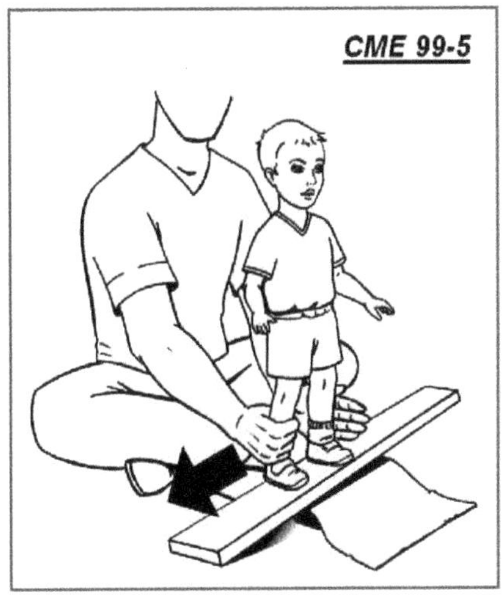

CME 99-5

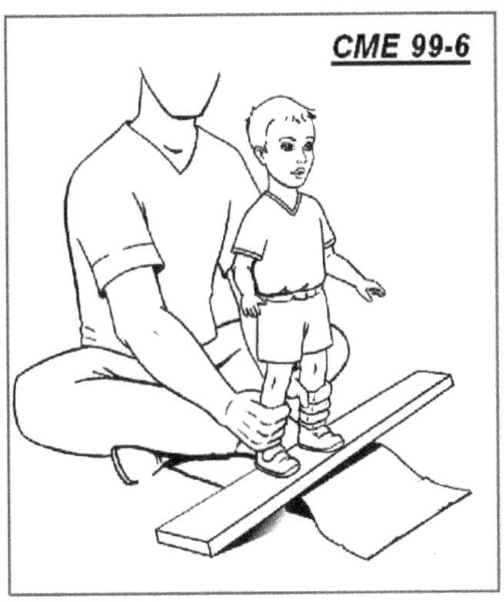

CME 99-6

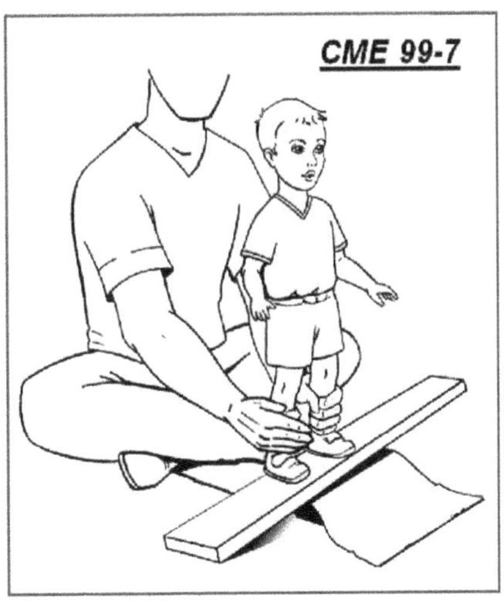

CME 99-7

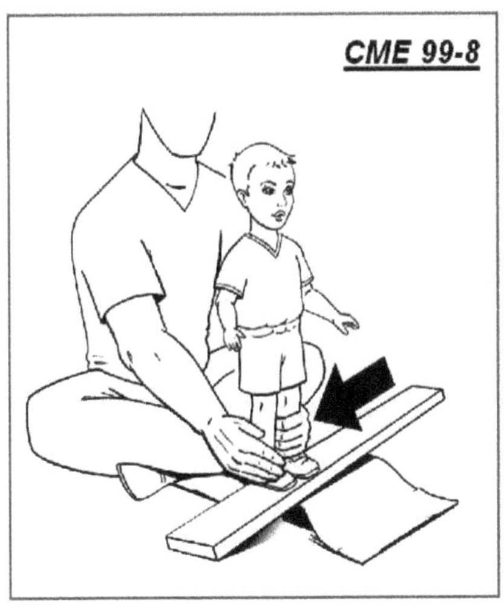

CME 99-8

CME THERAPY CASES

In this section of the manual a few examples were chosen among the successful cases treated with CME therapy by Ramon along 35 years of continuous practice.

CASE ONE: RACHEL

To keep privacy, the real name of the child was changed, and her personal information will not be mentioned, only the motor level at the beginning and at the end of the 4 days of CME therapy. Rachel was 5 years old when she came to CME therapy for the first time in July 1996 by the recommendation of other satisfied parents. Her diagnosis was developmental delay with motor and mental impairment detected in the post natal period. She began conventional stimulation immediately after diagnosis and the physical therapy intervention when Rachel was younger than 1 year old.

The CME initial assessment results showed a functional motor level equivalent to 10/11 months of a standard motor development performance. Her head and trunk control were appropriate and the highest motor function was "***walking supported by 2 hands***". Rachel was unable to maintain a free standing posture, neither was she able to initiate free steps, furthermore, she was taking the maximum of standing control from the hands of the parents and holding her trunk inclined forward. Her parents brought her for an 8 session program and the following inserted pictures illustrate Rachel's evolution during the 4 days of CME therapy.

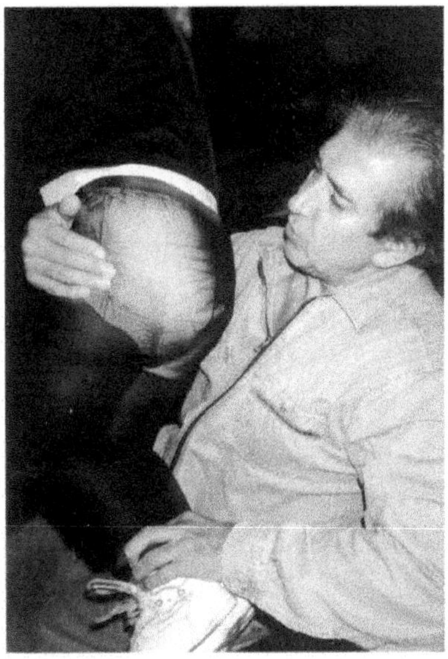

In this exercise Ramon is trying to place Rachel in a standing position by the ankles and hips, with no hand support. Appreciate the negative reaction of the girl by keeping the knees bent and making toes support and inclining the trunk forward to the right side. This is the typical reaction of a child having no previous experience in a standing position with out hand support. Remember that she was in motor therapy 4 years ago, without succeeding to control independent standing.

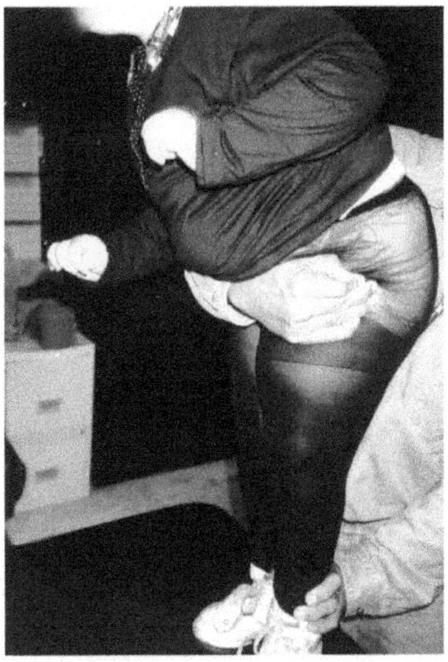

*(Ramon quote: I do remember that during the first session of therapy Rachel was very irritated and was heavily scratching my forearms, crying non stop, and making strong random movements of trunk and legs to avoid standing. Furthermore, after finishing the session the parents made this comment to another person: "**we are scared that Ramon will send us back home because our daughter is fighting him so much**").*

In this figure we can now appreciate the feet flat on the table and the knees straighter, but still Rachel is trying to avoid standing by making trunk flexion, which is contained by Ramon with right hand pushing the pelvis against his chest very strongly.

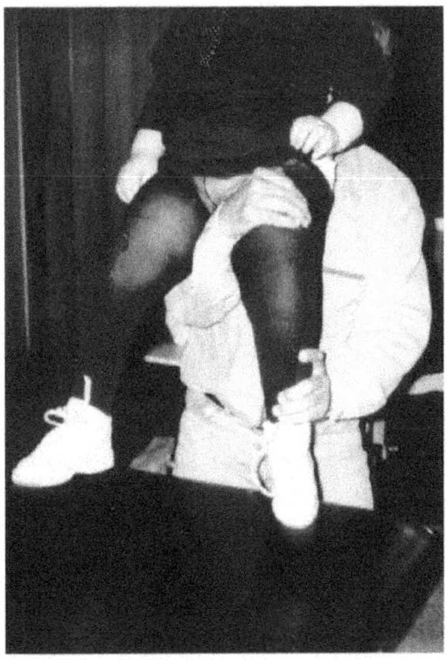

This figure illustrates how Ramon decreases the amount of support by only holding the left leg but taking significant amounts of weight on the chest and right shoulder because she is still trying to avoid the standing position by bending the knees.

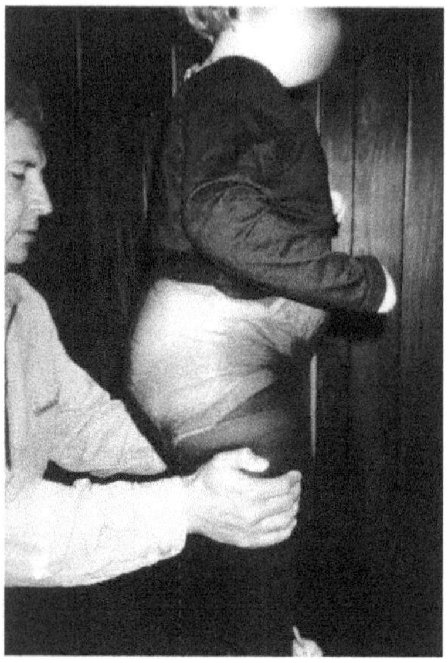

By end of first treatment session, Ramon succeeded to place the girl in a standing position by holding her by above the knees.

FOR THE FIRST TIME IN HER LIFE RACHEL CAN EXPERIMENT THE SENSATION OF STANDING WITHOUT HAND SUPPORT!!

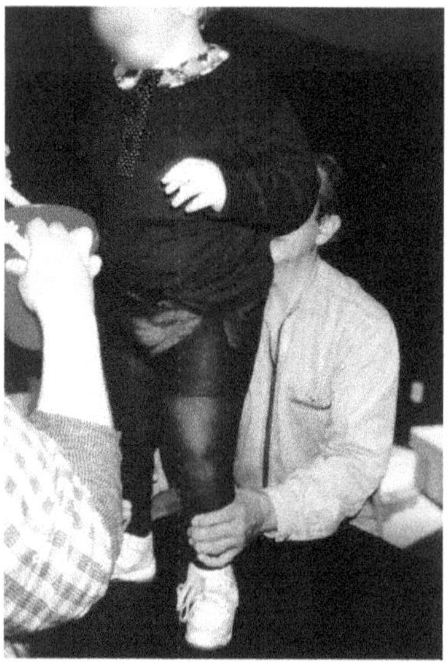

By end of the second session of CME therapy Rachel was able to tolerate the standing position by distal support at the level of the ankles.

Accordingly to CME characteristics, the child needs to be exposed to the influence of the force of gravity using the most distal grasp as possible, in order to provoke the appearance of absent motor functions and no cooperation, neither motivation are needed to provoke the automatic postural responses.

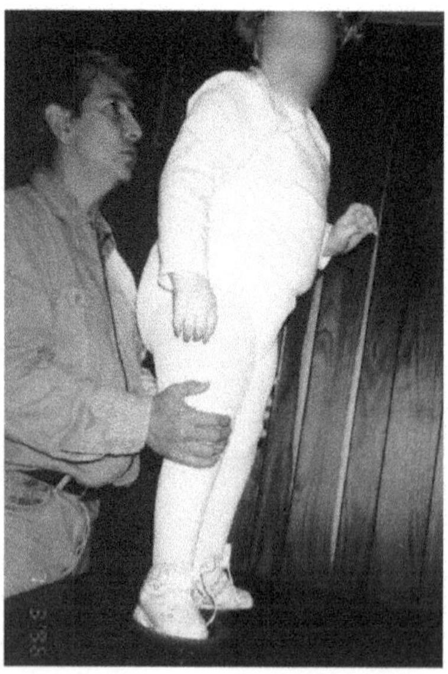

Second day of therapy sessions, Rachel is kept from the beginning in the standing position with distal support with normal acceptance and no more fighting.

At this point the therapy on table is finished and the child continues CME therapy on the floor.

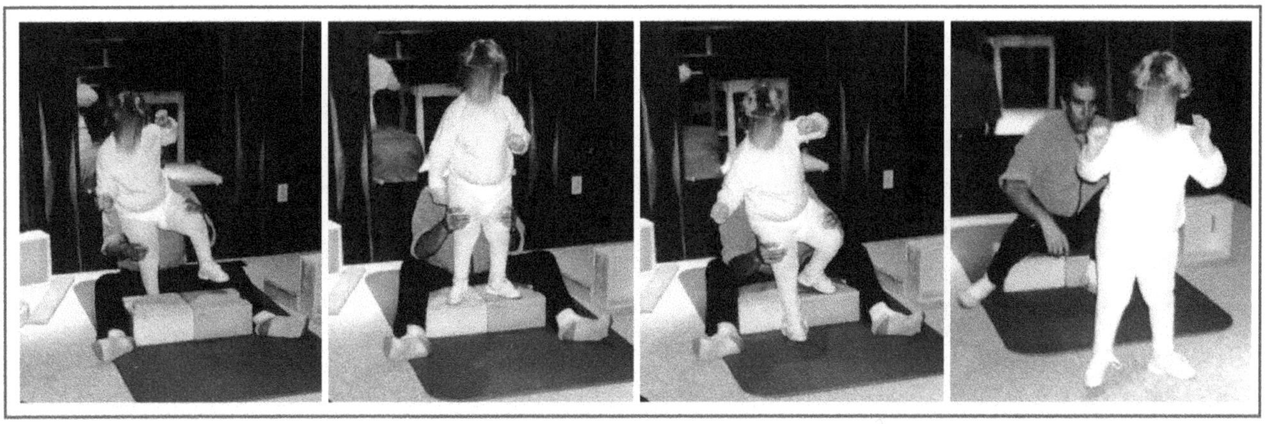

By the second day the afternoon session was dedicated to working balance reactions with the CME set of boxes, and the target was the free standing with beginning of spontaneous free steps. Rachel is no longer crying.

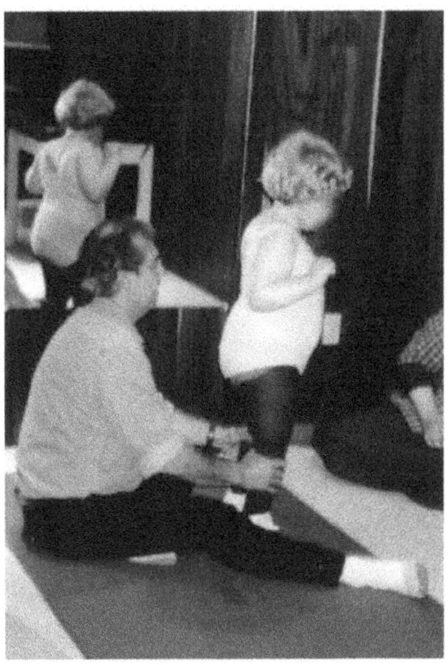

The third day of treatment was focused on preparing Rachel for the free steps experience, initiating steps by ankles on the floor.

Rachel begins to accept the challenge of walking.
With her arms in position in the typical first steps attitude.
At five years old!!

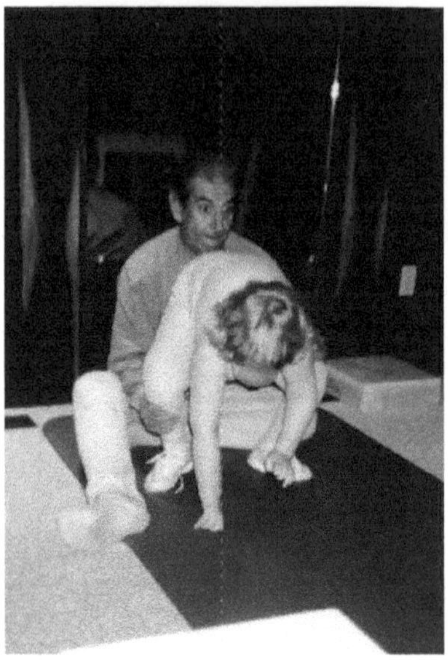

The first therapy session of the last day.
The goal was to provoke self standing from the floor, providing support by below the right knee and the left thigh.

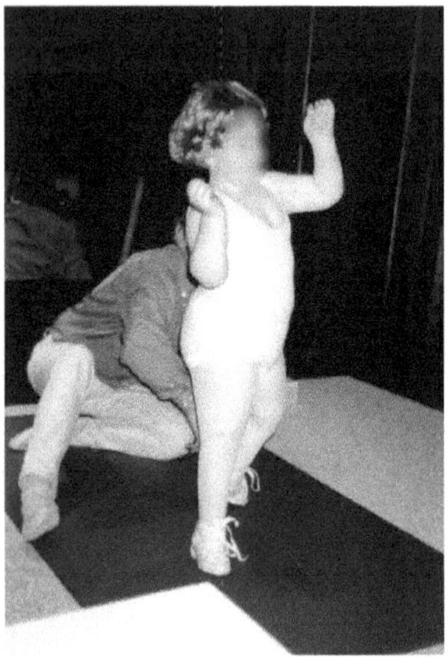

Rachel is completing a standing response and because her stability and balance control are "connected", she begins to take steps right away.

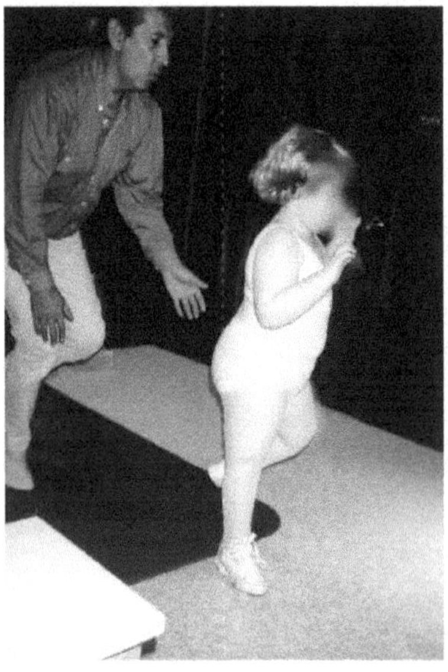

This picture was taken by Rachel's father 4 days after beginning to receive CME therapy.

She came to CME when she was 5 years old unable to do free standing or steps, and she went back home, being motor independent after 8 sessions of CME therapy!!

This is the result of giving efficacious stimulation to a powerful but neglected motor recovery potential.

This type of quick motor progress is not the norm in to the CME therapy interventions, furthermore, to obtain positive results constant effort from the families and therapist is necessary and perseverance with no dismay to stimulate the child until achieving the wanted motor response under only one condition: the brain must assimilate and organize the postural-functional motor information. The results of CME therapy in Rachel's case was exactly the aim of the parents, and a very important issue for Rachel's family because the achievement of independent walking was the missing part that conventional motor therapy did not achieve in four years of regular intervention.

There are many cases like Rachel in which the CME therapy intervention by Ramon or by other CME practitioners, was the turning point in the life of many children and their families, maybe taking more time to achieve the positive motor responses, but marking an objective advantage when compared to the former motor therapy that those children were receiving.

CASE 2: GUITI

The following pictures illustrate the case of Guiti P. who was born premature from a twin pregnancy, with low weight (less than 4 pounds), microcephaly and other difficult conditions; her twin sister enjoyed total indemnity. Muscle tone was abnormal with combination of spastic conditions plus functional dyskinesia with athetosis and ataxic components which affected more the right side of the body. She began CME therapy when she was two and a half months old in 1990, in Caracas, Venezuela, and Guiti's mother took pictures on April 1991, when Guiti was 14 months old. The photos illustrate a portion of CME therapy session focused on the provocation of Guitis's motor recovery potential concentrating the effort on the standing position and the steps.

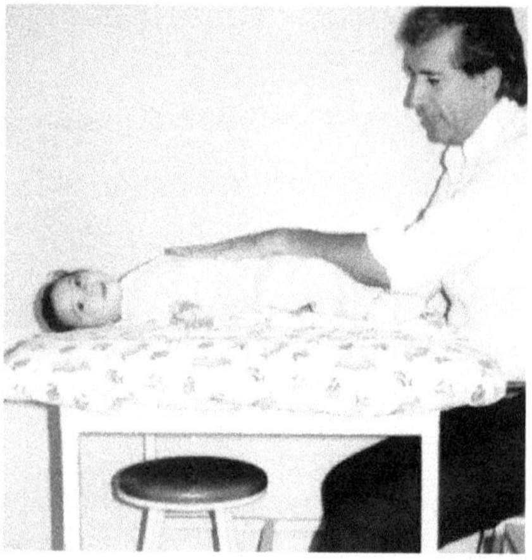

This picture illustrates the preparation of the trunk control exercise called Low Sandwich (because of the position of the therapist hands on the child).

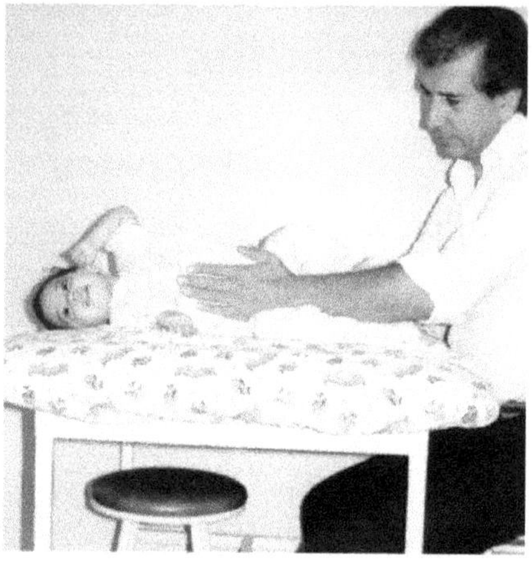

At this moment Guiti was placed sideways over on her right side, and the therapist maintained the position and waited for the child spontaneous and automatic sitting reaction.

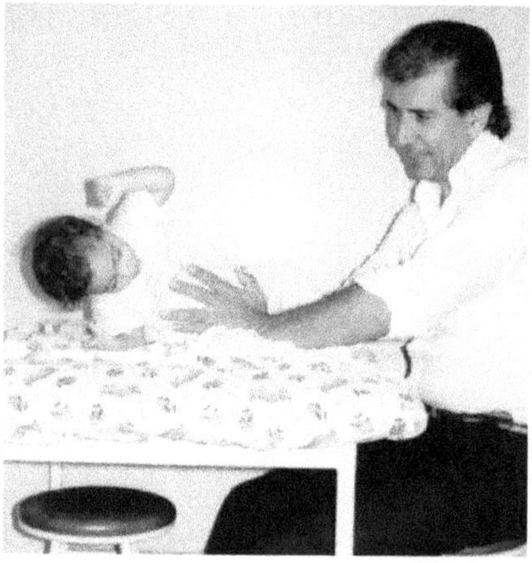

Guiti began by lifting her head from the table and trying to move the trunk upwards by the main command of the automatic reaction, provoked by the steady side ways position under the therapist's control.
The Moro pattern of the arms is noticeable, as a collateral reaction to the sitting effort.

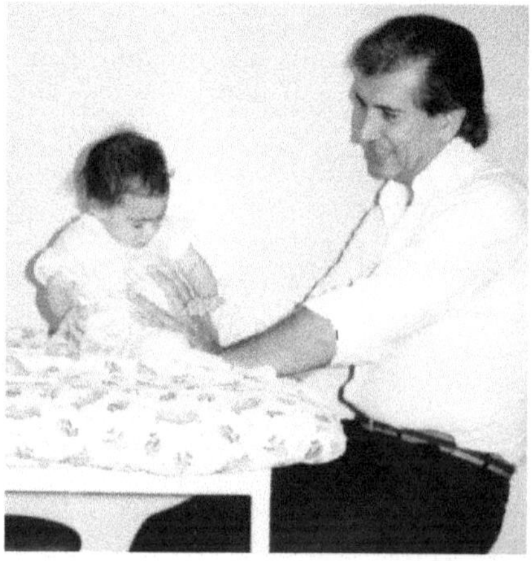

Final position of the exercise when Guiti reached the vertical posture of trunk.
We can notice that the arms are more relaxed and lack of head control remains.

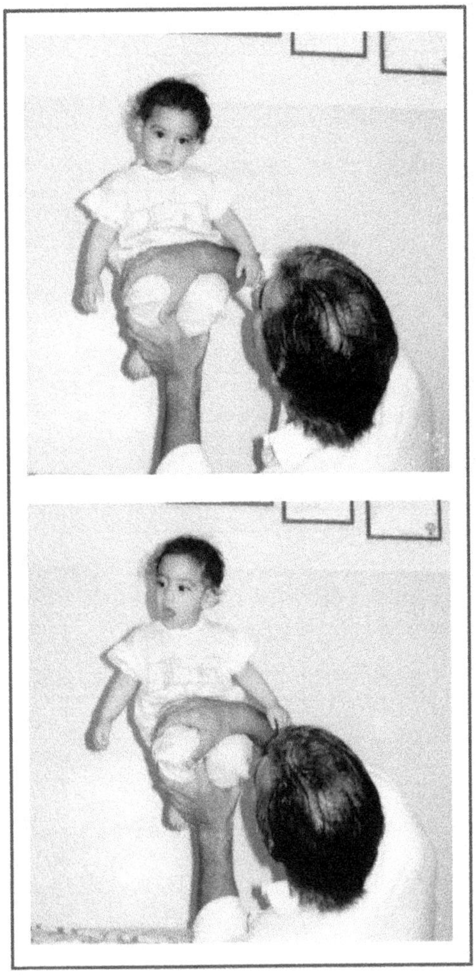

The pictures show Guiti in the final position of the exercise called Aerial Sitting, conceived to provoke automatic vertical trunk control response with no direct support on the trunk, as a consequence the "brain" is compelled to respond to the challenge of the free vertical experience.

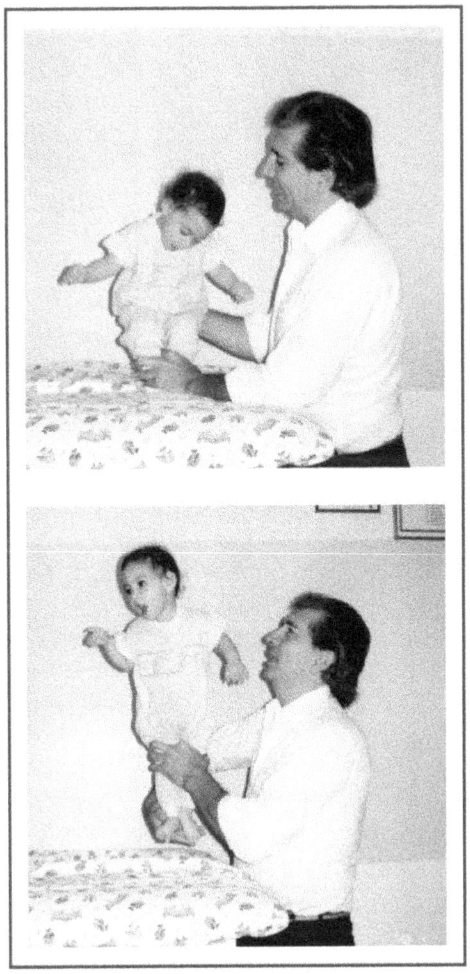

Here is illustrated the beginning of the Standing on palms exercise, stimulating the standing reaction with left hands on ankles and right hand under buttocks.

At this moment Guiti went up to a standing position, and the therapist placed the right hand under her soles with the girl suspended air borne, and still controlling the standing stability with left hand above the knees.

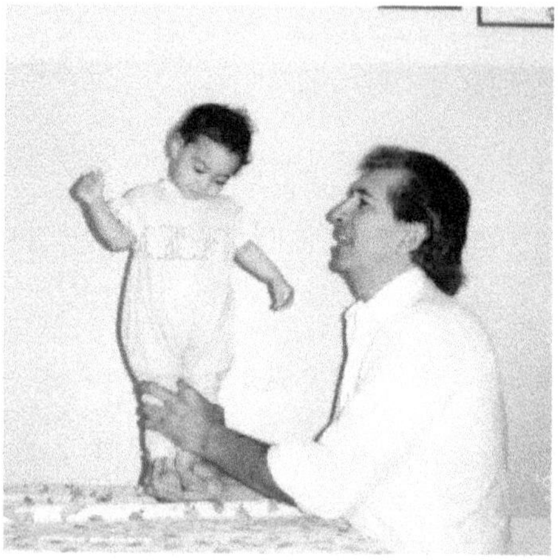

Preparing to move his left hand from the knees level to under feet.

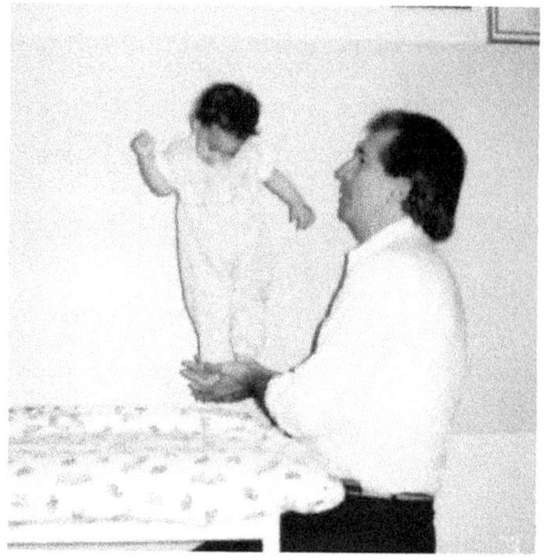

The therapist placed the left hand under the right one, and Guiti is in total control of the stability in a free standing position, (with collateral TNR), the therapist only needs to provide balance.

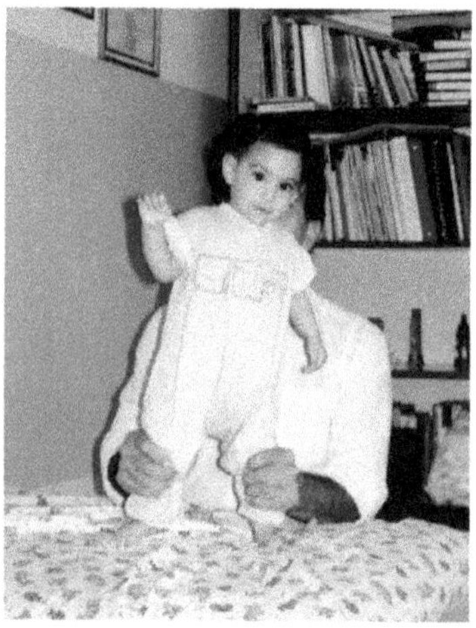

In this picture the therapist is preparing Guiti for the experience of walking supported by below knees. So, the child needs to respond with automatic control of the standing posture, shifting weight to left side in order to take the step with her right leg.

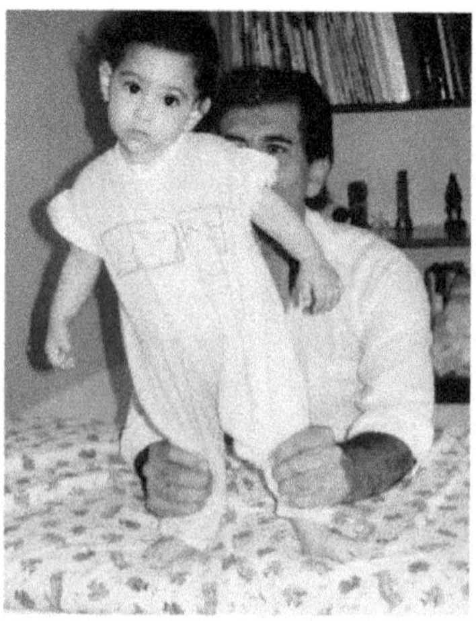

Almost at the end of the successful sequence of steps reaching the edge of the table. Her arms are totally relaxed (no more TNR) and the head is still inclined to the left side.These two pictures illustrated clearly two CME intrinsic characteristics:

- *Provoke absent motor functions.*
- *Expose the child to the force of gravity using distal support.*

Guiti is currently totally independent.

CASE 3: TEYA

She began CME therapy at the age of four and a half years old (in 2002), with the diagnosis of congenital cerebellar hypoplasia and deafness. To understand better Teya's case we will briefly explain the cerebellum characteristics, information adapted from the NINDS web page:

"The cerebellum is composed by two large hemispheres, and is located in the inferior posterior portion of the head, directly dorsal to the brainstem and pons, inferior to the occipital lobe. Cerebellum who means "little brain" in Latin is a region of the brain that plays an important role in the integration of sensory perception and motor output. Many neural pathways link the cerebellum with the motor cortex which sends information to the muscles causing them to move, and the spinocerebellar tract which provides feedback on the position of the body in the space or proprioception. The cerebellum integrates these two functions, using the constant feedback on body position to fine-tune motor functions. Because of this "updating" function of the cerebellum, lesions within it are not so debilitating as to cause paralysis, but rather present as feedback deficits resulting in disorders in fine movement, equilibrium, posture, and motor learning".

Teya was under the health care system in her country since a very early age, receiving all the necessary services including conventional physical therapy, and during the previous months before the beginning CME therapy she was receiving training to use a walker frame, furthermore the mother was advised to apply for a wheel chair, because none of the professionals involved in Teya's care considered her to be a candidate for motor independence, probably due to the severe diagnosis and poor aging-developmental-evolution. But the mother had the courage to disregard standard motor therapy and to find CME therapy, and decided to make a big effort to bring her daughter to Santiago. Chile, to take intensive CME therapy with Ramon Cuevas.

In the initial assessment of Teya in August 2002, she stopped getting 3 or complete response in item "***28: Standing posture by hands on wall***". She got a "1 or minimal response"; because she was afraid to keep standing position only by hands on the wall. This item needed the basic balance control response of keeping the center of gravity inside the base of support without the "***grasping***" compensation, which is the first immature reaction that motor delayed children do in order to keep standing posture and taking steps. Teya was never before placed in a standing position without grasping a solid support. In item "32: Steps by one hand-arm horizontal" she got a zero and then also a zero in item "34: Standing posture free". Her final motor level from assessment was equivalent of 10 month of age, being 4 ½ years old.

Luckily, since the first set of intensive therapy sessions Teya showed good tolerance to the new therapy responding very well to the exercises, but due to her size and total lack of control of balanced reactions, it was very hard for the mother and for the therapist to provoke functional responses holding such a tall girl by the ankles. However, it was a worthy effort because slowly but steadily Teya never stopped progressing since the beginning of CME therapy, and currently she can stand up from the floor by herself, walk freely on the street almost an unlimited distance turning to any necessary direction with functional balance control and step up and down small stair cases. Teya is still dealing with severe cerebellum atrophy and deafness and is not completely independent yet, but considering the way in which she manages strolling around after the last set of intensive therapy (March 2006), the option to become totally motor independent despite the diagnosis is increasing every single day.

Teya at the 2 years old, with her mother. She was receiving standard physiotherapy at that time.

Teya's mother quote about this picture: "the one of Teya sitting on the steps is the way her posture was until about 3 and a half years old. In that picture I was sitting beside her, I couldn't leave her or she'd topple over.

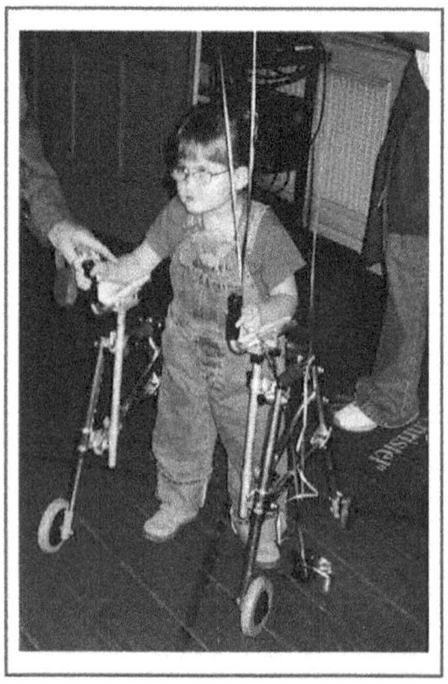

Teya standing with her adapted walker when she was 4 years old, before beginning CME therapy program.

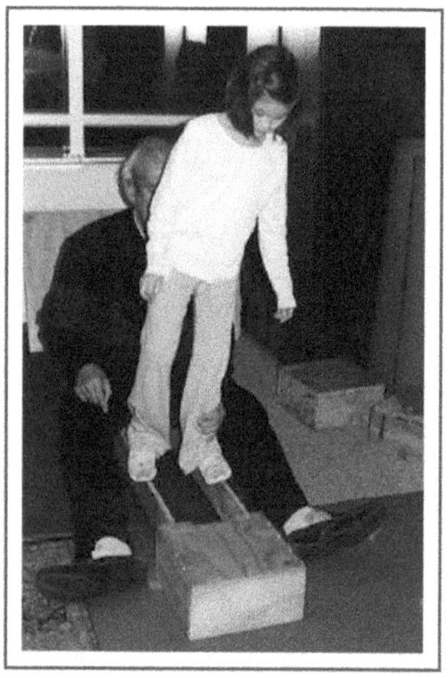

This picture shows Teya walking on a narrow parallel balance bars supported by stance leg at ankles level, in 2005.

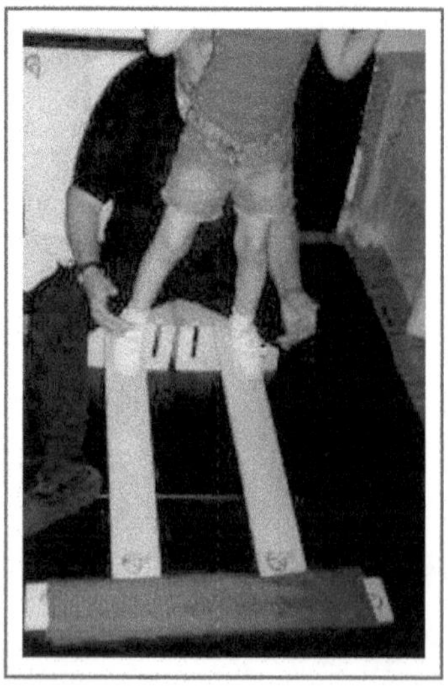

This picture from 2004, illustrates Teya while taking free steps on down sloped parallel balance bars, induced by light touching on each foot.

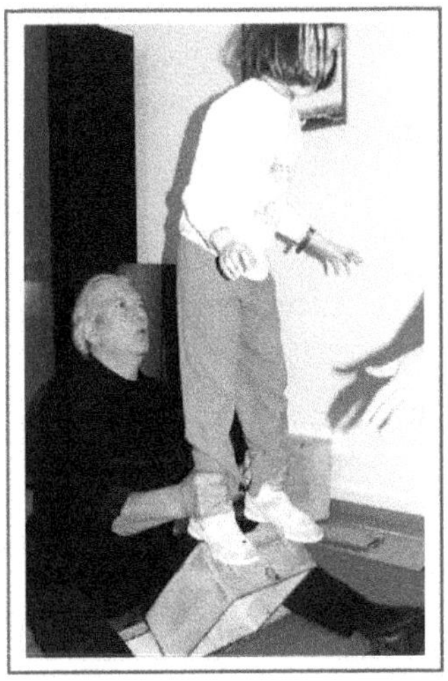

In this picture from May 2005, Teya is doing the exercises "Collapse on 30 centimeters box", which is very demanding in balance reactions. The mother is in front to prevent any risky situation.

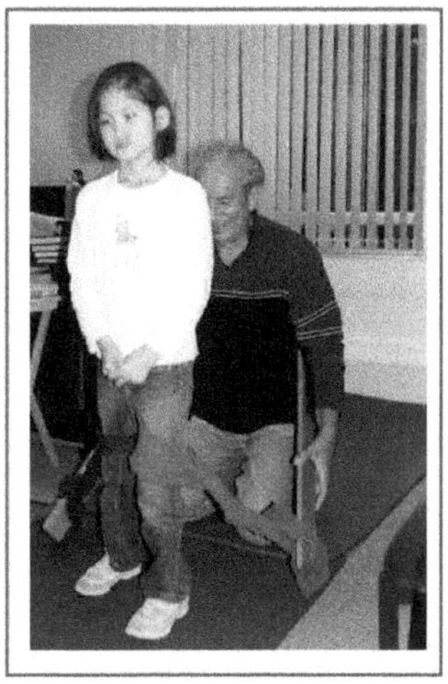

This photo illustrates the exercise called "Steps by straps above knees, remote control", May 2005.

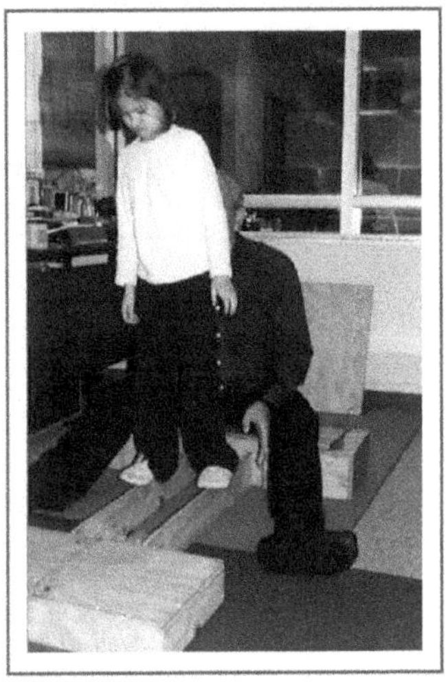

The picture shows Teya taking free steps on the narrow parallel balance bars in May 2005.

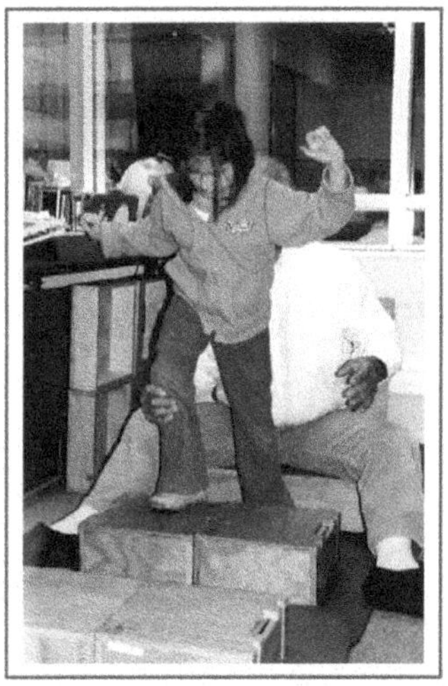

This picture from March 2006, shows Teya taking steps upon 15 centimeters boxes by support on the stance leg below the knee. After a few days of practice she was able to begin taking the step on top of 15 cm. box up without below knee support with a lot of effort and also she succeeded to step down the box freely. Nevertheless this was a very new function, which she would I need a lot of practice to achieve functional control.

The upper sequence of pictures were taken in Toronto at the beginning of April 2006, four weeks after she returned from Chile, from last intensive CME therapy program, shows Teya walking totally independent on the side walk near her home. This is the best reward for the perseverant mother who did not believe that her daughter's fate was to own the best wheel chair that the health system was willing to provide.

CASE 4: MACKAYLA

Mackayla was born in April 2001, with severe perinatal hypoxia as a consequence of quadriplegic spastic-athetoid cerebral palsy. She received conventional therapy treatment from birth until mid 2004 when she began to receive CME therapy with PT Catherine Meehan certified CME II practitioner, (who successfully passed examination for CME III certification in March 2006, in Santiago, Chile). From the beginning the parents noticed small but significant improvements in Mackayla motor functions as a result of the new therapy.

The father brought Mackayla to Chile in April 2006 for an intensive CME program of 7 days (21 sessions) with Ramon. At the initial assessment, she presented the characteristic signals of athetoid neck movements and posture as well in upper extremities. Mackayla experienced significant improvement in two basic functions during the 7 days of CME therapy; the first was trunk control reaction when sitting on the sofa with no trunk support (pictures below), function which was totally absent 4 days before.

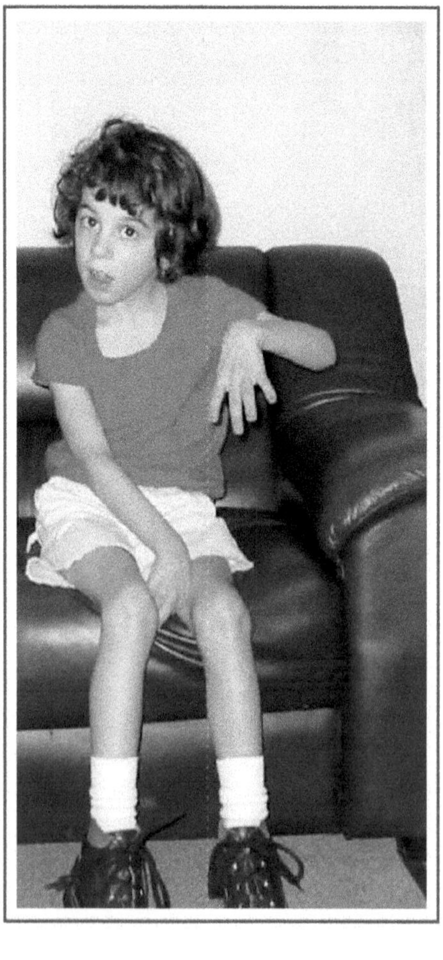

From that moment Mackayla become consistently able to control the trunk in an unsupported sitting posture for more than two minutes and she could also stay with the other hand down between the legs or without hands touching sofa.

The other wonderful improvement was the endurance to keep the standing position and take steps supported by the ankles. Mackayla only began to control the standing position by the ankles after the family incorporated CME therapy as a basic routine. The average amount of guided steps by the ankles, before the intensive therapy in Chile was 11 steps followed by collapsing sitting down and surprisingly, during the therapy session number 12 on the fourth day of the program Mackayla was able to take 246 consecutive steps guided by the ankles. She became excited and collapsed sitting 10 inches before touching the elevator door in front of Ramon's office; the father and the grandfather were witnesses of this amazing performance. This response to the exercise means that Mackayla has a great potential for postural control.

This picture illustrates the way in which the therapist is holding Mackayla in a standing position supported by the ankles. Here she is working in steps in/out of 15 cm box by ankles.

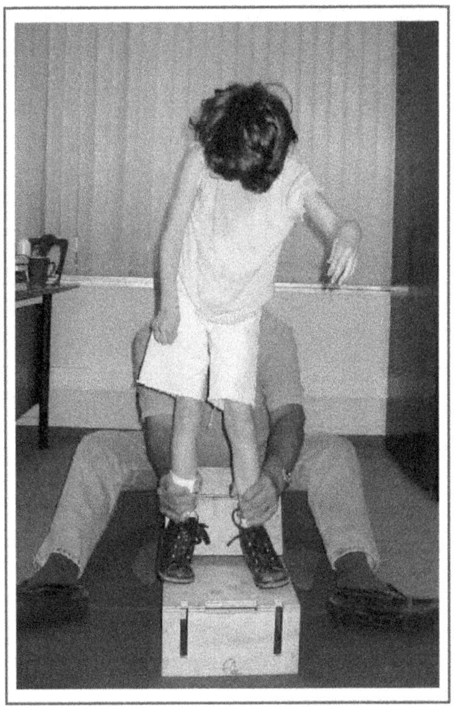

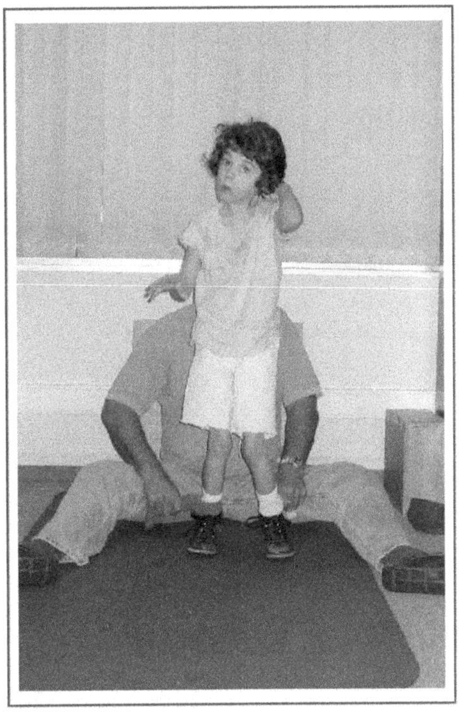

This picture shows a very challenging exercise called "Standing position by straps on ankles", in which Mackayla keeps the standing position by her self plus the effort of the therapist to maintain at all times the center of gravity on to the supporting base by only maneuvering on the straps placed around ankles.

CME RESULTS IN DOWN SYNDROME

The next graphic illustrates the results obtained with 16 Down syndrome children stimulated with CME therapy by Ramon in the Institute INVEDIN in Caracas, Venezuela, between 1973 and 1975, in which all of them began motor therapy from the ages of two to three months old. None of the children were affected by any heart correlated condition or any other complication, except for the Down syndrome clinic symptoms. The average frequency of the CME session was of 3 times per week (30 minutes per session) plus the home program exercises for the parents, recommended to be done once daily.

The collected data was done retrospectively in 1975, and it is not part of any scientific study to corroborate the efficacy of CME therapy, nevertheless it represents the objective and veritable results obtained in a small group of Down syndrome population treated with CME during three years. Not all of the children began the treatment at the same time, but all of them started therapy between January 1973 and March 1974.

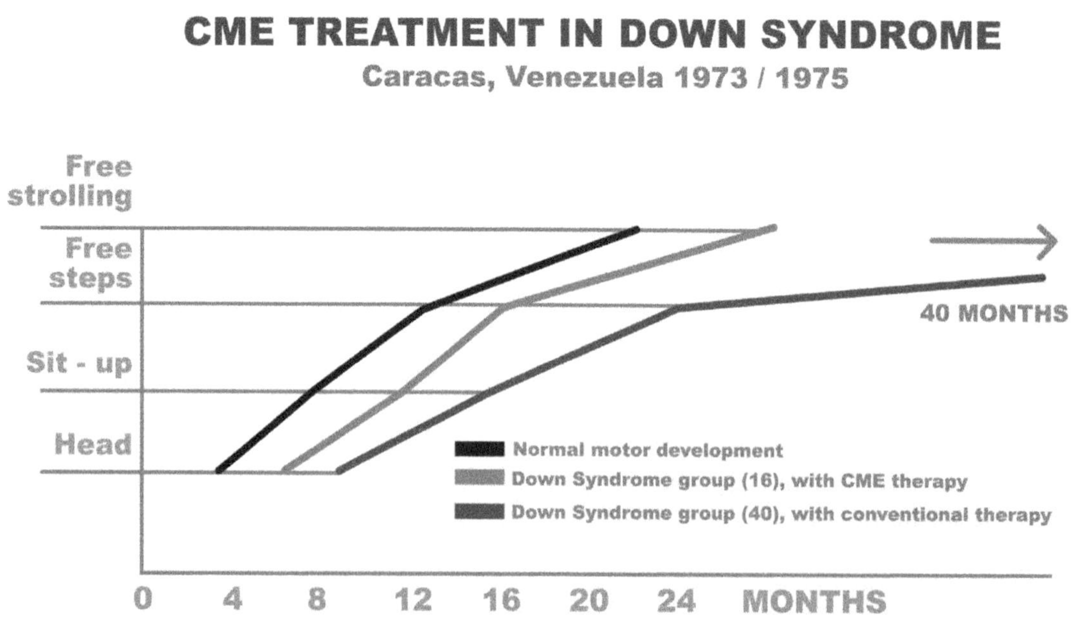

Figure 8.

The dark line shows the standard normal curve of motor evolution along the first 2 years of age, which is universally accepted and fully demonstrated by different authors of developmental motor scales. The blue line illustrates the average motor evolution of a population of 40 Down syndrome children, which were treated with the conventional motor therapy approaches but not under the early stimulation concept. This information for obtaining the average curve was collected from the files of the 40 children assisting to the special school of the Institute and all of them were older than 6.

The gray curve is the average motor evolution of the Down syndrome group treated with CME therapy and is clearly better than the curve of motor progress shown by the older Down syndrome children treated in the Institute before the implementation of CME therapy. The final conclusion from this retrospective study was that early treatment produces a substantial motor progress in children with Down syndrome with positives effects in the other areas of the development process, and that the CME therapy also contributes to push the motor evolution curve closer to the normal one.

NEUROLOGICAL BASES OF MOTOR IMPROVEMENT

Motor impaired children will progress only in direct relation with the neurological recovery potential of the brain plus quality of the motor therapy and in the third place, the earliest age of therapy beginning. In consequence the physical therapy approach chosen to treat motor impaired children will only act as catalyzing agent for the inherent organization process of the brain. To achieve optimal motor progress, the most important factor is the internal component because we cannot infuse nor create recovery potential, but only we can stimulate the brain trying to restart (if possible) the organization process.

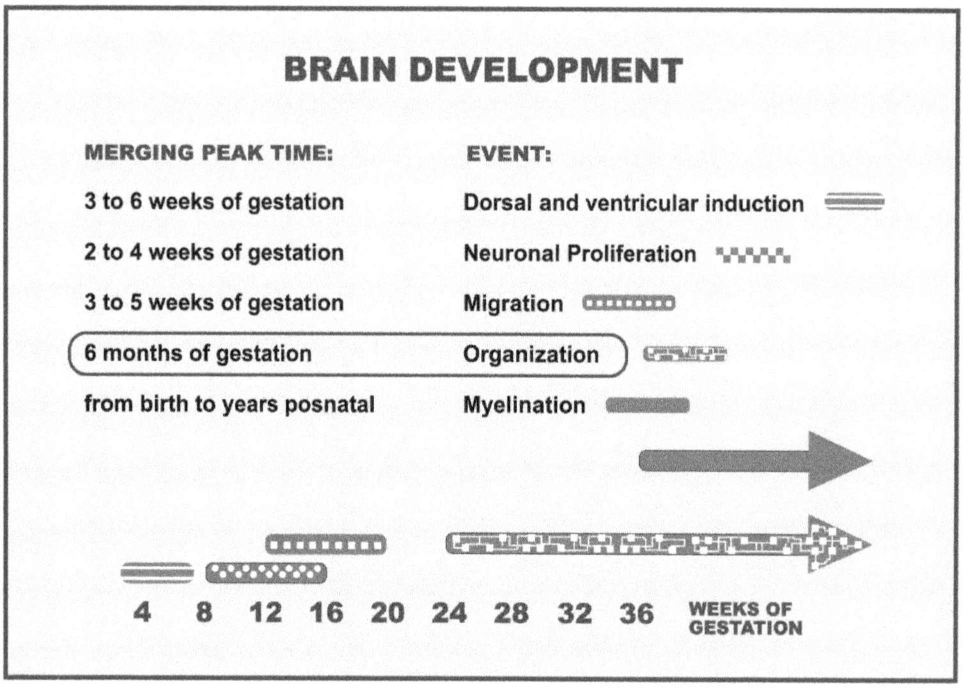

Figure 9.

The figure 10 illustrates the period during gestation in which brain formation begins to occur in the human embryo, and within the five main events its seems that the one called ***"organization"*** is the clue process for further evolution of the brain functions.

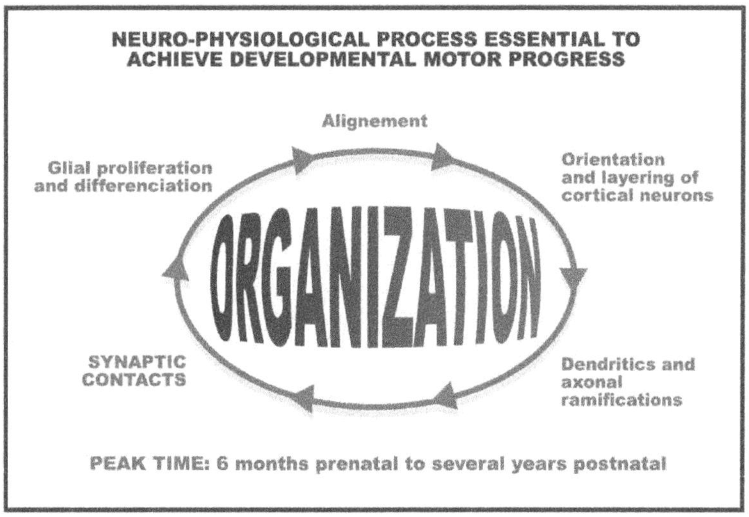

Figure 10.

This event begins around 6 months of gestational age lasting for many years after birth, and the most relevant neuro-physiological component of the "organization" is the one called "synaptic contacts" which is the origin of the nervous system network. The synaptic contacts are the physical connections between the axons terminals and the dendrites receptors which allow the nerve impulses to travel at 100 meters per second in 2 basic directions: efferent direction (from brain to organs) and afferent (from all body ton the brain).

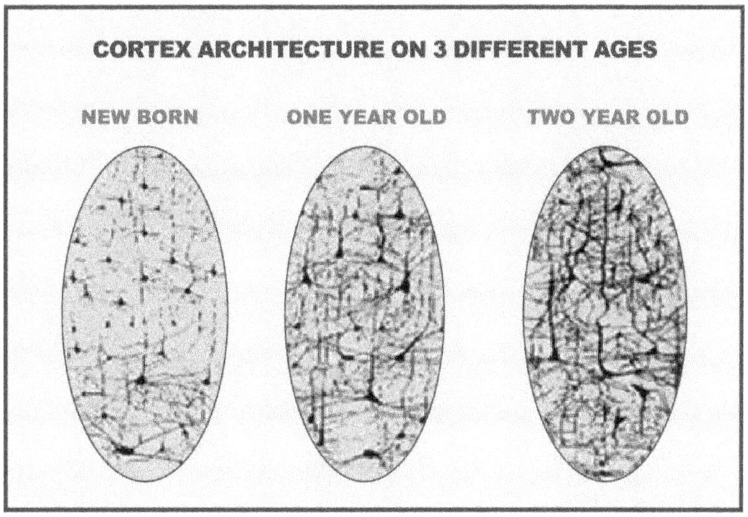

Figure 11.

The figure 11 illustrates 3 different chronological moments in human life in relation to the anatomical configuration of the brain cortex. The first oval corresponds to a new born, and is noticeable the neurons units and soft traces of connections of axons and dendrites. The middle oval shows the cortex of a one year old child, with more marked traces of network connections and less neurons units, and the last oval corresponds to a 2 years old child, with much more density in wiring network and no more neurons units, even less of them.

Observing this graphic we can conclude that improvement in developmental functions in the growing child is in some way correlated with the increasing of inter neuronal connections. Maybe the experience of motor postural challenges provoked by CME therapy can trigger the production of synaptic contacts in immature brains.

FINAL REFLECTION

The parents need honest and unbiased information from all the physical therapists, companies or entities working in pediatric rehabilitation. The most reliable therapist is the one who is able to explain to the parents in simple words the basic characteristics of the chosen intervention and the limitations imposed by the clinical characteristics of any given child. Also the professional must commit himself to established specific goals of motor progress, to be achieved after a limited time of intervention in order to demonstrate the alleged benefits of the therapeutic approach. If positive results are achieved in a short term, it is a demonstration that those results are consequence of the therapy intervention and not by the aging process alone; otherwise the therapist should recognize the failure and recommend the parents to search for better therapy or better hands. Nevertheless, some children are affected by severe brain damage caused by known or unknown causes and these children will not progress to motor independence no matter what type of physical therapy intervention is intended at an early or late moment.

(*) "*However, scientists from Harvard, Johns Hopkins, and other top institutions have recently begun to use tools such as stem cells and genetic modification to repair and regenerate brain tissue in their laboratories. For the first time in history, leading researchers see opportunity to treat the underlying causes, not just the symptoms, of these devastating conditions.*

Stem cells can divide into different types of cells. Although there are many different types of stem cells, the ones that interest scientists trying to cure childhood neurological disorders are neural stem cells. These amazing cells have a remarkable capability to become new, healthy neurons and glia that can be used to replace diseased ones and perhaps treat the underlying causes of pediatric neurological disorders, not just the symptoms.

Neural stem cells can be derived from both adult and embryonic sources, replicated in cultures to create enough tissue for transplant, and then injected into damaged regions of the brain to create new neurons and glia to potentially restore function to children disabled by neurological disorders.

Not only do injected stem cells produce new neurons and glia, but they also enhance the brain's self-repair process by communicating with their host. Scientists are studying how stem cells "talk" to each other and their environment through signaling molecules and proteins. Deciphering this language will help them learn exactly how to best harness the brain's capacity for self-healing in developing treatments for children with neurological disorders".

(*) Quoted from CNS Foundation web site.

FLASHBACK

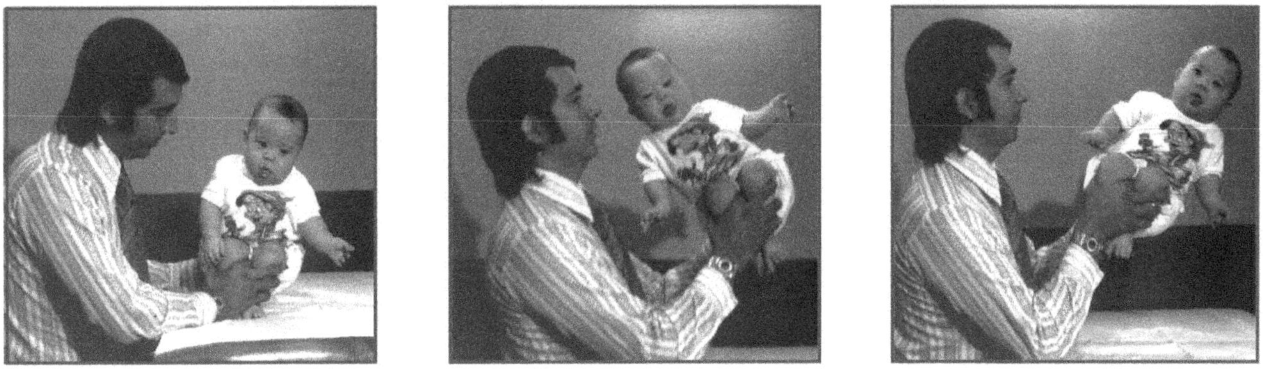

The upper 3 figures show the therapist working with a Down syndrome child of 6 ½ months old, in 1971 with an exercise for trunk control which was the precursor of the "aerial sitting" exercise. Here the child is supported by the left hand below the knees and the other hand on the sacrum area; currently this maneuver is called "aerial squatting", belonging to the head control area.

These 4 pictures from 1989, in Caracas, show Ramon doing the exercise called "Lotus sitting in suspension", for trunk control improvement. In the pictures below is possible to visualize the great effort of the child to reach automatic vertical posture of head and trunk. A clear proof of the child's active reaction is that the angle of Ramon's elbows stays unchanged in order to provoke the maximum dynamic postural response.

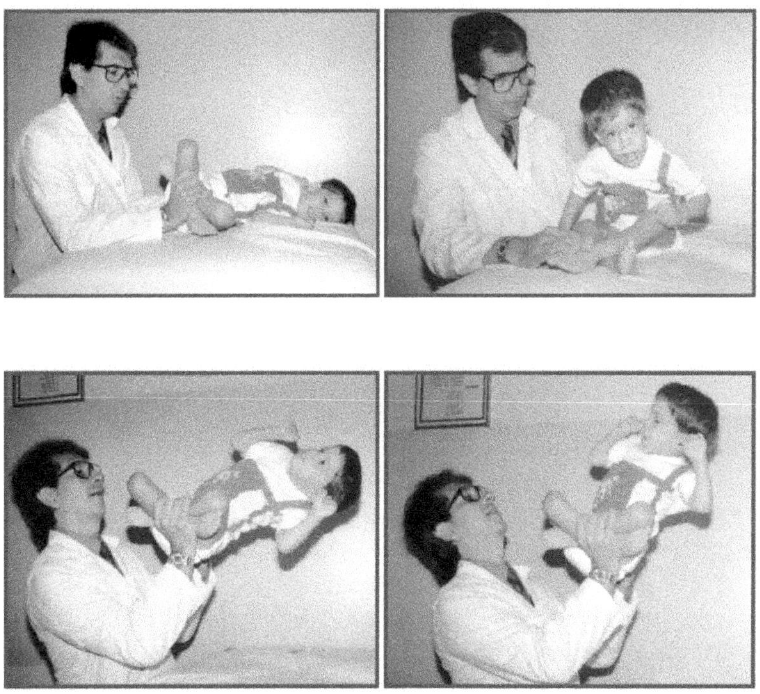

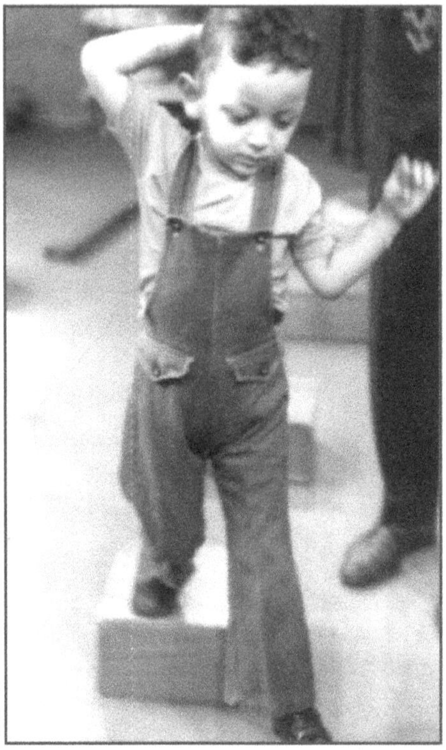

This picture shows Ramon in 1972, with a child 2 ½ years old, doing a sequence of up and down steps on 15 cm. boxes held by the neck.

This picture shows Ramon in 1996, with a girl 5 ½ years old, doing a sequence of up and down steps on 15 cm. boxes held by the neck.

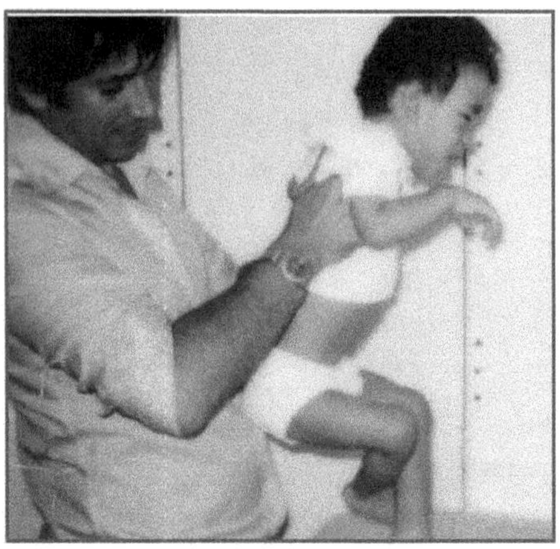

Ramon working in 1975 with a child of two and a half years old, affected by congenital developmental delay, macro-cephaly, etc., coming from traditional physical therapy since birth, and presenting no spontaneous weight bearing.

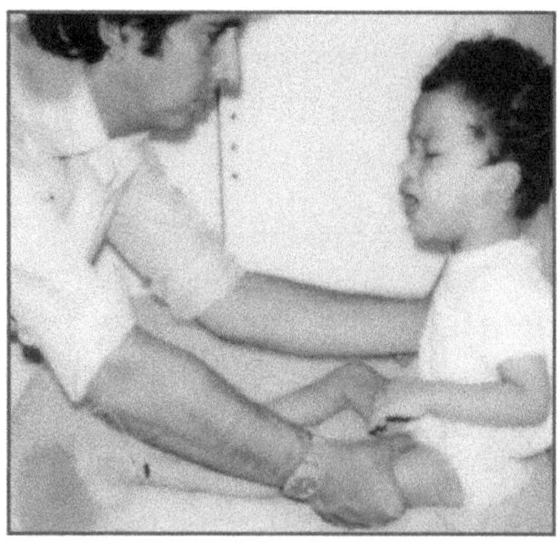

Beginning of the exercise called "Standing back to wall by below knees", one hand held above the knee and the other hand on the opposite arm pit. Simple maneuver to provoke the weight bearing reaction immediately with or without the child's cooperation.

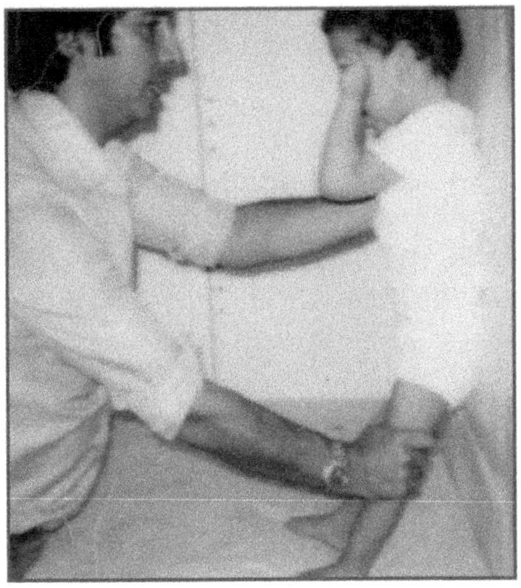

Lift back up against the wall, waiting for the acceptance of weight bearing and active trunk extension. The child is still crying.

Now Ramon moves both hands below the knees, and slides the child's heels closer to wall. At this time the child can feel his own weight and begins to get use to it.

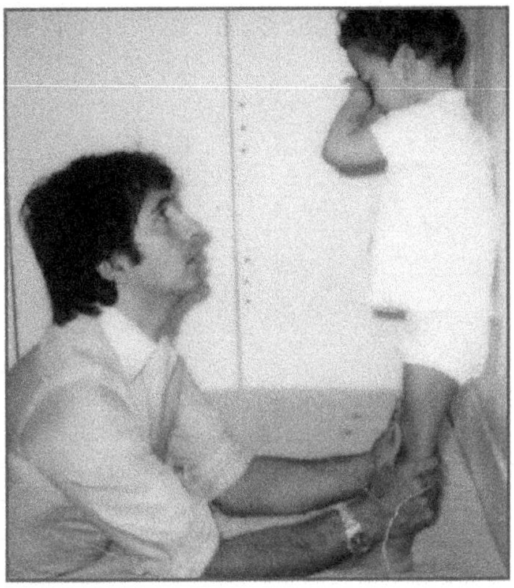

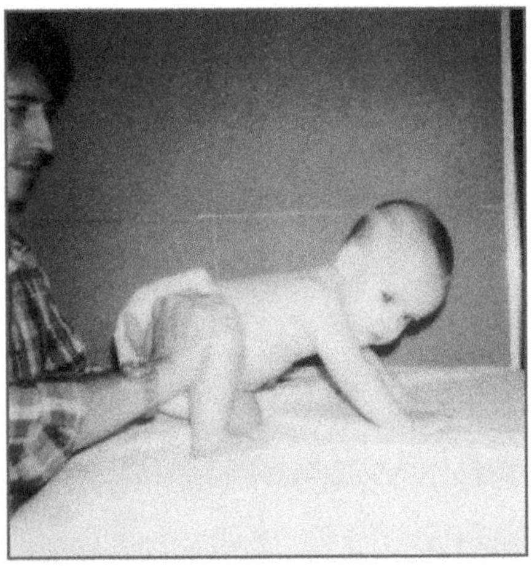

Caracas 1980, Ramon was working with a High Risk condition one year old child, having CME since 6 months old. This is the exercise called "4 points to standing by one thigh and abdomen push". Placing right leg on table.

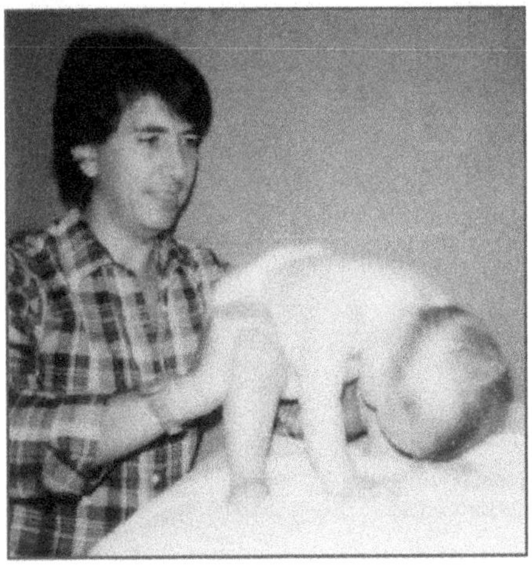

The left leg was placed and now the therapist applied soft pressure on the lower abdomen with the left hand in order to provoke the standing reaction.

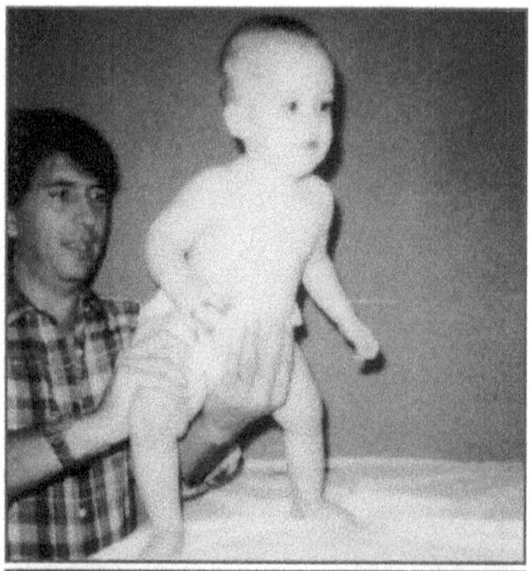

Here is ¾ up to standing position, which is a function not controlled yet by the child, who also wasn't able to control free standing or freesteps.

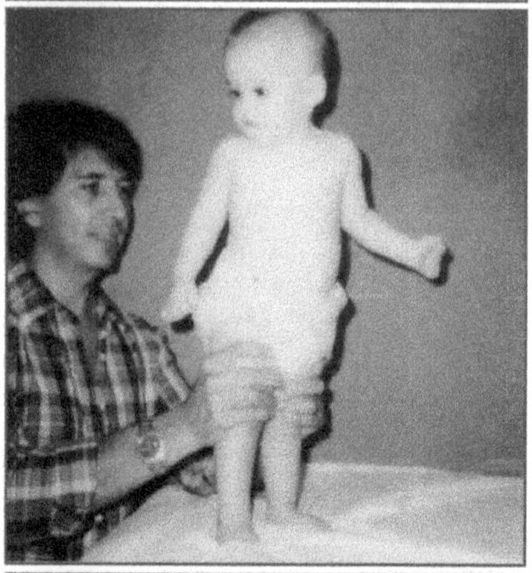

This picture illustrates a perfect final standing reaction; the therapist is holding the child by the thighs with legs together.

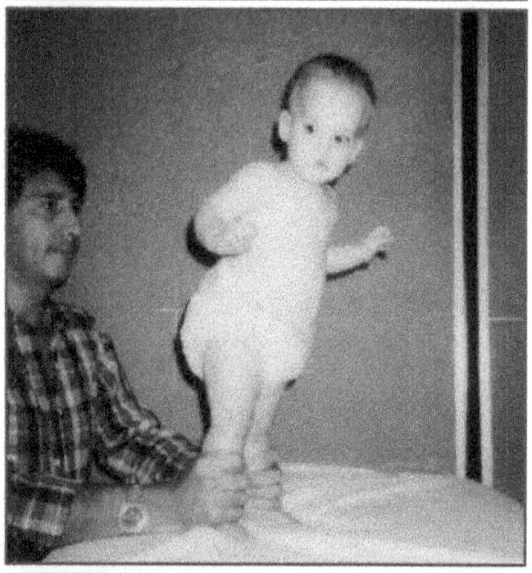

At that time Ramon began the idea of sequential exercises, which means to continue with upgraded functional challenges as long as the child responses are positive. In this case the child starts the sequence in a prone position, placed in a crawling position and stimulated to take a standing posture supported by below the knees, then the support is finally provided by ankles.

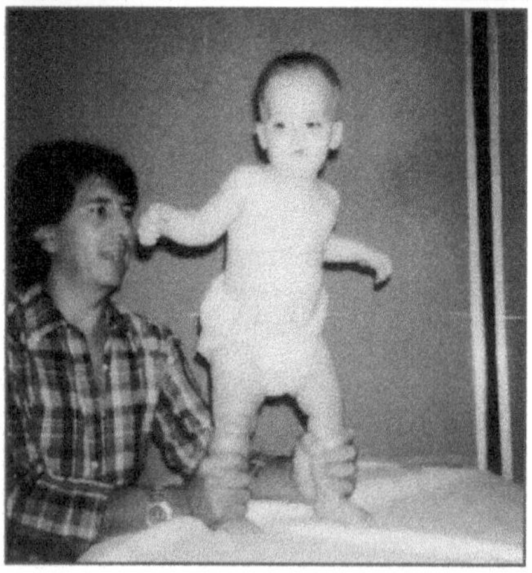

Carefully the child is induced to take steps by below knees as illustrated in the picture.

This picture shows Ramon in 1980, doing the exercise called "Aerial standing by one leg" to a developmentally delayed child of 16 months old. In this exercise it is very appropriate to do stretching of any necessary angle on the ankle, furthermore, provoke weight bearing response, hip and knee stability and balance reactions. The top upgrading situation occurs when the therapist moves right hand support off the left thigh of the child.

Ramon in 1972, trying to provoke head control responses with the exercise called "The clock". The little girl was 6 months old, with spastic quadriplegia. This was the beginning of CME head control exercises.

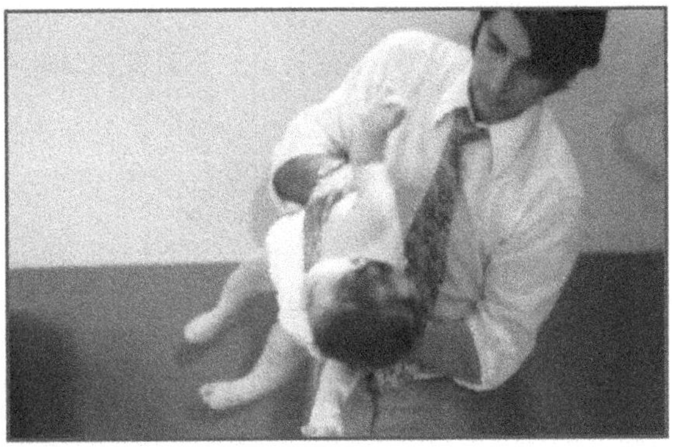

Trying to stimulate standing position in Down syndrome children in 1971 in Caracas, was for Ramon a crucial experience to develop one of the key characteristics of CME, which is "to provoke" instead of to facilitate motor responses.

At that particular time the concept of challenging the child's brain recovery potential using distal support was not part of the former theoretical framework of conventional therapy, nor of the current traditional approach assumptions in 2006. Furthermore, because official physical therapy was oriented towards treating children with spastic cerebral palsy, the Down syndrome infant population was out of any decent global policy of early motor stimulation.

The practical conflict afflicting pediatric physical therapy in the sixties and the beginning of the seventies was quite clear for Ramon, because from its origins physical therapy for infants was adapted from the approaches for adult rehabilitation in which the cooperation and motivation of the patient is extremely important, but when working with babies or young children it is not fair to expect from them cooperation or motivation.

Then it was Ramon's challenge to redefine the goals of the therapy and abandon the archaic model based on the idea that "***physical therapy helps the child to develop stronger muscles such as those in legs and trunk***".

This statement was and still is, totally wrong because functional motor handicap in developmental delayed child is not caused by a weakness in the muscle condition, furthermore the muscle weakness is a consequence of the lack of brain maturity to control basic motor-postural reactions.

So , Ramon began trying to provoke the absents automatics postural responses with different types of exercises using the most distal support that the child could tolerate instead of proximal support , and also the repetition of current controlled motor responses were also eliminated from the new therapy approach and replaced with exercises provoking absent motor responses. For instance, delayed children were commonly stimulated to stay in a standing position with hands supporting on the therapist hands or on different objects, and the final result was a standing position conditioned to hand contact but which did not increase free standing, because the automatic stability and balance reactions were not stimulated to develop when hand contact was active.

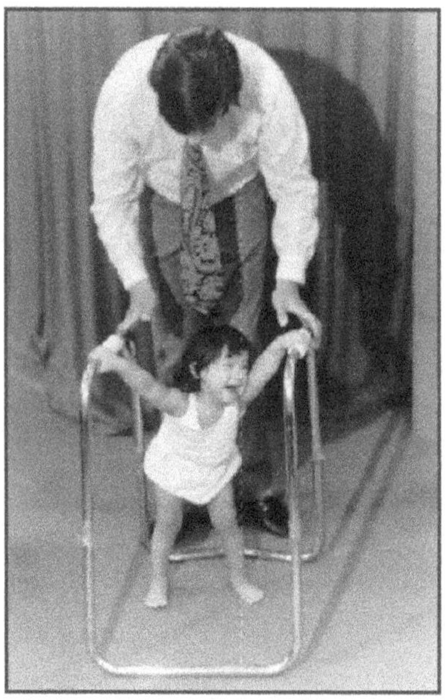

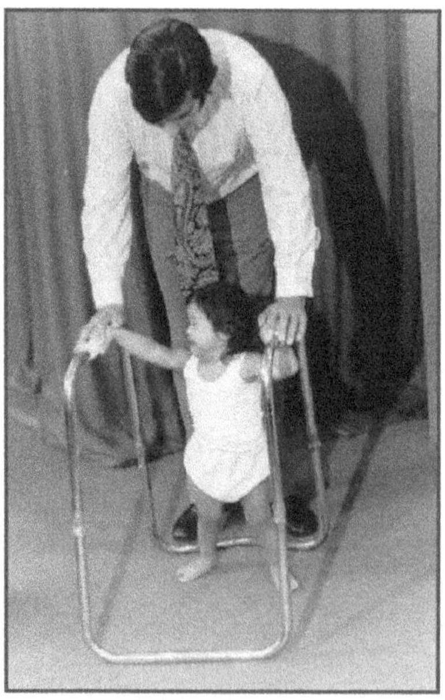

The left picture shows the little Down syndrome girl of 2 ½ years old in a standing position holding the frame with both hands plus the hands of the therapist placed on top of her hands. The next picture illustrates the reaction of the child after the therapist removes his hands and tries to slide the frame to provoke steps, so, the conclusion is that holding the child by the hands, does not lead to motor progress.

The next sequence of pictures illustrate a good example of Ramon's quest for new exercises (1972) to provoke and control the standing position without touching the child's hands. The case is a Down syndrome child of 16 months old; the first picture shows his reaction to standing with hand support (very inclined), the following pictures illustrate the child's reaction after Ramon releases his hands and provide support by both legs. The child slowly begins to extend the trunk by himself until reaching total vertical standing as shown in the last picture.

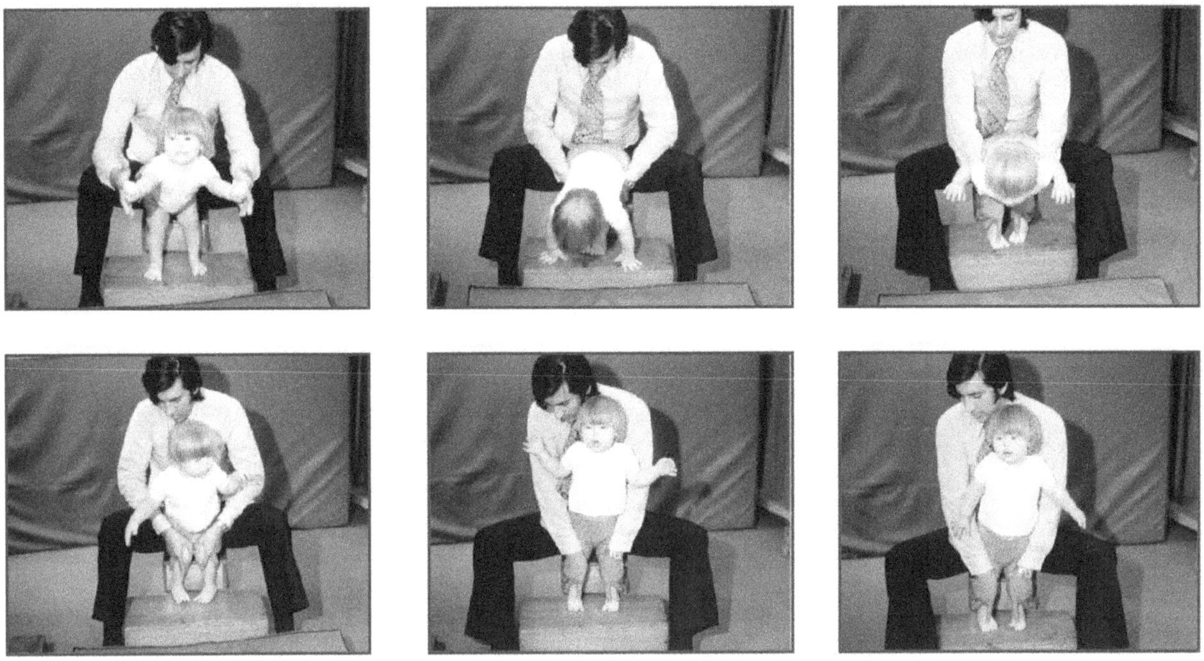

Once again, the child's reaction to the new maneuver demonstrates that distal support provokesautomatic responses, thus the child can experience the real sensation of being erect with full weight bearing and hands free, with the final result of increasing the main human postural reaction without promoting wrong compensation of hand support.

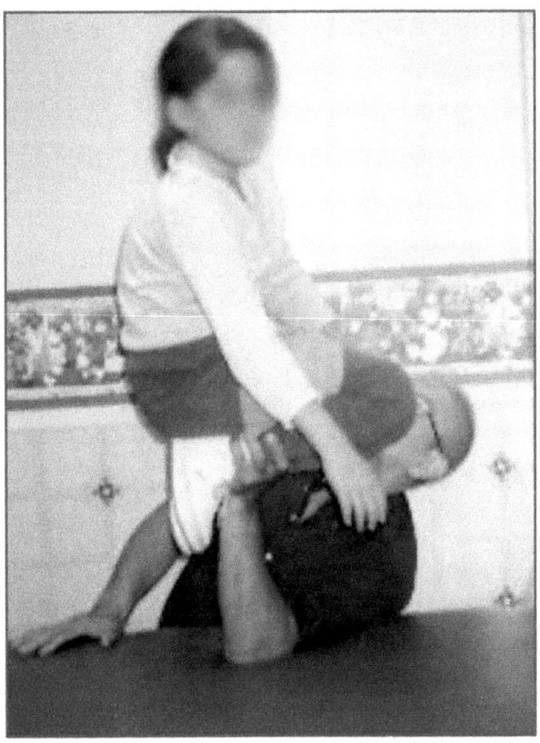

This picture illustrates the biomechanical final stage of the exercise called "Aerial sitting", performed by Ramon with a 72 pound child.

CME CONTINUING EDUCATION PROGRAM

CMECEP

MISSION STATEMENT

TO PROVIDE THE HIGHEST QUALITY AND UPDATED INFORMATION ABOUT THEORY AND PRACTICAL APPLICATIONS OF CUEVAS MEDEK EXERCISE THERAPY AND ITS PROGRESSIVE EVOLUTION AND CHANGES.

The Cuevas Medek Exercise Continuing Education Program was conceived by Ramon Cuevas to enhance the quality of CME teaching since the year 2000. Starting from that year Ramon reorganized the teaching team with new highly ethical assistants which accepted and supported from the beginning the evolution of the name of the therapy from the former MEDEK to the current CME therapy which stands for Cuevas Medek Exercise.

The unique goal of CMECEP is to provide the opportunity for physical and occupational therapists to progressively learn the state-of-the-art current curriculum of this motor therapy approach for pediatric rehabilitation.

The Cuevas Medek Exercises therapy continuing education program is integrated by four levels of learning:

Level one: The introductory course or CME I, regular (in groups) or tutorial (only one participant)

Level two: The intermediate course or CME II, regular (in groups) or tutorial (only one participant)

Level three: The advanced course or CME III, tutorial only

SOME MOMENTS FROM INTERACTIV CME III EXAMINATION

This picture illustrates Ramon doing CME called "Step up/down big Olympic podium by ankles" while Jo-Anne Weltman is attentively observing the maneuver.

This photo shows Bernie Connors doing CME called "Step down from 30 cm. balance bar by below knees", in front of Ramon.

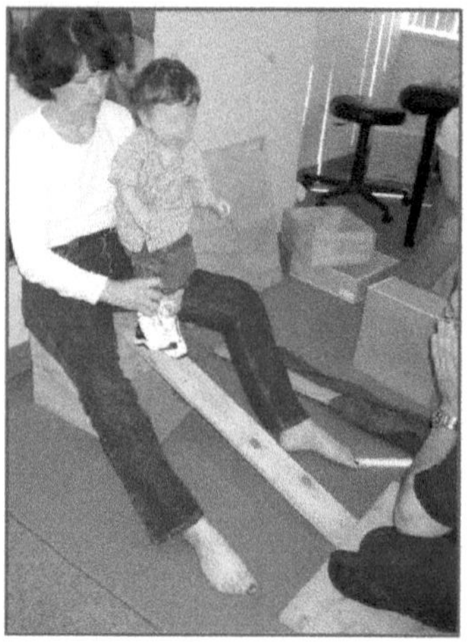

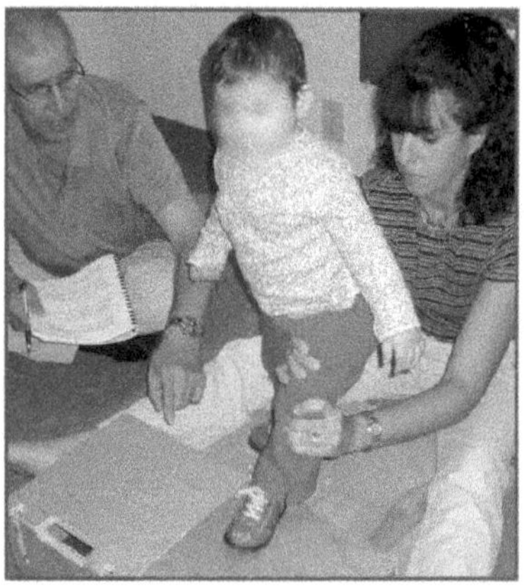

Ramon is observing Jo-Anne Weltman doing CME called " Step up 15 cm. box by one leg, swing leg free".

Simona DeMarchi and Ramon working together with CME called "Steps on 4 see-saw 30 cm boxes by ankles".

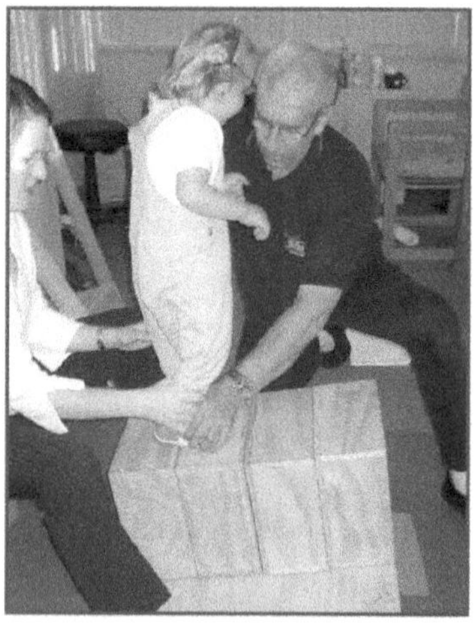

This picture shows Jo-Anne doing practicum of CME "Steps down big ramp by crossed straps". Simona, the girl's mother and Ramon follow attentively the manouver.

GLOSSARY

ADAPTATION: the process of adapting to something such as environmental conditions.

ADAPTIVE OR ADAPTATIVE: having a capacity of adaptation.

ADAPTATIVE-MOTOR CONTROL: classic definition from N. Bailey and A. Gessell, for developmental manual activities.

ANOXIA: Absence of oxygen in arterial blood or in the tissues.

ASSUMPTION: An Idea or statement assumed to be true without proof.

ATAXIA: Inability to coordinate voluntary muscle movements, unsteady movements and staggering gait.

AXON: nerve fiber that conducts impulses away from the body of a nerve cell.

BALANCE: equality of distribution.

BALANCE BAR: narrow piece of wood.

BALANCE BOARD: square flat piece of wood with a central cube.

BIOMECHANICS: the functional aspect of an activity.

BUFFER: device to reduce the shock due to contact

CENTIMETER: a metric unit of length equal to one hundredth of a meter.

CEREBRAL PALSY (*): group of chronic (non degenerative) conditions affecting the body movements and muscle coordination. Is caused by damage to one or more specific areas of the brain occurring during the pre, peri or post natal period.

CME: abbreviation of Cuevas Medek Exercises

COLLAPSE: the act of throwing yourself down, abrupt failure of function.

CONSCIOUS: having awareness of sensations and thoughts.

DENDRITE: the branching portion of a nerve cell that carries impulses toward the cell body.

DEVELOPMENTAL MOTOR DELAY: slow acquisition of motor functions.

DISTAL: directed away from fulcrum point.

DOWN SYNDROME: chromosomal abnormality, (produce global developmental delay) also call trisomy 21.

ECLECTIC: selecting what seems best of various styles or ideas.

ENCEPHALITIS: Inflammation of the brain usually caused by a virus.

EQUILIBRIUM: a sensory system located in structures of the inner ear that registers the orientation of the head.

FACILITATE: make easier.

FINE MOTOR CONTROL: archaic definition for functions involving manual activities.

FOUR POINTS: crawling position

FULCRUM: the pivot about which a lever turns.

FUNCTIONAL: performing or capable of performing

GLIA: Star-shaped cells that provide physical and nutritional support for neuron.

GRAVITY FORCE: attraction of the earth's mass.

GROSS MOTOR CONTROL: archaic definition for postural motor functions. *(CME concept: Indeed is no "gross motor functions" in human motor behavior and is a big mistake to titled the postural-functional activities with such name. For instance, to stand and walk required the complex interaction of many systems to cope with the biomechanical challenge to control the body vertical and functional using a reduced base of support as is the feet surface relateto the size of the body. Is no specie on earth matching the biomechanical characteristics of the human standing posture. Examples of the complex nature of standing and walking can be easily found in the technological field of robotics. For decades are available robots in the automotive industry (painting,soldering, assembling etc.), so manual activities can be replaced easily, but walking activities are only recently achieve in a very basic level by a robot made in Japan in 2000. Other example is the technology available for quadriplegic persons which allowed them to manage the motorized wheel chair by a device working with air pressure that the person apply with the mouth through a plastic tube. To name postural motor responses as "gross activities" and manual activities as "fine ones" does not honoredthe truth).*

HIGH RISK CONDITION: expose to a chance of damage.
HORIZONTAL: parallel to or in the plane of the horizon.
HYPERTONUS: abnormal high muscle tension to palpation or / and to stretching.
HYPOTONUS: abnormal low muscle tension to palpation and or to stretching.
IDIOPATHIC: arising from an unknown cause.
LATENT: potentially existing but not presently evident or realized.
LOCK A JOINT: to immobilize a joint to restrain flexion.
LOTUS POSITION: sitting position on the floor with legs crossed.
MATURATION: biological unfolding of events involved in an organism changing gradually from simple to more complex level of function.
MATURITY: state to being mature, full development.
MEDEK: Spanish acronym meaning Metodo Dinámico de Estimulación Kinésica, (Dynamic Method of Kinetic Stimulation).
MENINGITIS: Infectious disease characterized by inflammation of the meninges (the membranous tissues that surround the brain).
MIMIC: to imitate closely
MOTOR CONTROL: is the ability to regulate or direct the mechanisms essential to posture and movement.
MOTOR-POSTURAL ASSESSMENT: the act of evaluate the motor-postural reaction of a child, related to the chronological age.
MUSCLE TONE: the elastic tension of the resting muscle.
NEONATAL: related to the infant during the first month after birth.
NEURONE: any of the cells of nerve tissue, consisting of main portion or cell body and short fibers conducting impulses toward the cell, and one long fiber conducting the impulses away from the cell.
ORGANIZATION: is the neuro-physiological process essential to achieve developmental evolution.
PIVOT: the act of turning,
PER SE: intrinsically, with respect to its inherent nature.
POLICY: a plan of action adopted by an individual or social group.
POSTURE: position or arrangement of the body and the limbs
POTENTIAL: the inherent capacity for coming into being.
PREMATURE: born after a gestation period of less than the normal time.
PRONE: lying face downward.
PROGNOSIS: A prediction of the course of a disease.

PROGRESSIVE DISEASE: gradually advancing in severity.

PROPRIOCEPTION: The ability to sense the position, location, orientation and movement of the body and its parts.

PROVOKE: to incite to action.

PROXIMAL: situated nearest to fulcrum point.

ROBOT: a mechanism that can move automatically.

ROBOTICS: discipline concerned with the practical use of robots.

SANDWICH (by): position of hands, one in front of the other, containing the child's body in between.

SPECTRUM: broad range of related values or qualities or ideas or activities.

STABILITY: the quality or attribute of being firm and stable.

STIMULUS: acts to arouse action.

STROKE: A sudden loss of consciousness resulting when the rupture or occlusion of a blood vessel leads to oxygen lack in the brain.

SUPINE: lying face upward.

SYNAPTIC CONTACT: the point at which a nervous impulse is transmitted from an axon of one neuron to the dendrite of another.

TRAPEZE: a short horizontal bar suspended from 2 parallel ropes.

VECTOR: a straight line segment whose orientation in space is direction.

VERTICAL: at right angle to the plane of the horizon.

*(*) **From NINDS, National Institute of Neurological disorders and Stroke**: Cerebral palsy is an umbrella-like term used to describe a group of chronic disorders impairing control of movement that appear in the first few years of life and generally do not worsen over time. The disorders are caused by faulty development or damage to motor areas in the brain that disrupts the brain's ability to control movement and posture. Symptoms of cerebral palsy include difficulty with fine motor tasks (such as writing or using scissors), difficulty maintaining balance or walking, involuntary movements. The symptoms differ from person to person and may change over time. Some people with cerebral palsy are also affected by other medical disorders, including seizures or mental impairment, but cerebral palsy does not always cause profound handicap. Early signs of cerebral palsy usually appear before 3 years of age. Infants with cerebral palsy are frequently slow to reach developmental milestones such as learning to roll over, sit, crawl, smile, or walk. Cerebral palsy may be congenital or acquired after birth. Several of the causes of cerebral palsy that have been identified through research are preventable or treatable: head injury, jaundice, Rh incompatibility, and rubella (German measles). Doctors diagnose cerebral palsy by testing motor skills and reflexes, looking into medical history, and employing a variety of specialized tests. Although its symptoms may change over time, cerebral palsy by definition is not progressive, so if a patient shows increased impairment, the problem may be something other than cerebral palsy). All NINDS-prepared information is in the public domain and may be freely copied. Credit to the NINDS or the NIH is appreciated.*

REFERENCES

Bayley, Nancy: Scales of Infant Development, edited by The Psychological Corporation. N.Y. (1966)

Bergeron, M.: Les Manifestations Motrices Spontanees Chez L'enfant, edited by Herman & Cie. Paris (1947).

Birch, H. & Gussov, J.D.: Disadvantaged Children, edited by Harcourt Brace-world N.Y. (1970).

Brazelton, B.T.: Neonatal Behavioral Assessment Scale. Clinics in Developmental Medicine, Edited by Heinemann Medical N.Y. (1973).

Caldwell, B. & Bradley, R.: Early Stimulation in Mental Retardation and Developmental Disabilities, edited by Wortes J.M.D. US. (1975).

Connolly, K.: Sensory-Motor Coordination Mechanisms and Plans. Clinics in Developmental Medicine, edited by Wolf P.U.S. (1963).

Cuevas, Ramon. Early stimulation in high Risk Babies, presented at Fourth International Congress of IASSMD, Washington D.C. (1976).

Cuevas, Ramon & Nascimento John.: Early Stimulation, edited by Mental Delay Collection, Venezuela (1979).

Gesell, Arnold: Diagnóstico del Desarrollo, edited by Paidos, Buenos Aires (1966).

Koop, C. & Parmelee, A: Prenatal and Perinatal Influence on Infancy Behavior, edited by Osofsky J. N.Y. (1977).

Koupernic, C.: Psycho-motor Development in Early Infancy, edited by L. Miracle, Spain (1969).

Lindemann, K.: The infantile Cerebral palsy, edited by E.C. Medical, Spain (1968).

Mitchel, R.: Child's Growing and Development, edited by Editorial Pediátrica, Spain (1975).

Orozco, Aleida, & Cuevas R.: Medek Therapy in Psychomotor Retardation, degree thesis, Rehabilitation School, Venezuela (1981).

Parmelee, A.: Who is the Risk Infant ?, edited by Clinical Obstetric and Gynecology, US (1973).

Painter, G.: Teach your Baby, edited by Simon & Schuster, N.Y. (1970).

Ramey, C.. & Campbell, F.: Prevention in Developmental Retardation in High Risk Children, edited by University P.P., Baltimore (1977).

Schwartz, J & Begley, S.: The Mind and the Brain, edited by Regan Books, division of Harpers Collins, USA, (2002).

Shumway-Cook Anne & Woollacott Marjorie.: Motor Control, edited by Lippincott Williams & Wilkins, Baltimore, USA, (2001),

Town, B.: Neurological Development in Infancy, edited by William H. London (1980).

www.ingramcontent.com/pod-product-compliance
Ingram Content Group UK Ltd.
Pitfield, Milton Keynes, MK11 3LW, UK
UKHW052301280926
15471UKWH00006B/1051

9 789563 459197